Practical Guide to
Surgical and Endovascular Hemodialysis Access Management

Case Based Illustration

Practical Guide to
Surgical and Endovascular Hemodialysis Access Management
Case Based Illustration

Jackie Pei Ho
National University Health System, Singapore

Kyung J Cho
University of Michigan Health System, USA

Po-Jen Ko
Chang Gung Memorial Hospital, Taiwan

Sung-Yu Chu
Chang Gung Memorial Hospital, Taiwan

Anil Gopinathan
National University Health System, Singapore

World Scientific

NEW JERSEY · LONDON · SINGAPORE · BEIJING · SHANGHAI · HONG KONG · TAIPEI · CHENNAI · TOKYO

Published by

World Scientific Publishing Co. Pte. Ltd.

5 Toh Tuck Link, Singapore 596224

USA office: 27 Warren Street, Suite 401-402, Hackensack, NJ 07601

UK office: 57 Shelton Street, Covent Garden, London WC2H 9HE

Library of Congress Cataloging-in-Publication Data

Ho, Jackie Pei, author.

    Practical guide to surgical and endovascular hemodialysis access management : case based illustration / Jackie Pei Ho, Kyung J. Cho, Po-Jen Ko, Sung-Yu Chu, Anil Gopinathan.

       p. ; cm.

    Includes bibliographical references and index.

    ISBN 978-9814675345 (hardcover : alk. paper) -- ISBN 978-981-4725-30-9 (pbk : alk. paper)

    I. Cho, Kyung J., author. II. Ko, Po-Jen, author. III. Chu, Sung-Yu, author. IV. Gopinathan, Anil, author. V. Title.

    [DNLM: 1. Renal Dialysis--methods--Case Reports. 2. Arteriovenous Shunt, Surgical--methods--Case Reports. 3. Catheterization, Central Venous--methods--Case Reports. 4. Kidney Failure, Chronic--surgery--Case Reports. WJ 378]

    RC901.7.P48

    617.4'61059--dc23

                               2015021590

British Library Cataloguing-in-Publication Data

A catalogue record for this book is available from the British Library.

Typeset by Stallion Press

Email: enquiries@stallionpress.com

I dedicate this work to my beloved parents,
Hsueh Mei and Chung Chi,
for their gift of ever enduring love and life-long lessons
that have taught me all the ideals and true values
I cherish today and forever.

Preface

My training in vascular and endovascular surgery began in 1999 in Hong Kong. However, my exposure to hemodialysis access work started only from 2006. Teaching and training in hemodialysis access are not included in either undergraduate or postgraduate curricula. Neither was it emphasized during my vascular and endovascular training. Yet, ever since I took up hemodialysis access work, I have found this clinical area to be a highly specialized field, extremely fascinating and at the same time very demanding.

Hemodialysis access work is very different from the vast proportion of vascular and general surgeries. Unlike most surgeries, which are one-time procedures, hemodialysis access creation and management usually involve repeated surgical and endovascular procedures. Hemodialysis access work is also unique in that after creation or salvage, the accesses are constantly interacting with the environment through cannulations. In many situations, the clinician has more than one option when considering creation or salvaging an access. Most of the devices and technologies used in this field are borrowed from peripheral arterial disease and, hence, many unmet needs still exist. New hemodialysis-specific devices have emerged in recent years that hopefully will improve the current clinical picture. The vascular access is an access to maintain the survival of hemodialysis patients. A successful access strategy not only provides an access to the patient during a single time frame but can also sustain the patient's hemodialysis over a long period of time with a minimal number of surgeries and interventions.

Providing a high quality of service confers a huge difference to both the patients as well as the healthcare system. It is beyond any single clinician's capability to provide an optimal hemodialysis service. A multi-disciplinary team approach with a mutual management goal, consensual treatment strategy as well as efficient and effective communication is the recipe for success. Although hemodialysis access service has been relatively neglected in many countries, more and more hospitals and healthcare authorities are now giving priority to this field.

This book aims to provide practical knowledge, evidence-based guidance and tips on key aspects of hemodialysis access creation, maintenance and salvage. And I hope the cases included in this book will illustrate some of the treatment principles as well as demonstrate, in part, the complexity and variability of hemodialysis access work.

Acknowledgement

I am extremely grateful to many of my seniors, friends, contemporaries, co-workers, trainees and students. Without them, this book will never have come to fruition the way it has been written now. I would like to thank Dr. Siow Woei Yun for inspiring me to write this book; Prof. Sydney Chung for his invaluable advice on book writing and content planning; A/Prof. Davide Lomanto, Dr. Roger Ho, Dr. Malvyn Zhang, Prof. Aileen Wee and Ms. Carven Tam for their advice on publishing; Prof. Lee Chuen Neng and Prof. Yeoh Khay Guan for their support to me to pursue this project; Mr. Lee Cheuk Hung, Dr. Danny Cho, Dr. Alfred Wong, Dr. Cheng Shin Chuen, Dr. Ye Zhidong, and Dr. Sujith Wijerathne for editing the chapters; Ms. Iris Yuet, Mr. Han Young-Rok and A/Prof. Jimmy So for their encouragement especially in times of frustration; Ms. Kyi Zin Thant and Ms. Abdul Majeeth Salimdeen Razia for their contributions to the flow charts and diagrams; Dr. Wong Weng Kin and Ms. Candy Wu for their expertise on hemodialysis; Ms. Adeline Teo, Ms. Florence Ang and many of my trainees, Dr. Lynette Loo, Dr. Arunesh Majumder, Dr. Ryan Yak, Dr Amritpal Singh, Dr. Thng Yong Xian, and Dr. Shum Jia Yi for collecting many of the clinical images; Ms. Melody Hee Hui Shi, Ms. Julia Hee Loo Chin and Mr. Chia Yong Qing for the cover illustrations. Last but not least, Mr. Chua Hong Koon, Ms. Darilyn Yap and Ms Joy Quek from World Scientific for their professional advice and editing works on this book. All of them have made this odyssey of book writing a very pleasurable one.

Jackie P. Ho

Authors

Jackie Pei Ho 何蓓
Associate Professor
University Surgical Cluster
Yong Loo Lin School of Medicine
National University of Singapore
Consultant
Department of Cardiac, Thoracic & Vascular Surgery
National University Health System
Singapore

Kyung Jae Cho
Professor of Radiology
Department of Radiology
University of Michigan Medical School
Ann Arbor, MI, USA

Po-Jen Ko 柯博仁
Section Chief of Vascular Surgery
Chang Gung Memorial Hospital
Linkou, Taiwan

Sung-Yu Chu 朱崧毓
Lecturer
Department of Medical Imaging and Intervention
Chang Gung Memorial Hospital
Linkou, Taiwan

Anil Gopinathan
Consultant (Interventional Radiology)
Department of Diagnostic Imaging
National University Health System
Singapore

Contents

Abbreviation

AV	Arteriovenous
AVF	Arteriovenous fistula
AVG	Arteriovenous fistula graft
BA	Brachio-axillary
BAM	Balloon angioplasty maturation
BB	Brachio-basilic
BC	Brachio-cephalic
BBT	Brachio-basilic fistula transposition
CIN	Contrast induced nephropathy
CKD	Chronic kidney disease
CT	Computed tomography
CTO	Chronic total occlusion
CVA	Cerebrovascular accident
CVC	Central venous catheter
DASS	Dialysis associated steal syndrome
DBI	Digital brachial index
DP	Dorsalis pedis artery
DRIL	Distal Revascularization and Interval Ligation
DSA	Digital subraction scan
ECG	Electrocardiography
EIV	External iliac vein
ePTFE	Expanded Polytetrafluoroethylene
FTM	Failure to mature
GSV	Great saphenous vein
IJV	Internal jugular vein
IV	Intravenous

KDOQI	The National Kidney Foundation Kidney Disease Outcomes Quality Initiative
NKF	National Kidney Foundation
PAI	Proximalization of arterial inflow
PRA	Proximal radial artery
PTA	Balloon angioplasty
PT	Posterior tibial artery
Qa	Access flow
Qb	Blood flow set by the hemodialysis machine
RC	Radio-cephalic
RCT	Randomized controlled trial
RUDI	Revision using distal inflow
SCV	Subclavian vein
SVC	Superior Vena Cava
VP	Venous pressure

CHAPTER 1

General Principles of Hemodialysis Access Creation

Jackie P. Ho

Overview on ESRF and Hemodialysis Access

The prevalence of end stage renal failure ESRF patients is increasing Worldwide.[1] Japan, Taiwan, USA, Portugal, Singapore and Mexico are some of the countries having the highest ESRF prevalence. Hemodialysis is the modality of renal replacement therapy adopted by the majority of patients.[2] The hemodialysis access has become the life-line of many ESRF patients.

Different from treatment of various vascular diseases, hemodialysis access management involves creation of an abnormal vascular passage that will constantly and repetitively interact with the external hemodialysis devices. This passage can be in the form of a tunnelled catheter inserted into a large sized central vein, CVC, or a new connection between an artery and a vein using either patient's native superficial vein (arteriovenous fistula, AVF) or synthetic graft (arteriovenous fistula graft, AVG).

On the other hand, many patients do have existing cardiovascular diseases which will interfere with the choice, success, risk and durability of hemodialysis accesses.

The creation and maintenance of hemodialysis access require both open vascular surgical technique and endovascular wire and catheter technique. The planning and monitoring of hemodialysis access also require specific knowledge and skill sets. This book aims to provide an all compassing coverage of these aspects. Before going into a specific area, let us take a closer look at the center of hemodialysis access service — the patient and the access users.

Understand Your Patients and the Access Users

Diversity: Renal failure hemodialysis-dependent patients have a wide spectrum of medical risk, cardiovascular conditions, psycho-social and financial status. Their age range from pediatric to octogenarians. Some have low surgical and anesthetic risk while some have extremely high risk. Some are young, mobile and working with long life expectancy. Some are old and frail, who require a caregiver to attend clinics and treatments. There are a variety of illnesses leading to ESRF. Chronic medical co-morbidities are common among the hemodialysis population. Because of the multiple long-term medical problems, psycho-social and financial problems are frequently seen among the patients, rendering hemodialysis access service delivery more difficult. Depression and depressive symptoms are not uncommon in hemodialysis patients.

Service goal: The hemodialysis access serves as the life-line of those patients. Patients may require hemodialysis access for a couple of years to few decades. Hemodialysis access program therefore should not look to providing a one-off service but a long-term committed care with strategic planning. Due to the wide variation of their medical, surgical and psycho-social issues,[3,4] hemodialysis access strategy also needs to be tailored to suit individual's problems and needs. The aim is to enable patients to have hemodialysis access with longest duration, least number of surgical and interventional procedures and least access-related morbidity. The service should be set to minimize the disturbance of patients' and their caregivers' daily activities. Day procedure therefore is a more preferable model. Clinicians have to be sensitive to patients' emotion and psychological status. A friendly and caring healthcare team is one of the key elements for a successful hemodialysis access service.

End users: The end user of the hemodialysis access is not the nephrologist, surgeon, or interventional radiologist. The end users are the patient and the nursing staff of the dialysis center who needle the access few times a week. Unfortunately, the healthcare team and dialysis nurses rarely work side-by-side in the same facility. Many patients are unable to understand their access condition thoroughly or to convey the message between the two parties. A clear and standardized format of communication between the healthcare team and dialysis nurses can greatly facilitate the hemodialysis access service.[5] Dedicated patient education on access assessment and daily care will engage patients well in maintaining a functional access. In the long journey of maintaining a hemodialysis access, the patient himself/herself is an important team player.

Patient expectation: Currently surgically created vascular accesses for hemodialysis are being addressed as "permanent access". The term "permanent" may cause some confusion. These accesses are permanently created onto the patient's body. However, most accesses degenerate and develop stenosis with time and may not work permanently. Patients will become disappointed and frustrated when vascular access become blocked, which is against their expectation of "permanent". Therefore, patients should be counselled that the vascular access is only of "long-term" and not "permanent" use. Vascular access, especially AVF, may be perceived as a minor and simple operation. Patients may expect a straight forward process from creation to needling. Again, emotional upset may arise if the access failed to mature or require secondary procedures to enhance the maturation. Proper patient counselling and patient education help manage the expectation of the patients and avoid unnecessary frustrations.

General Principles

There are benefits and disadvantages for both hemo- and peritoneal dialysis. Make sure the patient has been well counselled on both dialysis modalities and had made an informed decision for themselves before creating a vascular access.

The three essential components of a hemodialysis access (except tunnelled CVC) are:

⇨ **Good inflow** — clinician has to find an artery which is easily accessible, reasonable in size, good flow and ensure creating a fistula on it will not compromise the end organ blood supply.

⇨ **Good outflow** — An outflow vein that drains well ultimately into right atrium.

⇨ **Good conduit** — Either native superficial vein or a synthetic graft that can easily be cannulated.

There are several general principles on surgical dialysis access:

(1) **Patient's native vein arteriovenous fistula (AVF) in general is preferable to synthetic graft arteriovenous fistula graft (AVG)**[6–8]

Usually, AVF has better patency and fewer septic complication than AVG, provided the native vein size and quality is satisfactory.

(2) **Use the most distal native vein first**[6]

This is to preserve more proximal vein for future use when the distal AVF failed. Distal forearm cephalic vein will be considered first, followed by mid-forearm cephalic vein and antecubital vein. Basilic vein of the arm, situated medially, usually deeper in sub-cutaneous tissue and next to medial antebrachial cutaneous nerve, will be chosen when cephalic vein is small or exhausted. Basilic vein of the forearm may also be used for AVF creation if the size is good.

(3) **Choose non-dominant or less functional upper limb for access creation**

Patients will have reduced functionality of the upper limb bearing the vascular access during dialysis for about 4 hours in 2–3 days every week. After dialysis, they also need to avoid exertion of that limb for half a day to minimize re-bleeding. It is more convenient for the patient if the vascular access is situated on the non-dominant or less functional limb. This is particularly true if the patient had suffered a previous stroke and there is weakness in one of the upper limbs. Bearing in mind the vein of the weaker limb may shrink in size and contracture may present, vascular access creation or subsequent cannulation in the weaker limb could be more challenging.

(4) **It is more preferable to place dialysis access on upper limb than lower limb**

There are both medical[9] and social disadvantages of placing the vascular access in the lower limb compare to upper limb. Venous stenosis and deep vein thrombosis symptom in the lower limb are more likely to cause problems than in the upper limb. Hygiene in general, is better for the upper limb and therefore will have less septic complications. The social embarrassment of exposing the upper limb for cannulation is also less than the lower limb, especially for female patients.

(5) **Avoid or minimize the duration of percutaneous tunnelled CVC**

Disadvantages of percutaneous tunnelled CVC include line sepsis, social inconvenience and induce central vein stenosis or thrombosis (risk increase with duration). Certainly, tunnelled CVC has to be avoided or its duration minimized for patients with reasonable life expectancy. On the other hand, the advantages of tunnelled CVC are no needling pain on every dialysis session, no risk of steal syndrome and no surgery required. A long-term tunnelled catheter may be the best hemodialysis access option for some patients, e.g. patients with very limited life expectancy (terminal malignancy or terminal medical conditions), patients with extensive arterial atherosclerosis involving both upper and lower limbs, patients with hematological diseases whose bleeding risk will worsen in future.

In some clinical situations, the principles may conflict with each other. Clinicians have to weigh the importance of these principles based on individual patient's specific conditions. Some examples are discussed below.

A 63-year-old petite lady, with tunnelled CVC inserted via the right internal jugular vein IJV eight months ago, attends your clinic for vascular access creation. Besides diabetes and hypertension, she also suffers from depression. She has marginal sized left forearm cephalic vein clinically as well as on ultrasound study (average 2.2 mm). Antecubital vein over left elbow is prominent and measures 2.8 mm on ultrasound. Brachial, radial and ulnar pulses are all palpable. Diameter of her radial artery measures 1.8mm on ultrasound study. A small radial artery together with a marginal sized vein, the chance of maturation failure of a left RC AVF is high. To avoid prolonging the tunnelled CVC duration and inducing disappointment, a more reasonable approach would be to use the more proximal antecubital vein for BC AVF creation as the first attempt.

A 57-year-old morbidly obese lady (Fig. 1) with history of diabetes, sleep apnoea and heart failure (ejection fraction 35%) had tunnelled CVC via right IJV two months ago and then blocked one week ago required change of the catheter to the right femoral vein. On examination, left forearm cephalic vein is not visible, antecubital vein is palpable and of reasonable size over the cubital fossa. Ultrasound study confirmed the left forearm cephalic vein is small (average 1.5 mm). The left arm cephalic vein size averaged 2.9 mm but the subcutaneous fat above the arm cephalic vein measured 15–20 mm. All the pulses over left upper limb are palpable and strong.

The option for this lady would either be left BC AVF and a subsequent superficialization procedure of the cephalic vein fistula, or a forearm loop BC AVG. There are pros and cons for either strategy. Left BC AVF and superficialization bears a higher risk of wound complication and subsequent difficulty of cannulation. It will take a longer time to mature for cannulation and the tunnelled CVC blockage might recur again. The superficialization operation will likely require general anesthesia although one may try regional block first.

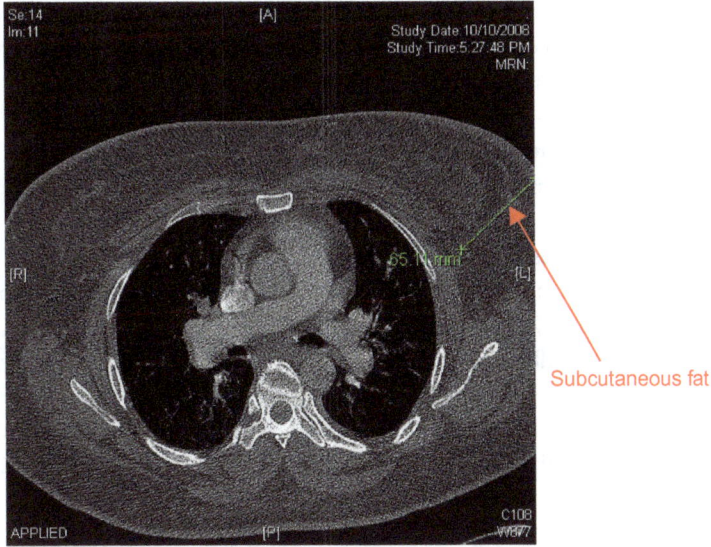

Fig. 1. A cross section of patient's thoracic computed tomography done for evaluation of pulmonary pathology few months before the referral for hemodialysis access.

On the other hand, forearm loop BC AVG enables earlier and easier needling, the patency may be shorter than BC AVF. The AVG creation procedure can be performed under general anesthesia or brachial plexus block even though a good regional block is going to be challenging. The subsequent septic complications of AVG would also be higher. There is no absolutely right or wrong strategy. A thorough discussion with the patient about the benefits and risks of both strategies should be conducted to understand the patient's preference and concern. If the patient has no strong preference over either strategy, I would prefer to proceed with a forearm loop BC AVG, making the anastomosis of both artery and vein below the elbow skin crease. By this, we provide a secure vascular access early for the patient. In the future, if the loop BC AVG fails and cannot be salvaged, there is a chance that the cephalic vein over the arm would have pumped up. As such, be ready to perform a single stage BC AVF and superficialization.

References

1. United State Renal Data System. *International Comparisons*, Atlas of ESRD.
2. Fresenius Medical Care. *ESRD Patients in 2011: A Global Perspective*.
3. Sridharan S, Berdeprado J, Vilar E, *et al.* A self-report comorbidity questionnaire for haemo dialysis patients. *BMC Nephrol.* 2014; **15**: 134.
4. Farrokhi F, Abedi N, Beyene J, *et al.* Association between depression and mortality in patients receiving long-term dialysis: A systematic review and meta-analysis. *Am J Kidney Dis.* 2014; **63**(4): 623–635.

5. Berdud I, Arenas MD, Bernat A, *et al.* Appendix to dialysis centre guidelines: recommendations for the relationship between outpatient haemodialysis centres and reference hospitals. Opinions from the Outpatient Dialysis Group. Grupo de Trabajo de Hemodiálisis Extrahospitalaria. *Nefrologia.* 2011; **31**(6): 664–669.
6. National Kidney Foundation. KDOQI clinical practice guidelines and clinical practice recommendationsfor 2006 updates: hemodialysis adequacy, peritoneal dialysis adequacy and vascular access. *Am J Kidney Dis.* 2006; **48** (Suppl 1): S1–S322.
7. Sidawy AN, Spergel LM, Besarab A, *et al.* The Society for Vascular Surgery: Clinical practice guidelines for the surgical placement and maintenance of arteriovenous hemodialysis access. *J Vasc Surg.* 2008; **48**(suppl):2S–25S.
8. Dhingra RK, Young EW, Hulbert-Shearon TE, *et al.* Type of vascular access and mortality in US hemodialysis patients. *Kidney Int.* 2001; **60**:1443–1451.
9. Miller CD, Robbin ML, Barker J, *et al.* Comparison of arteriovenous grafts in the thigh and upper extremities in hemodialysis patients. *J Am Soc Nephrol.* 2003; **14**(11): 2942.

CHAPTER 2

Algorithm of Assessment and Planning of Hemodialysis Access Creation in Challenging Conditions

Jackie P. Ho

Please remember the three essential components of a hemodialysis access:

⇨ Good inflow
⇨ Good outflow
⇨ Good conduit

Evaluation for First Vascular Access Creation

History:

- Cause of renal failure (e.g. diabetic nephropathy may associate with peripheral arterial disease, lupus nephropathy may associate with thrombophilic tendency etc.);
- Pre-emptive (current creatinine and glomerular filtration rate GFR) or already started hemodialysis with tunnelled CVC (duration of tunnelled CVC and any problem associated with it);
- Any significant medical condition (malignancy, symptomatic ischemic heart disease, stroke that causes limb weakness etc.);
- Which side is the dominant hand (preferably create the vascular access over the non-dominant hand to minimize inconvenience for patient during hemodialysis).

Physical examination:

- Skin condition;
- Position and condition of tunnelled CVC if any;
- Visibility, size, texture of cephalic vein of the target limb with tourniquet applied;
- Radial, ulnar, brachial pulses and Allen's test;
- Brachial blood pressure of both upper limbs (if it feels weak over the target limb).

Duplex ultrasound vein and artery study:

- As part of the assessment for the decision of hemodialysis access creation;[1]
- Either performed by the attending clinician who is going to perform the vascular access creation, or by a specialized vascular sonographer.

The Duplex ultrasound study of the limb for vascular access creation should cover both arterial and venous systems. Arterial system evaluates the size, flow velocity, waveform, any anatomical anomaly and degree of calcification of the subclavian, brachial, radial and ulnar artery. Venous assessment evaluates the size and patency of cephalic, basilic and axillary vein. For venous assessment, the position of patient and the environmental condition have to be standardized[2,3] because of the possibility of huge variation in vein size with patient position, tourniquet application, room temperature and anxiety of patient. In our institution, we have the patient sit in a semi-recumbent position (about 60°) without tourniquet, keeping the room temperature 24 °C or above.

The ideal situation is that a specialized vascular sonographer performs the arterial and venous assessment for the patient,[4] while working closely with the access creation clinician so that the anatomical features relevant to surgical procedures are understood well. The sonographer not only routinely follows a scanning protocol but is also able to pick up conditions that might affect the access creation surgery. For example: circumferential heavy calcification of radial artery near the usual RC AVF anastomosis site, remarks on focal area of superficial vein significant stenosis even though the average vein size is quite satisfactory. Another way is that the operating clinician performs the Duplex ultrasound study for the patient during the clinical assessment and draws down the plan, provided the clinician has proper training on vascular ultrasound and adequate clinic time is assigned for each patient. Alternatively, the vascular sonographer will scan patients using a standard protocol and the operating clinician will perform a quick scan again to look for specific conditions in the operating theatre just before the surgery.

If there is a long period of tunnelled CVC inserted over the limb where the vascular access is to be created, additional central vein studies like central venogram, CT or MR venogram might be warranted as pre-operative investigation.

Evaluation for Subsequent Vascular Access Creation

In patients with previous failed single or multiple vascular access(es), or those patients running out of access option, a more detailed assessment is required in addition to the basic evaluation discussed above. As mentioned at the beginning, we would like to know:

(1) Where is an ideal location for a good inflow?
(2) Is there any good superficial native vein or does one need to use a synthetic graft as conduit?
(3) Is there a good outflow vein that drains into the right atrium?

In addition, a new access may not be immediately available for patients with complicated access situation. We need to find out if the patient currently has a temporary tunnelled catheter *in situ* or we need to decide where to place the temporary dialysis access. So the additional question is:

(4) Where should the temporary hemodialysis access be placed which will cause minimal interference for future vascular access? If a temporary access is already *in situ*, where should the next access be? Do we need to change the temporary access site?

I would suggest the following steps in evaluation and planning:

Step 1: Thorough evaluation of patient's vascular history relating to hemodialysis access is an essential part. This will be more challenging if the patient's previous vascular access procedures were done in other countries, institutions or if the handling clinicians have already left the facility and proper records are not available. This information include:

- all previous access procedures;
- any specific difficulty encountered in previous procedures, e.g. high bifurcation of radial and ulnar artery, small sized brachial artery, steal syndrome etc.
- site and duration of previous tunnelled CVC to evaluate risk of central vein stenosis;
- previous investigations for central vein status;
- previous history of central vein stenosis required intervention;
- review any previous fistulogram study to understand the vessel condition.

I suggest to use a pictorial form to help the evaluation process (Fig. 1).

Step 2: Detailed clinical and ultrasound study[5] of the arterial and venous system of the limbs by the operating clinician for possible next vascular access. Do not take the old assessment result for granted. Patient's arterial and venous condition may change with time. It is advisable to repeat the clinical and ultrasound study thoroughly if the previous study was performed more than 4–6 months ago. If lower limb vascular access is planned, all lower limb pulses and any abnormal ankle edema should be thoroughly assessed for presence of peripheral arterial disease and deep vein problem respectively.

Step 3: Evaluate patient's medical risk for any surgical procedure, regional or general anesthesia. Some specific health issue may also significantly affect the decision of access creation. For example, patients with implantable pacemaker (and defibrillator). Usually, it is preferable not to place the vascular access on the same side of pacemaker (or defibrillator) to minimize the risk of any infection of vascular access spreading to involve pacemaker wire. Moreover, pacemaker wire situated for a long time in the central vein may cause central vein obstruction.

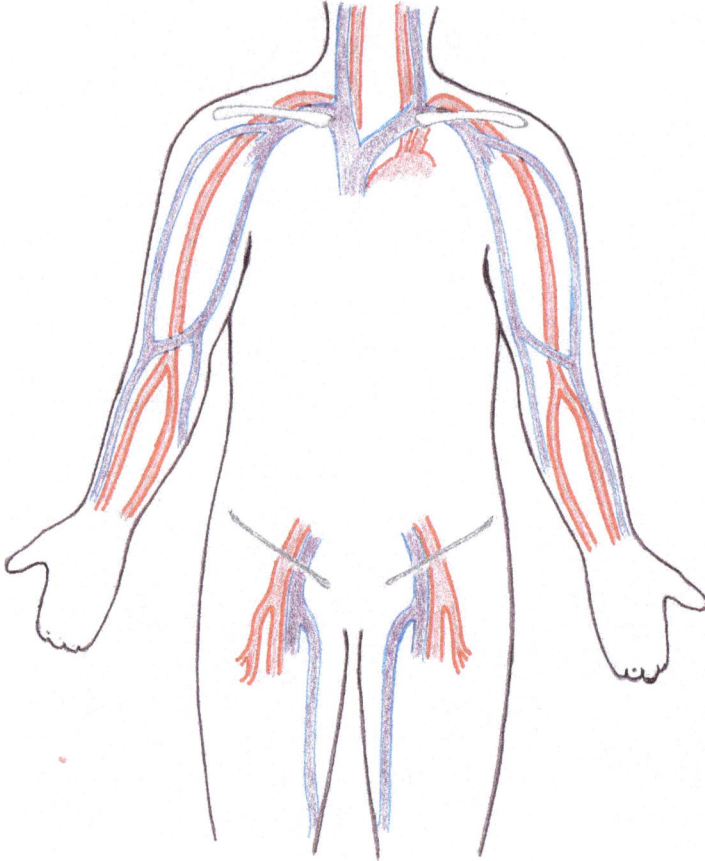

Fig. 1.   Pictorial form for evaluation of hemodialysis access creation.

Step 4: Further investigation if necessary. If central vein obstruction is known or suspected, one can either choose the side unlikely to have obstruction or perform central vein imaging and intervention prior to access creation. I prefer conventional venogram for diagnosis and proceed to therapeutic intervention if obstruction is identified before or in the same session of the vascular access surgery (depending on facility and expertise availability). Occassionally, an arterial angioplasty may be needed to improve the arterial inflow before a vascular access creation to avoid steal syndrome or to facilitate maturation.

Although the basic principle is fistula first, AVG does offer benefit for patients[6] with specific conditions especially for those already on tunnelled CVC hemodialysis for long duration. The maturation time for AVG is much shorter than AVF and does not require stage procedures at all body locations. Deciding between AVF and AVG, the main questions to consider and to address are:

- Medical conditions and life expectancy of patient;
- Duration of CVC and problems related to CVC;

- Presence or potential presence of central vein obstruction;
- Risk of steal syndrome;
- Track history of patient's previous AVF and AVG;
- Patient's preference and expectation.

Sometimes there is more than one option of vascular access and each has its advantages and risks. Thorough discussion with the patient on all options to reach a mutual access plan is the way to ensure patient's satisfaction.

Vascular access clinicians have to remember that in occasional situations, tunnelled CVC or conversion to peritoneal dialysis could be a better solution. For example, patients with very poor life expectancy like disseminated malignancy, severe thrombocytopenia or coagulopathy with high bleeding risk, etc.

Case 1

Mr E, 60 years old, had a stroke event resulting in right upper limb weakness 10 years ago. He also had DM nephropathy and ESRF started hemodialysis six years ago. He had used his right RC AVF for 5.5 years which was blocked recently during his admission for a severe pneumonia. A tunnelled CVC was inserted via his right IJV during the hospital stay. He was referred to my clinic after discharge.

During the clinic consultation, Mr E expressed strong reluctance for any vascular access created over his left upper limb (functional limb) and lower limb. The right brachial, radial and ulnar pulses were palpable and strong. There was a thrombosed aneurysmal fistula near the RC anastomosis and the forearm cephalic vein was thrombosed. Clinically, the right arm cephalic vein was not visible or palpable. A visible prominent basilic vein running over the medial-posterior side of the right forearm was noted. Ultrasound study showed small sized right arm cephalic vein (average <1.5 mm), while the basilic vein over the forearm measured 2.8–3.0 mm diameter and the arm basilic vein about 3.5–4.0 mm. The right IJV tunnelled CVC was *in situ* for four weeks.

Three options were discussed with Mr E:

1. Right BB AVF over the elbow level followed by second stage BBT, or single stage BB AVF and BBT;
2. Mobilize the forearm basilic vein to form a loop and connect to distal brachial artery;
3. Right forearm loop BB AVG.

Since the patient is relatively young, native vein AVF is preferred over AVG due to its better patency and lower infective complication. On the other hand, there is also a worry that the longer maturation time of AVF will prolong the tunnelled CVC time and increase the chance of central vein obstruction in the future.

Among the two options of standard arm BB AVF and forearm loop BB AVF, the success rate of arm BB AVF is higher as the vein size is bigger. However, the forearm

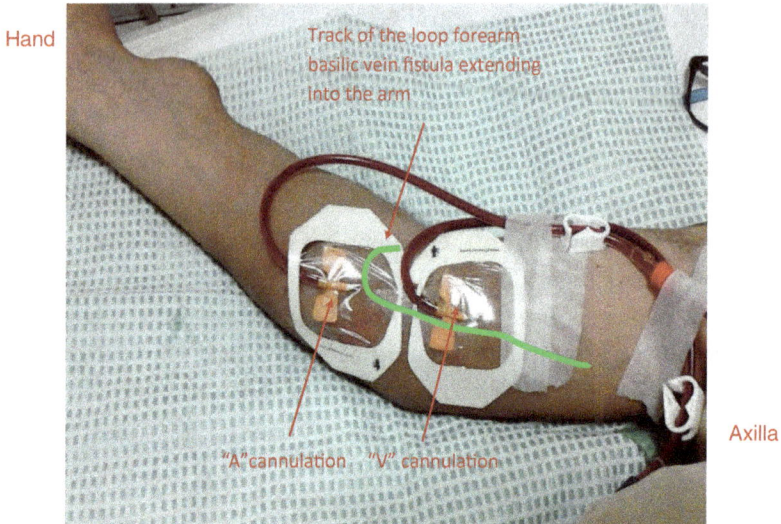

Fig. 2. Clinical photo of the positioning of the right upper limb and the two cannulation sites over Mr E's right forearm loop BB AVF.

basilic vein will be wasted. Whereas if we use the forearm basilic vein first, the arm basilic vein could be preserved for future BB AVF and BBT when the forearm basilic vein fails.

Eventually, the consensus decision was to mobilize the right forearm basilic vein and form a loop to connect to the distal brachial artery (forearm BB AVF). As the basilic vein was far medial, a long segment was mobilized out over a long length over the mid- and proximal forearm to place it into a more lateral position for ease of cannulation in the future. After the AVF creation, good thrill was felt over the loop forearm basilic vein as well as the distal arm. Ten weeks after AVF creation, the diameter of the forearm loop basilic vein ranged between 5.3 mm and 6.2 mm. The basilic vein over the elbow region was 6 mm. Mr E was brought back to the hospital dialysis unit for initial cannulations. Due to the location of the forearm loop BB AVF, Mr E needed to abduct and externally rotate the shoulder joint to facilitate the cannulation (Fig. 2). Subsequently, his cannulation was transferred back to the community hemodialysis center. Difficulty was encountered and he was required to return twice to the hospital dialysis center for cannulation. Eventually, the cannulation was successful and his right IJV tunnelled catheter was removed four months after the access creation.

Case 2

Mr T, 59 years old, DM, HT, diabetic nephropathy. Renal failure since November 2008. He was admitted with thrombosed right arm brachial artery to proximal basilic vein AVG in October 2013.

Hemodialysis access history as follows (Fig. 3):

Nov 08 — Tunnelled CVC through right IJV.

Jun 09 — Left BC AVF created.

Aug 09 — Left BC AVF matured and was used. Tunnelled CVC was then removed.

Jun 11, Feb 11, Aug 11 — Stenosis of the left BC AVF required fistuloplasty.

Sep 12 — Thrombosis of left BC AVF. Right IJV was found to be thrombosed and a tunnelled CVC was inserted into the left IJV.

Oct 12 — Central venogram showed total occlusion of segment of right brachiocephalic vein. Angioplasty and stenting of the right brachiocephalic vein were performed, followed immediately with right BC AVF creation.

Dec 12 — Poor maturation of right BC AVF. Fistulogram performed showed long segment stenosis with balloon angioplasty performed.

Feb 13 — Right BC AVF failed to mature, with long segment restenosis detected. Right upper limb basilic vein size was small over the elbow region and more reasonable over the proximal arm (3.8 mm). Proceeded to right brachial artery to proximal arm basilic vein AVG. Right central venogram performed showed patent in-stent right brachiocephalic vein. Right arm BB AVG starts needling two weeks after operation. Tunnelled catheter was removed.

Sept 13 — Acute thrombosis of right BB AVG. Graft thrombectomy was performed and fistulogram showed stenosis present over vein-graft anastomosis. Balloon angioplasty was performed. Patency was restored.

Sept 13 — One week later, readmitted for acute thrombosis of the right BB AVG again. Graft thrombectomy and fistulogram performed showed re-stenosis over the same site. Surgical revision of the vein-graft anastomosis with patch angioplasty performed.

Oct 13 — Admitted again for right BB AVG thrombosis.

Clinical examination showed good left brachial, radial and ulnar pulses.

Ultrasound study showed left brachial artery 3 mm diameter over elbow region, left brachial vein 3.2 mm, basilic vein only 2.5 mm over elbow, 3.8 mm over upper arm.

Need to rule out left central vein stenosis with previous five months' history of tunnelled catheter.

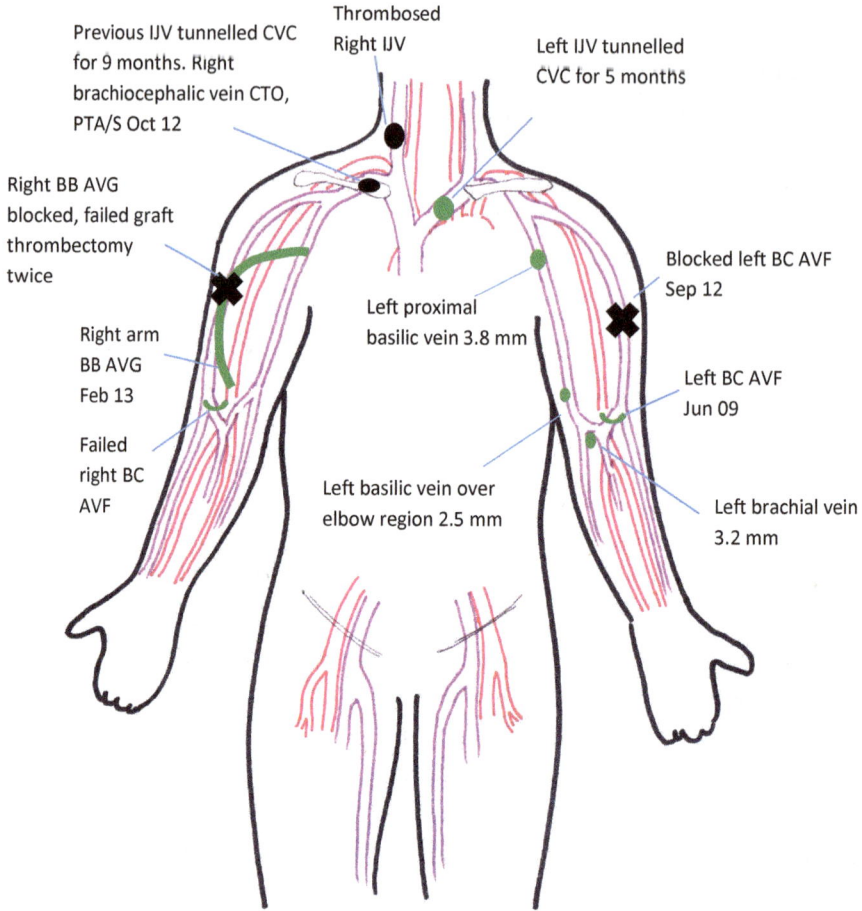

Fig. 3.   Pictorial illustration of Mr T's vascular conditions relating to hemodialysis access.

Where will the next vascular access be?

Right BA AVG is not a good option as the left brachiocephalic vein stenosis might narrow down again and cause trouble. Furthermore, surgical dissection is difficult over right elbow region because of previous multiple surgeries.

Patient is relatively young and mobile and he does not prefer a vascular access over the thigh or groin region.

For the left upper limb, there are a few options of vascular access: Left BB AVF followed by BBT, forearm loop AVG either implanting on brachial vein or basilic vein as outflow, left arm brachial artery to proximal basilic vein AVG.

Using native vein vascular access BB AVF and BBT would provide a better patency if successfully matured. The downside is the long maturation process for BB AVF and BBT, which takes at least 3–4 months. Patient has right central vein stenosis and history of left tunnelled CVC for five months, it may not be desirable to keep tunnelled catheter for another 3–4 months over the left central vein.

Left forearm loop AVG is preferred so that when this forearm AVG fails in the future, the patient still has the arm basilic vein fistula option. The elbow basilic vein is 2.5 mm, which is marginal for AVG creation. Based on the patient's previous history of frequent outflow vein stenosis, the patency of using elbow basilic vein as outflow might be rather short. Left brachial vein is of better size and more favourable as outflow vein for AVG. However, it might induce more hand and forearm swelling when a graft is implanted into a deep vein. Left arm brachial artery to proximal basilic vein AVG will take away the potential future BB AVF and BBT option from a relatively young patient.

Where to put the temporary hemodialysis access?

Right IJV is blocked and the planned next access is over the left upper limb. One can still use the left IJV for CVC provided the new vascular access can be used rapidly. Alternatively, the tunnelled CVC could be moved to femoral vein and enable proper interrogation of the left central vein before new access creation. In this patient, the tunnelled CVC was moved from left IJV to his right femoral vein.

Outcome

Left side central venogram performed in the operating theatre showed absence of central vein obstruction. Left forearm loop brachial artery to brachial vein AVG was implanted in the same session (Fig. 4). After the operation, left hand and forearm swelling were obvious. Elastic bandage was applied to left hand and forearm. The swelling subsided two weeks later and cannulation commenced. The femoral tunnelled catheter was removed one month post-operation.

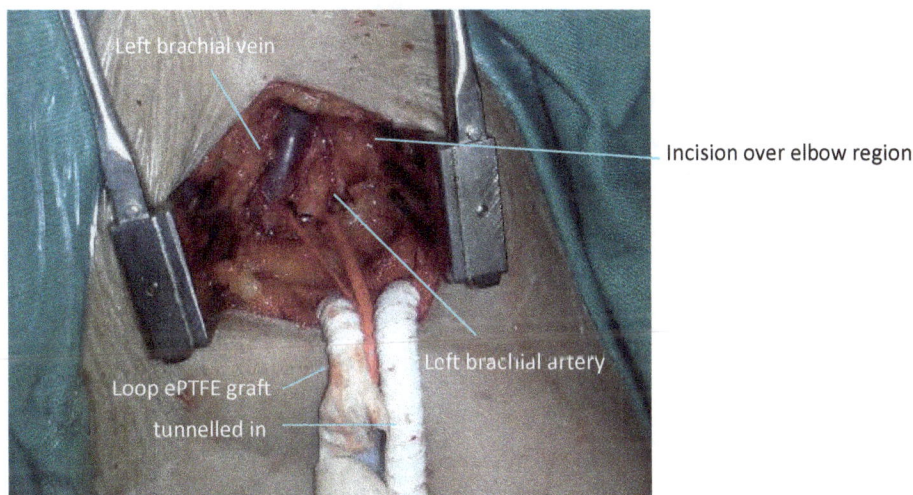

Fig. 4.   Clinical photo of left brachial artery and vein for AVG creation.

Case 3

Mr G, 71 years old, DM, HT, IHD, diabetic nephropathy with ESRF developed two years ago. His previous vascular access was created and managed in another country and no record was available. Based on patient's memory, he had left IJV tunnelled catheter for a few months since commencement of hemodialysis. He was told his superficial veins of both upper limbs were small. A right forearm loop AVG was created two years ago. Subsequently the right forearm AVG thrombosed and a left forearm loop AVG was created a year ago. Fistuloplasty for the left AVG was performed twice during the last year in his home country. He was then being admitted because of blocked left forearm AVG. The patient was dominant in the right hand. He had no fluid overload feature and the serum potassium was 4.9 mmol/L.

Clinical examination showed good brachial, radial and ulnar pulses of the left upper limb. Focused ultrasound study showed that the left forearm loop AVG was connected

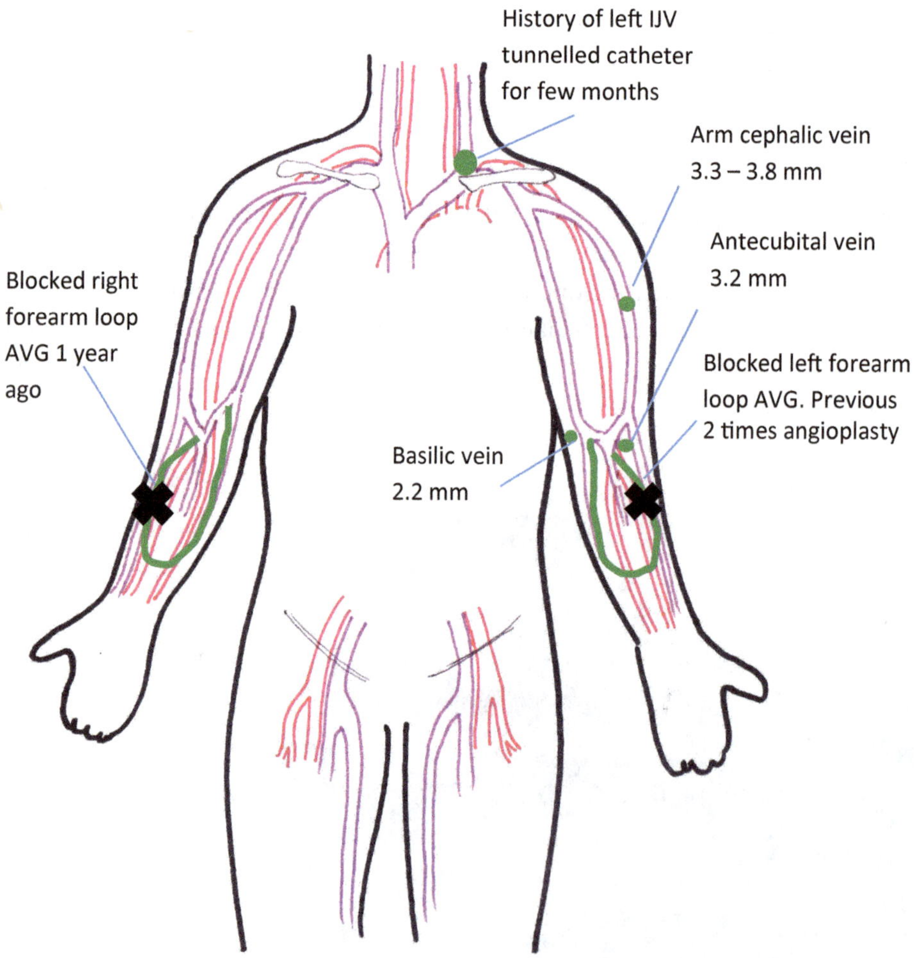

History of left IJV tunnelled catheter for few months

Arm cephalic vein 3.3 – 3.8 mm

Antecubital vein 3.2 mm

Blocked right forearm loop AVG 1 year ago

Blocked left forearm loop AVG. Previous 2 times angioplasty

Basilic vein 2.2 mm

Fig. 5.   Pictorial illustration of Mr G's vascular conditions relating to hemodialysis access.

between the brachial artery and brachial vein. Over the elbow region, the median cephalic vein was 3.2 mm and basilic vein was 2.2 mm. The left arm cephalic vein connecting to the antecubital vein was ≥3.3 mm throughout its length. The distance between left brachial artery and antecubital vein was about 3 cm just above cubital fossa skin crease and about 2 cm below the skin crease. The right arm cephalic and basilic vein size were marginal around 2 mm (Fig. 5).

Management options

The viable options for this patient would be:

1. Emergency left forearm AVG thrombectomy, fistulogram and fistuloplasty;
2. Temporary tunnelled right IJV CVC and left BC (brachial artery to antecubital vein) AVF creation;
3. Emergency left forearm AVG thrombectomy together with pre-emptive left BC AVF creation.

Emergency graft thrombectomy may salvage the current graft and avoid a tunnelled catheter for the patient. However, in view of frequent development of stenosis over a short period of time and lack of previous fistuloplasty information, there may be very resistant lesion(s) present along the graft or outflow vein. Thus, the risk of re-thrombosis after salvage thrombectomy and fistuloplasty is high. Patient is also likely to require frequent repeated interventions to maintain the graft patency in the future after successful salvage.

The second option of creation of left BC (brachial artery to antecubital vein), AVF might provide a more durable vascular access as the size of antecubital and arm cephalic vein was good. The downside is that the patient will require a temporary tunnelled CVC for 6–8 weeks. Another challenge in Mr G's condition is the presence of surgical scar and thrombosed graft around the elbow region. The thrombosed graft is situated between the brachial artery and the antecubital vein. The distance between the antecubital vein and brachial artery is also wide.

For the third option, the major risk will be steal syndrome if both the loop forearm AVG and the newly created BC AVF flow well. There is no absolute right or wrong option. Most of the time, with thorough counselling, the patient will be able to choose the better option for himself/herself. In this case, the pros and cons of all management options were explained and Mr G preferred to have a new vascular access (left BC AVF) creation.

Outcome

Right IJV tunnelled catheter was inserted. He then underwent left BC AVF under brachial plexus block. The antecubital vein distal to the cubital fossa skin crease was dissected and the thrombosed graft in between the brachial artery and antecubital vein was divided and ligated to facilitate AV anastomosis. Adhesion was encountered and dissected. Left BC AVF was created. The thrill was good at six weeks post-operation and the cephalic vein fistula reached 6 mm. The access was successfully used for needling and the tunnelled catheter was removed two months after insertion.

Case 4

Madam N, 66 years old, DM, HT, IHD, diabetic nephropathy with ESRF developed in 2004. She practiced peritoneal dialysis between 2004 and 2006, and suffered complications of peritonitis thrice. She was converted to hemodialysis in 2006. Her vascular access was initially managed in another hospital.

Jun 06 — Left BC AVF, failed to mature. Hemodialysis via right IJV tunnelled CVC.

Nov 06 — Left forearm loop BB AVG created but developed steal syndrome, which required ligation of the AVG.

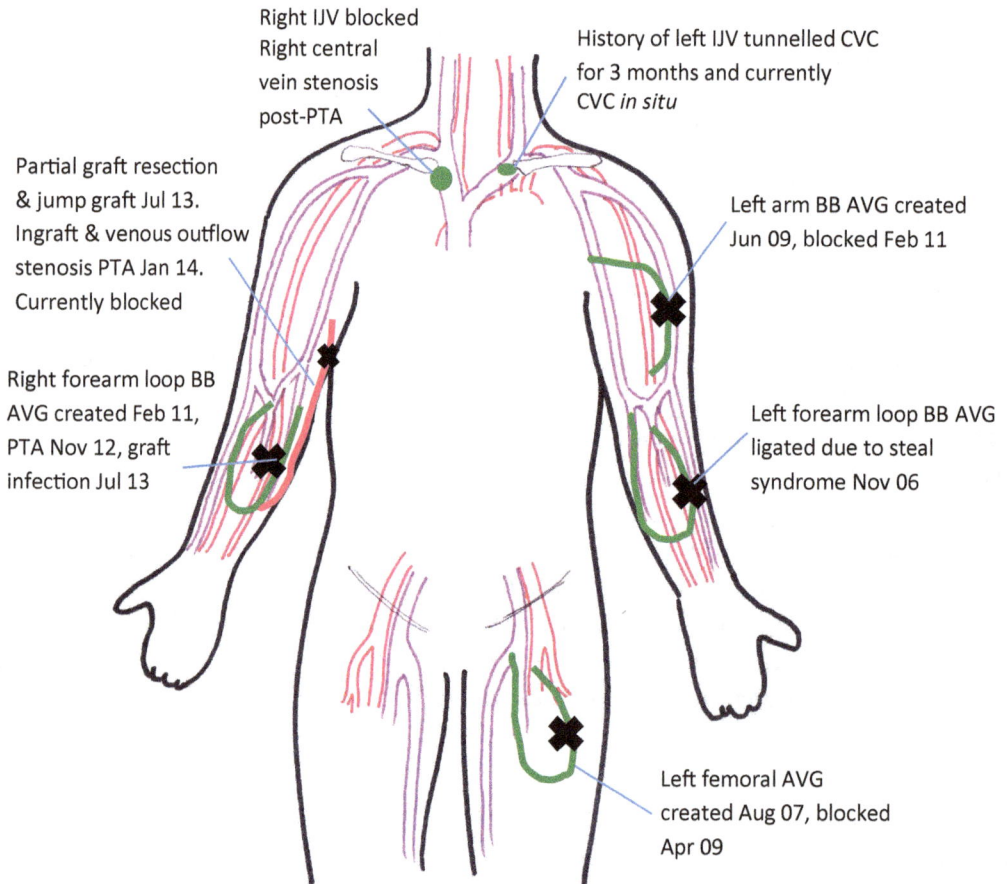

Fig. 6.   Pictorial illustration of Madam M's vascular conditions relating to hemodialysis access.

Aug 07 — Left thigh femoral artery to saphenous vein AVG. IJV tunnelled CVC was removed one month later.

Apr 09 — Repeated AVG stenosis required fistuloplasty twice and eventually blocked. Right IJV was found to be occluded. Tunnelled CVC was inserted via left IJV.

Jun 09 — Left arm BB AVG (proximal basilic) created. Left IJV CVC was removed one month later.

Feb 11 — Left arm BB AVG was blocked and failed salvage procedure (graft thrombectomy and fistuloplasty). Right forearm BB AVG was created.

Oct 12 — Referred to my clinic for raised venous pressure.

Nov 12 — Fistuloplasty for right BB AVG vein-graft anastomotic stenosis.

Jul 13 — Admitted to another hospital for infected AVG and abscess. Partial graft resection and jump graft to mid-arm basilic vein was performed.

Jan 14 — Back to my clinic. VP raised to 180 mmHg, fistulogram showed high grade in-graft stenosis (70%), vein-graft anastomosis stenosis (90%) and central vein stenosis (90%) between subclavian to brachiocephalic vein junction. Balloon angioplasty was performed to all lesions. Residual stenosis in central vein was about 30%.

Apr 14 — Admitted with blocked upper limb AVG. There was mild redness over the right upper limb AVG. Patient's white cell count was 10.7×10^9/L. She was in fluid overload on admission. Temporary dialysis catheter insertion was attempted by the on-call nephrologist trainee through right femoral vein but failed. A temporary CVC was inserted through her left IJV (Fig. 6).

Clinical examination showed good bilateral brachial and radial pulses. Bilateral ulnar pulses were not palpable. Bilateral femoral and popliteal pulses were good but both side ankle pulses were absent. Mild swelling and bruising was noted over right groin. Left IJV temporary CVC was *in situ*. Ultrasound study showed patent bilateral axillary vein size about 1.2 cm. Right proximal basilic vein size was 6 mm. Bilateral arm cephalic veins were small in size (<1.5 mm).

Management options

With a history of graft infection and presence of redness over her right upper limb AVG, leucocytosis, possibility of low grade infection of the AVG is high. The risk of progression of graft infection after emergency graft thrombectomy is substantial. Together with known history of right central vein obstruction and residual stenosis after angioplasty, emergency graft thrombectomy is not a good choice of management.

In view of absent lower limb distal pulses and multiple cardiovascular risk factors, right femoral AVG carries a non-negligible risk of lower limb steal syndrome. Therefore this is also not a good option.

There was no option of native vein AVF for the upper limbs.

A new AVG can be created between brachial artery and axillary vein on both sides. However, the right side had a confirmed central vein obstruction issue and thrombosed graft with possibility of harbouring infective pathogen. On the left side, with history of tunnelled CVC for three months, a small risk of central vein obstruction existed and the temporary catheter was on the same side occupying the surgical field for BA AVG creation.

Outcome

The initial management for this lady was antibiotic treatment for the potential right AVG infection. The tunnelled catheter was transferred to right femoral vein under ultrasound and fluoroscopy guidance in the intervention suite. The redness over the blocked AVG subsided (over one week) with antibiotic therapy and her white cell count and C-reactive protein returned to normal level. Left central vein venogram and creation of new vascular access were then carried out in the same session. Inside the operating theatre, central vein venogram confirmed absence of stenosis over the left side. There were heavy scarring and multiple thrombosed AVG *in situ* around the left cubital fossa region. Together with history of steal syndrome after the forearm loop AVG in year 2006, the proximal brachial artery was used for inflow anastomosis with the graft. A 6 mm loop left upper arm AVG using a ring supported ePTFE graft was created. No steal syndrome was detected after the AVG creation. The first cannulation (Figs. 7 and 8) of the graft was performed as in-patient at day 7 post-operation. Subsequently, the community dialysis centre took five weeks to get use to her new AVG. Her right femoral tunnelled CVC was removed two months after the new AVG creation.

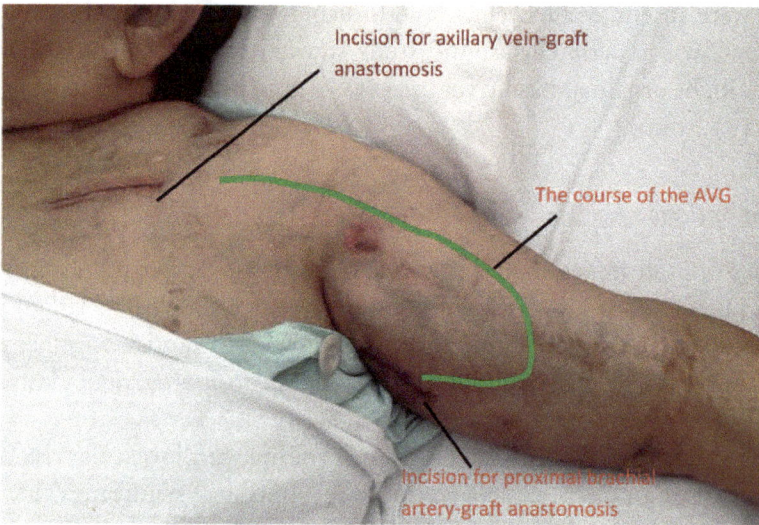

Fig. 7.   Clinical photo of patient one week after arm proximal brachial artery to axillary vein AVG creation.

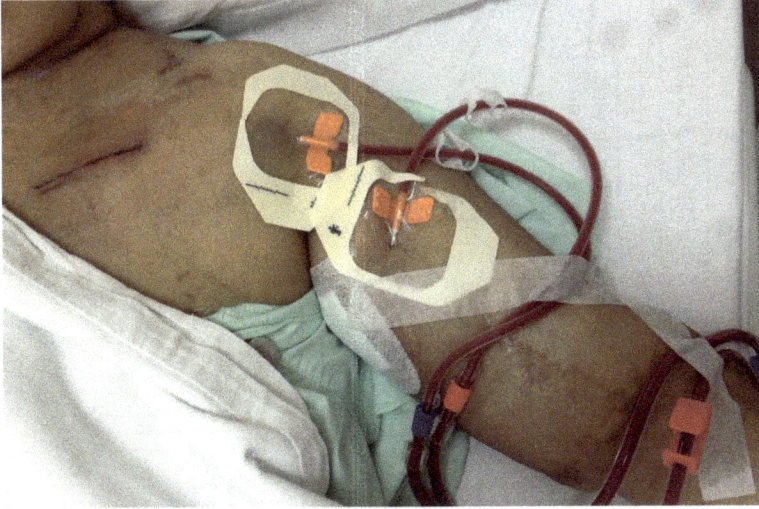

Fig. 8. Cannulation of left arm loop BA AVG.

Please also see Chapter 14 Case 3 for another illustration.

Figure 9 and Table 1 shows the strategies to create pre-emptive AVF.

Strategy for Creation of Pre-emptive AVF

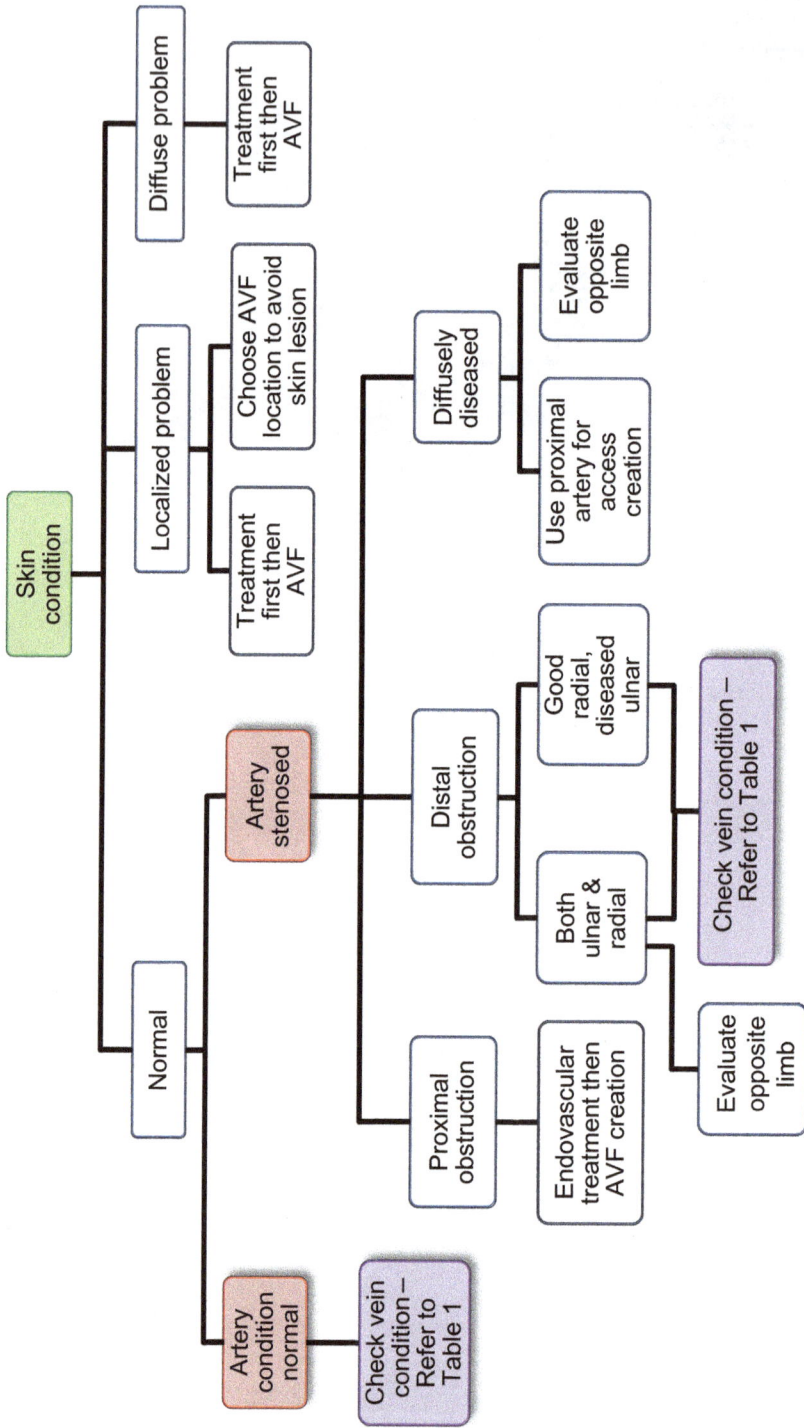

Fig. 9.   Pre-emptive AVF strategy.

Table 1.   Preferred pre-emptive AVF with different arterial and venous condition of upper limb.

Artery condition	Vein condition					
	Good distal forearm cephalic	Good mid forearm cephalic	Good cubital fossa cephalic, median cephalic	Good basilic	Good forearm basilic	No good superficial vein
Normal	RC AVF	PRA AVF	BC AVF	BB AVF + BBT	Loop forearm basilic vein to radial artery AVF	Brachial artery brachial vein AVF or PRA[2] radial vein AVF
Good radial, diseased ulnar	CO_2 PTA UA[1] then RC AVF	CO_2 PTA UA[1] then RC AVF, or Cephalic vein loop to proximal artery[3]	BC AVF	BB AVF + BBT	Loop forearm basilic vein to proximal UA[1]	Consider opposite upper limb
Both radial and ulnar diseased	Cephalic vein loop to proximal artery[3] or consider opposite upper limb	Cephalic vein loop to proximal artery[3] or consider opposite upper limb	CO_2 PTA ulnar/radial artery then BC AVF	Basilic vein loop to proximal artery[3]	CO_2 PTA ulnar/radial artery then forearm loop basilic vein AVF	Consider opposite upper limb

[1] UA — Ulnar artery.

[2] PRA — Proximal radial artery.

[3] Proximal Artery — Axillary artery or proximal brachial artery.

References

1. Huber TS, Ozaki CK, Flynn TC, *et al.* Prospective validation of an algorithm to maximize native arteriovenous fistulae for chronic hemodialysis access. *J Vasc Surg.* 2002; **36**(3):452–459.
2. van Bemmelen PS, Kelly P, Blebea J. Improvement in the visualization of superficial arm veins being evaluated for access and bypass. *J Vasc Surg.* 2005; **42**(5): 957–962.
3. Korten E, Spronk S, Hoedt MT, *et al.* Distensibility of forearm veins in haemodialysis patients on duplex ultrasound testing using three provocation methods. *Eur J Vasc Endovasc Surg.* 2009; **38**(3): 375–380.
4. Allon M, Lockhart ME, Lilly RZ, *et al.* Effect of preoperative sonographic mapping on vascular access outcomes in hemodialysis patients. *Kidney Int.* 2001; **60**(5): 2013–2020.
5. Parmley MC, Broughan TA, Jennings WC. Vascular ultrasonography prior to dialysis access surgery. *Am J Surg.* 2002; **184**(6): 568–572, discussion 572.
6. Sgroi M, Patel MS, Wilson SE, *et al.* The optimal initial choice for permanent arteriovenous hemodialysis access. *J Vasc Surg.* 2013; **58**(2): 539–548.

CHAPTER 3

Creation of Vascular Access

Jackie P. Ho

Surgically created vascular accesses can be broadly classified as AVF (arterialize the native superficial vein as fistula/conduit) and AVG (artificial conduit to connect between an artery and an outflow vein). This chapter will not be able to cover all aspects of vascular access creation as the knowledge and skill involved are tremendous. We aim to illustrate the basic principle, commonly used access and stress the practical tips and tricks for successful vascular access creation.

As mentioned in Chapter 1, the basic requirements to get a functional vascular access are:

⇨ **Good inflow** — arterial supply
⇨ **Good outflow** — outflow vein, deep vein and central vein
⇨ **Good conduit** — native superficial vein or synthetic graft

Besides the above three elements, we also have to take care of the practical aspects of the vascular accesses,[1] thus repeated cannulation is easy and safe (Rules of 6s). The requirements of AVF and AVG are:

⇨ **Good luminal size** — usually 6 mm is the required minimal luminal diameter for easy cannulation
⇨ **Superficial** — Prefer <6 mm below skin level
⇨ **Straight pathway** — for easy cannulation
⇨ **Enough length for insertion of two needles** — a minimal 10 cm relatively straight segment or ≥2 separated straight segments of 4 cm each or longer

Many a time, an AVF or AVG may flow well but still cannot be used because it is tortuous, deep in the subcutaneous tissue or variation in depth in the subcutaneous tissue. Clinicians have to bear all these in mind during the pre-operation planning, access creation surgery and consideration for adjunct procedure to assist maturation if necessary.

Common Types of AVF

Radio-cephalic (RC) AVF

RC AVF (Cimino-Brescia) is the preferred first option of vascular access if conditions allow.

Location: RC AVF can be created over the (1) anatomical snuff box, (2) around wrist region or (3) distal to mid-forearm (Fig. 1) depending on several factors:

- Radial artery calibre, wall texture and flow;
- Cephalic vein size;
- Distance between the cephalic vein and the radial artery.

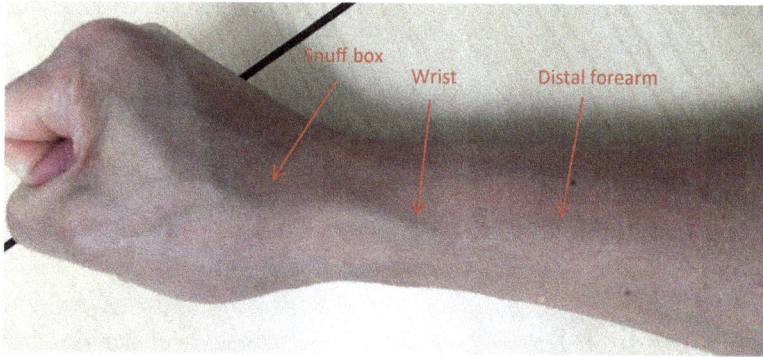

Fig. 1.   Different locations for RC AVF creation.

The more distal the RC anastomosis, the longer the cephalic vein fistula available in the forearm for needling. Using snuff box RC also has the advantage of preserving an option of wrist secondary RC AVF in the future if snuff box anastomosis narrows down or becomes blocked. However, both calibre of radial artery and cephalic vein become smaller with more distal locations (snuff box use dorsal branch of radial artery). Therefore only a small proportion of ESRF patients have good sized radial artery and cephalic vein able to go for snuff box RC AVF. Although it is not an absolute relation, radial artery is usually larger in men and associated with body height. It is rare to find a petite female patient suitable for snuffbox RC AVF.

Radial artery just proximal (2–4 cm) to the wrist joint is superficial and easy for surgical dissection. It becomes deeper and covered by the brachioradialis muscle and tendon further proximal in the forearm. Thus, the area just proximal to the wrist is a favorable location for RC AVF formation. One may consider making the AV anastomosis in a more proximal part of forearm if there is a (1) small and circumferentially calcified radial artery over wrist region, (2) fibrotic or small sized cephalic vein around wrist region, and

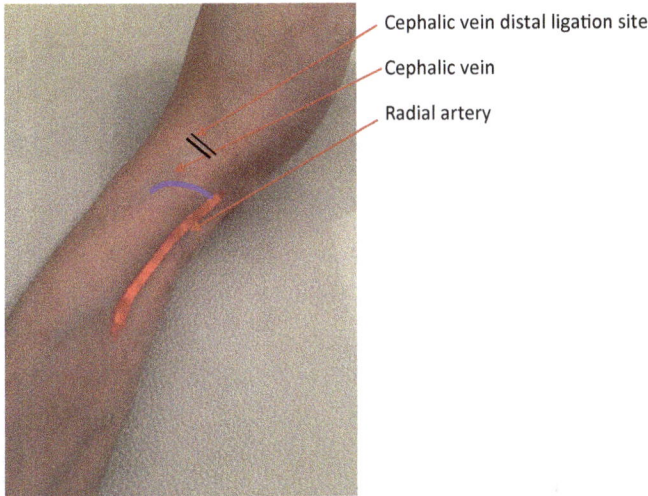

Fig. 2. Wrist cephalic vein mobilized medially (purple line) to form a gentle curve and join to the radial artery (red line).

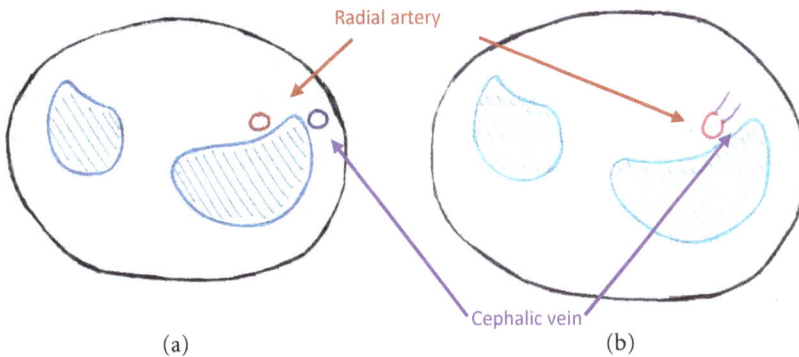

Fig. 3. (a) Cross section of wrist region showing the relative position of the left radial artery and left cephalic vein. To minimize the chance of kinking, the arteriotomy should be over 1 or 2 o'clock position (b).

the situation is better in the more proximal part of the forearm. When dissecting out radial artery over the mid forearm, the risk of hitting the cutaneous branch of radial nerve will increase. This risk has to be explained to the patient on pre-op counselling. Over distal forearm, the cephalic vein may run posteriorly and be farther away from radial artery. More caution is needed to ensure that the mobilized cephalic vein in all planes connect to the artery smoothly and avoid any twisting.

Anastomosis: The RC AVF can be made in end-to-end or end-to-side manner. I prefer end (cephalic)-to-side (radial) because it is technically easier. Without dividing the radial artery completely, it forms a stable base for anastomosis to take place (Figs. 2 to 4). Furthermore, there will be less spasm of the artery. The length of arteriotomy should be at least 4 mm, preferably 5 mm, and can be up to 6 or 7 mm depending on the alignment of the cephalic

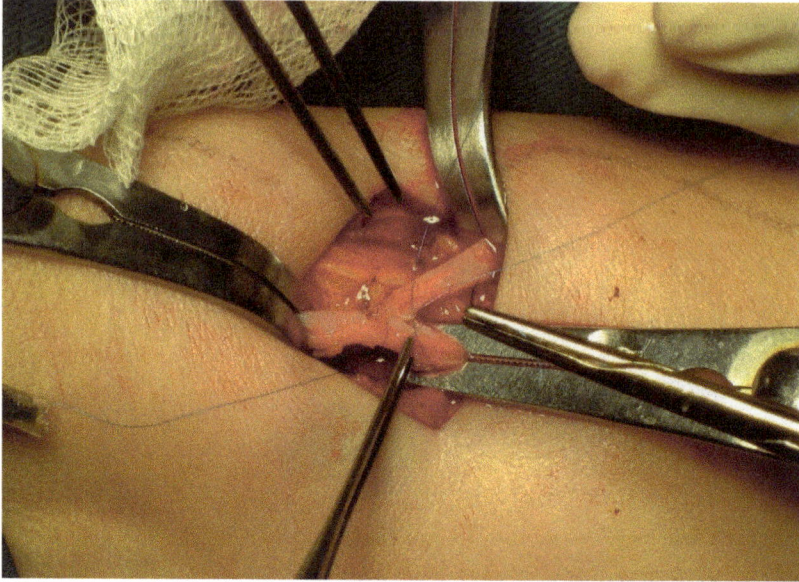

Fig. 4.   Creation of end-to-side anastomosis between forearm cephalic vein and radial artery (Courtesy of Dr. Ye Zhidong).

vein to the radial artery. The AV anastomosis is usually made with 7/0 non-absorbable monofilament suture (e.g. Prolene, Ethicon, NJ, US). The suture bite on both arterial and vein wall should be less than 1 mm. The suture bite on arterial wall may need to be increased if the wall is calcified and frail.

Heavy circumferential calcification is more frequently seen in distal radial artery of ESRF patients. Even with patent lumen, calcified artery is not a desirable inflow for AVF creation because (1) stiffness of the arterial wall makes anastomosis difficult, (2) clamp with stronger force is required to control bleeding during anastomosis which may induce stenosis in the future, and (3) after AVF creation, the calcified artery is unlikely to vasodilate to increase flow in response to reduced outflow resistance. Simple examination of the radial pulse could be misleading. If ultrasound machine is available in the theatre, it is best to do an immediate pre-operative quick scan to evaluate the condition over planned surgical site.

Vessel spasm: General anesthesia and brachial plexus block may reduce vessel spasm. However most patients have medical co-morbidities, therefore it is better to avoid general anesthesia. In the absence of anesthetic support, one may consider local anesthesia together with local immersion or IV injection of papaverin or glyceryl trinitrate (monitor blood pressure if IV injection given), or provide mild sedation to reduce the sympathetic output. The operator may also dilate the cephalic vein around the planned anastomotic region with heparin saline injection, with proximal and distal control applied (Fig. 5). Vein dilators are more reserved for structural focal stenosis rather than vasospasm.

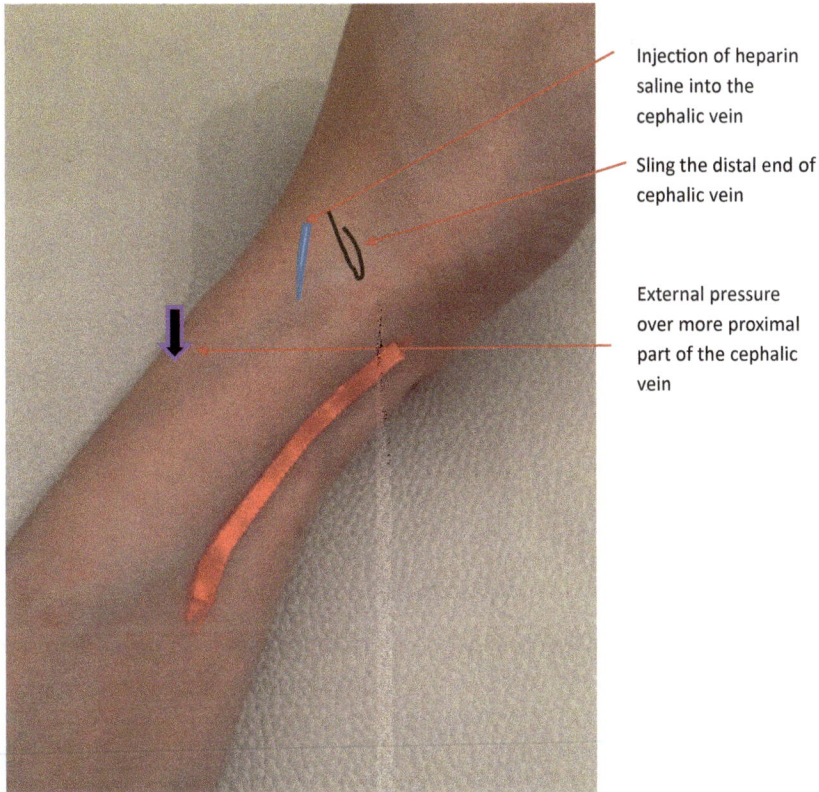

Fig. 5. Schematic diagram illustrating the method to dilate a vasocontricted cephalic vein.

Post-operation assessment: Upon skin closure and dressing applied, make it a routine step to check the thrill of the AVF over the forearm again and also finger perfusion condition. Occasionally, the closure of subcutaneous tissue may compress the AV anastomosis and affect the fistula.

Brachio-cephalic (BC) AVF

Location: Usually the cephalic vein of antecubital region or median cephalic vein is being mobilized to join to the brachial artery. Occasionally, median cubital vein is being used (Fig. 6). The incision can be made below the cubital fossa skin crease or just above the skin crease, depending on the size and distance between the brachial artery and the cephalic/median cephalic vein. Cephalic vein is often farther away from brachial artery and therefore, median cephalic vein is used to connect to brachial artery. However, if the median cephalic vein is small or absent, a longer segment of cephalic vein will be mobilized to connect to brachial artery. In some occasions, a good sized median antebrachial vein can be used for BC AVF over distal part of antecubital region. Both cephalic and basilic vein could be arterialized after the fistula creation.

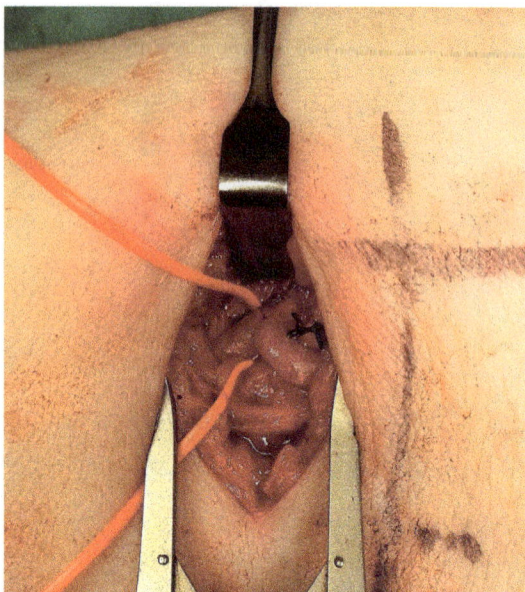

Fig. 6. Left median cephalic vein to distal brachial artery anastomosis in end-to-side manner.

Anastomosis: End-to-side anastomosis is used. Operatior has to make a careful judgement of the size of AV anastomosis based on the size of the brachial artery, any ipsilateral radial and ulnar artery disease, and the size of the median cephalic or cephalic vein. Usually around 4 mm would be a good size. Occasionally, 3 mm or less arteriotomy is made for small diameter brachial artery. Non-absorbable monofilament 7/0 suture is preferred. Side-to-side anastomosis can be used in some occasions to enable utilization of forearm vein in addition to arm cephalic vein when AVF matured.

Post-operation assessment: The radial and ulnar pulses and finger perfusion should be checked routinely after BC AVF creation. If the radial pulse is not palpable but the finger skin color looks alright, a simple method to assure perfusion is to check finger oxygen saturation with the pulse oximeter. For patients under local anesthesia, the hand sensation and motor power can be assessed easily.

Brachio-basilic (BB) AVF and basilic vein transposition (BBT)

The arm basilic vein is usually used for AVF creation when all cephalic vein options are exhausted. Because of its deeper location and close proximity to medial antebrachial cutaneous nerve, it has to be mobilised to a more superficial position to enable easy cannulation.

There are different methods to create BB AVF good for cannulation. It can be done as a two-stage or single stage procedure. Some opt to elevate the basilic vein and some prefer to transpose (BBT) the basilic vein to a more lateral and superficial plane.[2] During the

Table 1. Advantages and disadvantages of single and two stages brachio-basilic fistula creation.

	Two-stage	Single stage
Advantages	1. Start with small incision BB AVF under local anesthesia. Only commit to bigger magnitude operation when basilic vein is matured. 2. Less chance of twisting on transposition with arterialized vein.	1. Shorter process to get ready for use. 2. Less operation session required.
Disadvantages	1. Two procedures required, longer time to ready for use. 2. Patient may default for second procedure and the initial effort is wasted.	1. Patient would be upset if the AVF failed after a long incision. 2. More difficult procedure as vein size is small and not arterialized.

transposition, some would create a subcutaneous pocket laterally from the main wound and park the mobilized basilic vein in; some create a separate subcutaneous tunnel to track the basilic vein through; and some use endoscopic mobilization[3] together with subcutaneous tunnel. In a two-stage BB AVF and BBT procedure, some clinicians keep the existing AV anastomosis and some prefer to take it down and re-do the anastomosis during the second stage. There is no hard evidence as to which method gives more superior result than others.[4] Table 1 summarizes the advantages and disadvantages of single stage and two-stage procedure.

My personal approach is to use two-stage BB AVF and BBT approach for patients with basilic vein size between 2.5 to 3.9 mm over elbow region. Single stage BB AVF and BBT is a better option if basilic vein is already arterialized as in secondary AVF (e.g., failing loop forearm AVG with basilic vein already arterialized, BC AVF with branch communication to basilic vein), or occasionally in patients with large sized native basilic vein (≥4.0 mm).

Areas for Caution

- Ultrasound assessment of the basilic vein immediately before BB AVF surgery is important in selecting and locating the segment that is good for AV anastomosis. Over the antecubital region, there is frequently one or two tributaries joining to the basilic vein that makes its size vary significantly along the pathway. Occasionally, dural basilic veins are observed. Furthermore, the location of basilic vein could be variable and deep. Pre-operative localization minimizes unnecessary tissue dissection.
- Risk of steal syndrome in general is higher in AVF using brachial artery for inflow. Size of brachio-basilic anastomosis has to be judged based on the size of brachial artery, quality of brachial artery pulsation, wall quality and size of basilic vein. Again, around 4 mm would be a safe range but fine adjustment is needed for each individual condition.
- Be aware of any twisting of the basilic vein during transposition, especially through a newly created subcutaneous tunnel.
- Whether undermining a subcutaneous pocket or creating a separate subcutaneous tunnel to host the mobilized basilic vein, the operator has to ensure that the thickness

of subcutaneous tissue is <6 mm to enable easy cannulation by dialysis nurses in the future.

- Ensure the course of the transposed basilic vein is relatively straight and lateral enough to enable easy cannulation. The positioning of the arm during operation can be rather different from the position that patient assumes during dialysis and so as the position of the fistula.
- In patients with short and obese arm, the course of the superficialized basilic vein fistula can be rather short. If the basilic vein size is good, creating the BB AVF using basilic vein over proximal forearm will help extend the length of the usable fistula. Alternatively, maximize the length of basilic vein by mobilizing all the way up to axilla. This can be limited by poor skin condition around the axilla region of obese patients.
- Sometimes, the communicating branch between basilic vein and deep vein can be wide and short. Retraction and bleeding of those perforating branches can cause major bleeding complications. Extra caution needs to be given in tying off those branches. Plication with 7/0 polypropylene suture may be required.
- The relationship and degree of entanglement between basilic vein and the cutaneous nerve are highly variable. If there is too much entanglement, one may consider taking down the AV anastomosis and then moving the basilic vein away from the nerve before re-doing AV anasotmosis. There is also a risk of catching one of the cutaneous nerve fibre during wound closure (forming neuroma), which the operator needs to look out for.
- Generalized oozing can be troublesome over the transposition wound or subcutaneous tunnel and causes wound complication subsequently. Bleeding complication can be minimized by meticulous hemostasis over the wound, two layers continuously suture closure of the fascia and subcutaneous tissue, and placing a drain for short period of time. I do not give heparin routinely for second stage BBT procedure or straight forward BB anastomosis. Heparin will only be required if difficult anastomosis is expected. If heparin is given, reversal may be required before wound closure.

Other Potential Upper Limb AVF Locations

Connecting the proximal radial artery PRA with either cephalic vein or median cephalic or antebrachial vein over proximal forearm

Size of the proximal radial artery is usually bigger than the distal radial artery and less calcified. This AVF can be considered as the first option in patients without good sized distal forearm cephalic vein or heavily calcified small sized distal radial artery.[5]

The PRA AVF can also be a secondary AVF in patients with non-salvageable failing RC AVF.

Either the proximal forearm cephalic vein, median cephalic or antebrachial vein can be used to connect to the PRA, depending on their anatomical proximity. AV anastomosis can be done in end-to-side or side-to-side manner. In side-to-side anastomosis, both antegrade and retrograde flow of blood inside the venous fistula can be preserved.

In patients with diseased ulnar artery, creating RC AVF or PRA AVF may result in steal syndrome of the upper limb. Detailed arterial evaluation (arterial duplex and/or arteriogram) and also balloon angioplasty of the ulnar/radial artery may be needed prior to using marginally diseased upper limb arteries for AVF creation.

Utilizing forearm basilic vein

Forearm basilic vein, which is good in size, is present in a small proportion of patients. The posterior-medial position of the forearm basilic vein (Fig. 7) precludes its use as AVF in its original location. If both radial and ulnar pulses are strong, one can mobilize the basilic vein anteriorly and laterally and connect to either ulnar or radial artery over mid-forearm (Fig. 8). If the distal radial and ulnar arteries are small and calcified, the forearm basilic vein can be mobilized and looped over to form anastomosis with PRA or distal brachial artery.

Utilize deep vein transposition

This procedure is not commonly performed because it usually requires two-stage procedure, long wound, and longer maturation time, and potentially causes upper limb swelling. Brachial vein or proximal radial vein can be utilized to join the brachial artery or PRA respectively. One must ensure proximal and central veins are clear of obstruction.[6] A thorough assessment of the venous system and detailed planning are required. Nonetheless, this vascular access can be an option for some patients without good superficial upper limb vein.

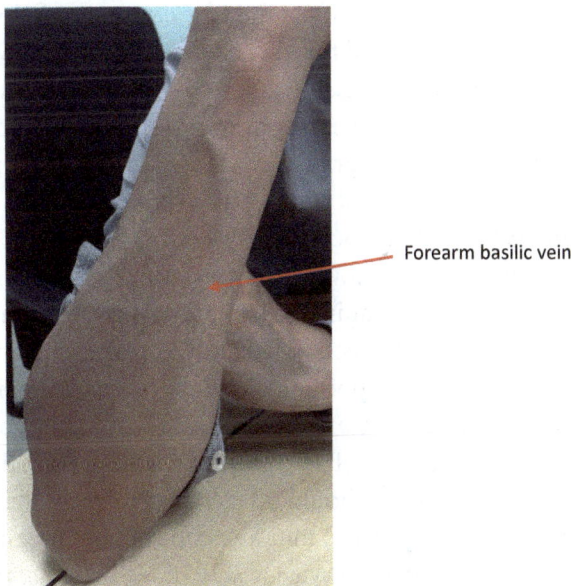

Forearm basilic vein

Fig. 7. Good sized forearm basilic vein in a healthy individual.

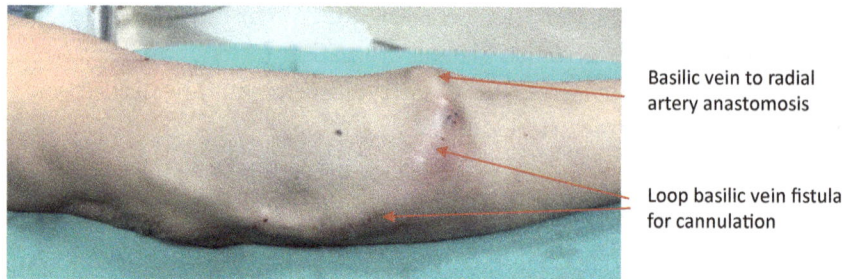

Basilic vein to radial artery anastomosis

Loop basilic vein fistula for cannulation

Fig. 8.   Left forearm basilic vein mobilized and form a loop connected to the mid-forearm radial artery.

Notes on AVG

When no good quality superficial vein is available in the upper limb or early vascular access is required to minimize the tunnelled CVC duration, AVG is considered. It consists of a conduit (synthetic or biological graft) connecting between a good arterial inflow and a good venous outflow (without central venous obstruction). There are several general considerations as discussed below:

- Currently, ePTFE grafts are the most common conduit used for hemodialysis access.[7] There are also polyurethane grafts with studies[8,9] showing similar patency as BBT and ePTFE graft. However, these polyurethane grafts are not as widely available worldwide as ePTFE ones. Biological and biosynthetic grafts are also used but they are more expansive and not commonly available.

- Usually 6 mm is the acceptable size of conduit for easy cannulation for hemodialysis. Therefore, 6 mm graft is usually used for AVG creation. If the inflow artery size is small, a tapered graft can be considered (e.g. 4–6 mm or 4–7 mm Gore-tex graft. W. L. Gore & Associates, Inc. Flagstaff, AZ, US). The dialysis nurses should be informed not to needle the smaller diameter segment of the graft towards the arterial end. In specific conditions like cross femoral AVG or collar AVG, 7 mm graft can be used.

- Steal syndrome in AVG tends to occur early compared to AVF due to the larger diameter of the graft, which offer a low resistance pathway and less adaptation time for the body. At the same time, flow requirement to prevent graft thrombosis is also higher for AVG than AVF. Careful vessel selection for graft implantation and decision on arterial anastomosis size plays an important role to prevent complication. In general, around 4 mm diameter arterial-graft anastomosis would be good for AVG using brachial inflow. When using a 6 mm ePTFE graft, there will be size discrepancy between the arteriotomy and the graft. Care should be exercised to space the discrepancy evenly. Depending on the actual artery size, 5 mm anastomosis for proximal brachial artery, and 6 mm for axillary or femoral artery generally should be safe. In diffusely calcified artery, a slightly bigger arteriotomy might be needed to facilitate anastomosis but one has to balance against the risk of steal syndrome.

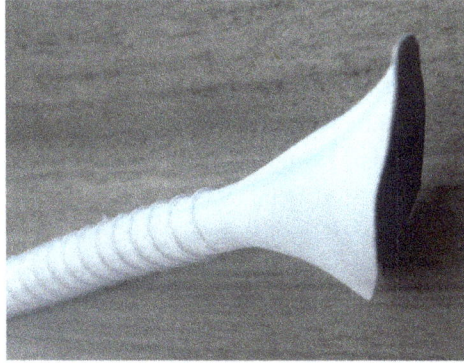

Fig. 9. Flare end of ePTFE Distaflo (BARD Peripheral Vascular Inc. Tempe, AZ, US).

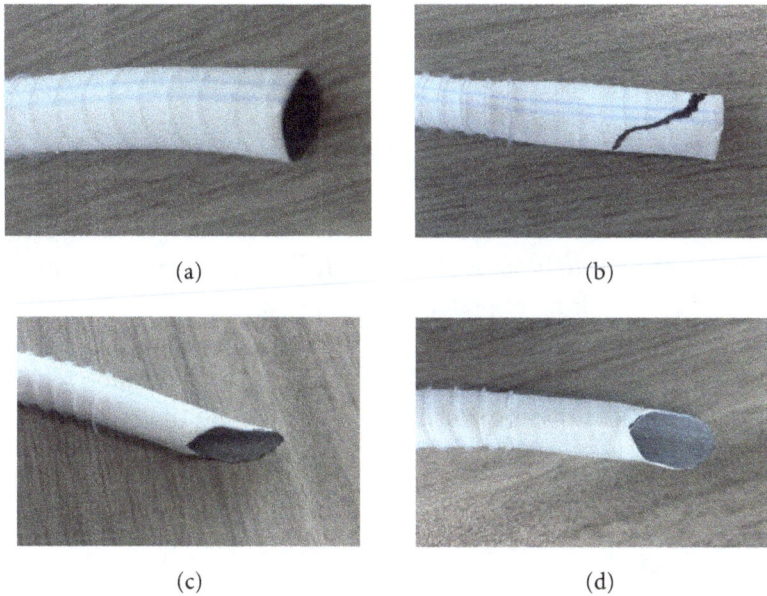

(a)

(b)

(c)

(d)

Fig. 10. (a) Straight end of ePTFE graft for arterial anastomosis in AVG. (b) "S" shape cutting line for venous end of the ePTFE graft. (c), (d) The opening of the graft after "S" shape cutting for venous anastomosis.

- Long-term AVG failure is usually due to vein-graft anastomosis stenosis. Therefore, clinicians can make a generous sized vein-graft anastomosis either using a cuffed ePTFE graft (e.g. Venaflo or Distaflo graft, BARD Peripheral Vascular Inc. Tempe, AZ, US) or make a "S" shape cut over the venous end of the graft (Figs. 9, 10a–d).
- 6/0 monofilament non-absorbable sutures are used for AVG anastomosis. Suture with 1:1 needle to thread ratio can be used to minimize suture hole bleeding (e.g. Gore-tex suture CV6, W. L. Gore & Associates, Inc. Flagstaff, AZ, US).
- Ring supported ePTFE graft is preferred if the AVG crosses a joint (e.g. Brachio-axillary AVG) to reduce the risk of kinking. Thrill usually cannot be felt over a ring supported ePTFE graft.

- Prospective randomized studies[10,11] showed better patency with cuffed ePTFE graft than non-cuffed ones. It is technically difficult to suture the cuffed graft onto a vein size <4 mm. The cuffed ePTFE grafts are more suitable for outflow veins that are larger in size (e.g. proximal basilic or axillary vein).
- To ensure the graft is superficial enough for easy cannulation, the subcutaneous tunnel for the graft is preferred to be superficial to the superficial fascia.
- During construction of the subcutaneous tunnel for graft placement, the configuration should ensure that at least one 10 cm segment or two 4–6 cm segments straight pathway are available for needling.
- Ultrasound assessment and localization of target vein for anastomosis immediately before surgery is recommended for basilic vein, proximal cephalic and axillary vein especially in obese patients to minimize extent of soft tissue dissection.

Common AVG Placement

Forearm loop AVG

In most situations, basilic vein is used as outflow vein (Fig. 11). Occasionally, cephalic and brachial vein will be used for AVG creation. It is preferred that both the arterial and venous anastomosis are below the cubital fossa skin crease so that the AVG does not cross the elbow joint. If the anastomosis has to be above the skin crease due to particular vessel situations, ring supported ePTFE graft is more preferred.

There are various methods to tunnel the graft and form a loop configuration. Most importantly, one should ensure there is no kinking of graft over the loop region. I prefer a transverse incision over the distal forearm and bury the loop in a small subcutaneous

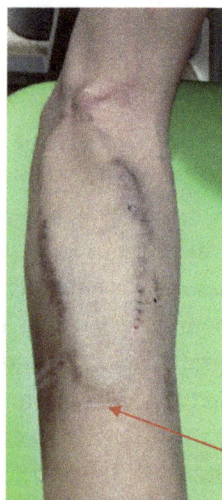

Distal forearm transverse incision for tunnelling the graft

Fig. 11.   Left forearm loop BB AVG.

pocket just proximal to the incision to minimize risk of wound break down and graft exposure.

There is no absolute rule whether to do arterial or venous anastomosis first. If a cuffed ePTFE graft is used, it is better to start with venous anastomosis first so that adjustment of the graft length can be made during arterial anastomosis.

Arm C shape AVG

Brachial artery to proximal basilic vein (Fig. 12) or brachial artery to proximal cephalic vein AVG. The brachial artery to graft anastomosis usually is located proximal to cubital fossa skin crease.

The advantage of this type of AVG is that it does not cross a joint and, at the same time, preserves the axillary vein for future AVG creation. However, the potential disadvantage is that the proximal incision may be close to the axilla and runs a higher risk of infection if the skin condition is not good.

In patients with short and obese arm, the AVG may assume a horizontal "V" shape pathway (Fig. 13) to have an adequate straight part for needling.

Arm loop AVG

A loop arm AVG can be used when proximal brachial artery is selected as arterial inflow due to extensive scarring around cubital fossa, atherosclerotic disease of distal upper limb artery or abnormally high brachial artery bifurcation in the upper arm (Fig. 14).

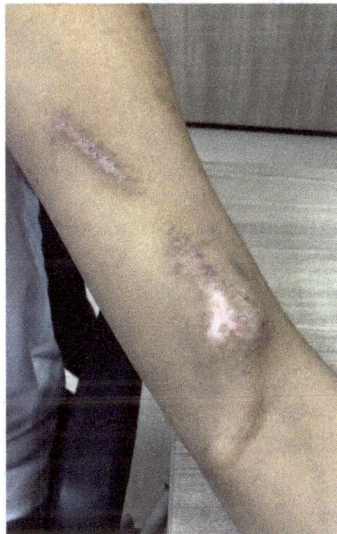

Fig. 12. Left arm brachial artery to proximal basilic vein AVG.

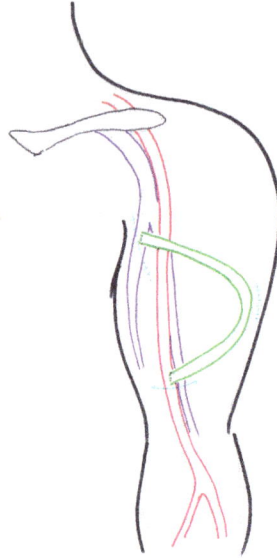

Fig. 13.   Horizontal "V" shape brachial artery to proximal basilic vein AVG provides relatively straight segment of graft for cannulation in obese patient.

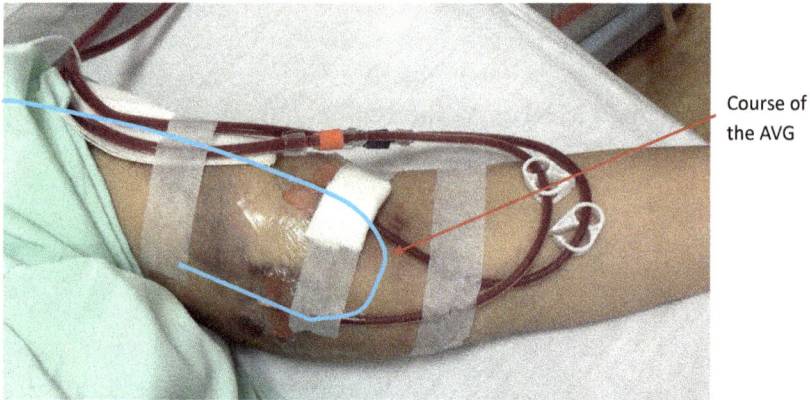

Course of the AVG

Fig. 14.   Left arm loop proximal brachial artery to axillary vein AVG was constructed for a patient with high bifurcation of brachial artery over the mid-arm.

Brachial artery to axillary vein BA AVG (C shape)

Ultrasound localization of axillary vein is preferred immediately before the surgery. Shoulder support facilitates surgical dissection of axillary vein. The axillary vein can be exposed using a subclavicular incision. It is underneath the pectoralis major muscle and costocoracoid membrane. Sometimes, part of pectoralis minor muscle has to be cut for adequate exposure.

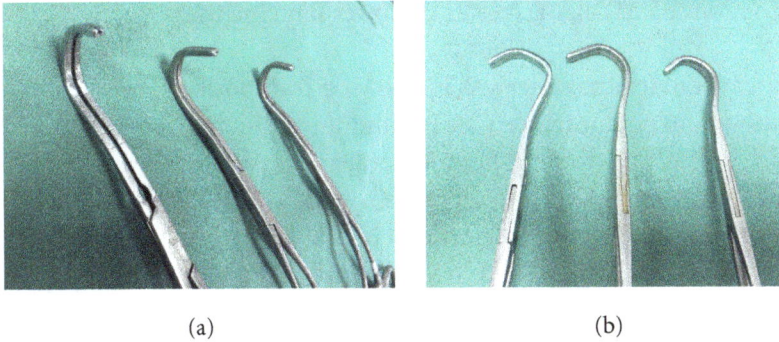

(a)   (b)

Fig. 15. Darra clamps (a) front view (b) side view for side clamping of the the axillary vein.

Axillary vein is usually big (>8 mm) with multiple tributaries. Side clamping (Fig. 15) of the axillary vein enables less extensive dissection of soft tissue away from the axillary vein. Cuffed and ring supported ePTFE graft (e.g. Distaflo, BARD Peripheral Vasculan Inc., Temple, AZ, US) is preferred for BA AVG.

In constructing the subcutaneous tunnel for BA AVG, ensure the majority part of the tract is superficial to the superficial fascia and close to the skin. The tunneller may go deep in the upper arm region, which renders subsequent cannulation challenging.

Other Potential Upper Limb AVG Placements

(1) Collar AVG (Axillary artery to contralateral axillary vein)
(2) Axillary artery to ipsilateral internal jugular vein AVG
(3) Axillary artery to femoral vein AVG

Lower Limb Vascular Access

Lower limb vascular access becomes a valuable option when upper limb accesses are exhausted or extensive bilateral brachiocephalic vein and SVC blockage is encountered. However, the complications related to deep vein obstruction and steal syndrome are less forgiving in lower limb compared to upper limb.[12] Therefore, a detailed assessment of the lower limb arterial and venous system is mandatory before proceeding to lower limb vascular access creation. Clinical examination of all lower limb pulses and toe pressure index are basic screening studies for arterial disease. Venous duplex of femoro-popliteal as well as iliac veins and IVC should be done for venous outflow assessment. If duplex quality for iliac vein and IVC is poor, conventional venogram has to be considered. Skin condition and skin hygiene need to be assessed. After lower limb access creation, vigilant follow-up is necessary to detect potential complications that develop with time; these include pseudoaneurysm, infection, venous hypertension, deep vein stenosis or thrombosis, and distal extremity ischemia.

Various options of vascular access in the lower limb include:

- Saphenous vein to femoral artery AVF with loop transposition;
- Saphenous vein to popliteal artery AVF with transposition;
- Superficial femoral vein to superficial femoral artery AVF with transposition;
- Femoral artery to saphenous or femoral vein AVG;
- Popliteal artery to saphenous or femoral vein AVG;
- Suprapublic cross femoral artery to femoral vein AVG.

The details of lower limb vascular access will be discussed in Chapter 14.

When there is No Suitable Cephalic Vein, Should Next Option be BB AVF/BBT or BB AVG?

There are several aspects we need to review to compare the two including:

- Success rate, any procedure required to assist maturation and success;
- Early post-operative complications;
- Time from creation to ready for cannulation;
- Primary patency;
- Assisted primary and secondary patency;
- Number of interventions required to maintain a functional access;
- Long-term complications, e.g. infection, pseudoaneurysm etc.

There are one prospective non-randomized[8] and two prospective randomized studies[13,14] addressing this question. One study uses Polyurethane graft (Vectra[8] BARD Peripheral Vasculan Inc., Temple, AZ, US) and one uses biosynthetic graft (Omniflow[14] Le Maitre Vasculan Inc., Burlington, MA, US). The general findings of the studies are that BB AVFs take longer time to mature, often require two-stage operation, more early post-operative complications but have better primary and assisted primary patency, less long-term complications and less secondary interventions required. However, BB AVG can be used earlier, have less post-operative complications but perform worse in longer terms. The secondary patency is comparable between BB AVF and BB AVG probably due to aggressive surveillance and widely available endovascular interventions. The success rate and steal syndrome for both access modalities are variable in the three studies.

In actual clinical practice, we also need to consider whether patient is for pre-emptive vascular access, or incident ESRF with tunnelled CVC *in situ*, duration of tunnelled CVC, any problem related to CVC, patient's medical condition and previous access infection history, patient's expectation and acceptance to more extensive wounds. For pre-emptive access, young and medically well patients, BB AVF would be more preferred. For older patients with limited life expectancy, patients already on long duration of tunnelled

CVC, or those with frequent CVC complications (recurrent blockage, infection), BB AVG is probably a better choice. There is no simple answer to this question. A thorough evaluation and discussion with individual patients about the pros and cons of BB AVF and BB AVG to reach a consensus clinical decision is the way to achieve better patient satisfaction and outcomes.

Hemodialysis Access in Elderly Patients

Many countries see an increasing percentage of incident ESRF patients with advanced age.[15] The definition of elderly patients in different studies varies. Some take 65 years old[16] as a cut off, some take 67[17] and some take 80 years old[18]. Nonetheless, elderly patients usually have more calcified peripheral arteries, higher incidence of various medical co-morbidities including diabetes, ischemic heart disease, heart failure, peripheral arterial disease and malignancies. They are also likely to face more social and financial difficulties including memory problem, caregiver's availability, locomotive problems and lack of constant income. A meta-analysis[19] showed elderly patients have a higher failure rate of AVF compared to non-elderly adults. Other studies identified female gender, diabetes, heart failure, specific ethnicity and short pre-ESRF nephrology care were particularly associated with poor outcome of AVF in elderly patients.[20] Nonetheless, survival benefit was shown even in octogenarian patients using AVF over AVG or tunnelled CVC.[21] Thus, native vein AVF should still be the first choice of hemodialysis access, provided reasonable upper limb artery and superficial vein are present. Furthermore, due to the more challenging medical, physical and social conditions in the majority of elderly patients, a thorough evaluation and detailed patient counselling to reach an individualized vascular access strategy is required. More flexible endorsement of various vascular access modalities to suit patient's unique clinical and social situations is preferred. Understanding the general success rate of different site AVF,[16,22] specific risk factors affecting maturation[20] and the usual time required for AVF to mature[17] in elderly patients would greatly facilitate patient counselling.

References

1. Shenoy S. Innovative surgical approaches to maximize arteriovenous fistula creation. *Semin Vasc Surg.* 2007; **20**: 141–147.
2. Dagher F, Gelber R, Ramos E, Sadler J. The use of basilic vein and brachial artery as an AV fistula for long-term haemodialysis. *J Surg Res.* 1976; **20**: 373–376.
3. Martinez BD, LeSar CJ, Fogarty TJ, *et al.* Transposition of the basilic vein for arteriovenous fistula: An endoscopic approach. *J Am Coll Surg.* 2001; **192**: 233–236.
4. Hossny A. Brachiobasilic arteriovenous fistula: Different surgical techniques and their effects on fistula patency and dialysis-related complications. *J Vasc Surg.* 2003; **37**: 821–826.
5. Jennings WC. Creating arteriovenous fistulas in 132 consecutive patients. Exploiting the proximal radial artery arteriovenous fistula: reliable, safe, and simple forearm and upper arm hemodialysis access. *Arch Surg.* 2006; **141**: 27–32.

6. Jennings WC, Taubman KE. Alternative autogenous arteriovenous hemodialysis access options. *Semin Vasc Surg.* 2011; **24**: 72–81.

7. Scott E, Glickman MH. Conduits for hemodialysis access. *Semin Vasc Surg.* 2007; **20**: 158–163.

8. Kakkos SK, Andrzejewski T, Haddad JA, *et al.* Equivalent secondary patency rates of upper extremity Vectra Vascular Access Grafts and transposed brachial-basilic fistulas with aggressive access surveillance and endovascular treatment. *J Vasc Surg.* 2008; **47**: 407–414.

9. Glickman MH, Stokes GK, Ross JR, *et al.* Multicenter evaluation of a polyurethaneurea vascular access graft as compared with the expanded polytetrafluoroethylene vascular access graft in hemodialysis applications. *J Vasc Surg.* 2001; **34**: 465–472; discussion 72–73.

10. Sorom AJ, Hughes CB, McCarthy JT, *et al.* Prospective, randomized evaluation of a cuffed expanded polytetrafluoroethylene graft for hemodialysis vascular access. *Surgery.* 2002; **132**(2): 135–140.

11. Ko PJ, Liu YH, Hung YN, *et al.* Patency rates of cuffed and noncuffed extended polytetrafluoroethylene grafts in dialysis access: a prospective, randomized study. *World J Surg.* 2009; **33**(4): 846–851.

12. Cull JD, Cull DL, Taylor SM, *et al.* Prosthetic thigh arteriovenous access: outcome with SVS/AAVS reporting standards. *J Vasc Surg.* 2004; **39**(2): 381–386.

13. Keuter XH, De Smet AA, Kessels AG, *et al.* A randomized multicenter study of the outcome of brachial-basilic arteriovenous fistula and prosthetic brachial-antecubital forearm loop as vascular access for hemodialysis. *J Vasc Surg.* 2008; **47**: 395–401.

14. Morosetti M, Cipriani S, Dominijanni S, *et al.* Basilic vein transposition versus biosynthetic prosthesis as vascular access for hemodialysis. *J Vasc Surg.* 2011; **54**: 1713–1719.

15. Lassalle M, Ayav C, Frimat L, *et al.* The essential of 2012 results from the French Renal Epidemiology and Information Network (REIN) ESRD registry. *Nephrol Ther.* 2014. pii: S1769-7255(14)00631-2.

16. Renaud CJ, Ho P, Lee EJ, *et al.* Comparative outcomes of primary autogenous fistulas in elderly, multiethnic Asian hemodialysis patients. *J Vasc Surg.* 2012; **56**: 433–439.

17. Hod T, Patibandla BK, Vin Y, *et al.* Arteriovenous fistula placement in the elderly: when is the optimal time? *J Am Soc Nephrol.* 2015; **26**(2): 448–456.

18. Nadeau-Fredette AC, Goupil R, Montreuil B, *et al.* Arteriovenous fistula for the 80 years and older patients on hemodialysis: Is it worth it? *Hemodial Int.* 2013; **17**(4): 594–601.

19. Lazarides MK, Georgiadis GS, Antoniou GA, *et al.* A meta-analysis of dialysis access outcome in elderly patients. *J Vasc Surg.* 2007; **45**(2): 420–426.

20. Hod T, Desilva RN, Patibandla BK, *et al.* Factors predicting failure of AV "fistula first" policy in the elderly. *Hemodial Int.* 2014; **18**(2): 507–515.

21. Hicks CW, Canner JK, Arhuidese I, *et al.* Mortality benefits of different hemodialysis access types are age dependent. *J Vasc Surg.* 2015; **61**(2): 449–456.

22. Weale AR, Bevis P, Neary WD, *et al.* Radiocephalic and brachiocephalic arteriovenous fistula outcomes in the elderly. *J Vasc Surg.* 2008; **47**(1): 144–150.

Suggested reading

Vascular Access: Principles and Practice by Samuel Eric Wilson.

Appendix

The anatomy of superficial veins over the forearm and cubital fossa has lots of variations. Figure 16 illustrates the common H-shape (Fig. 16a) and M-shape (Fig. 16b) of the antecubital veins.

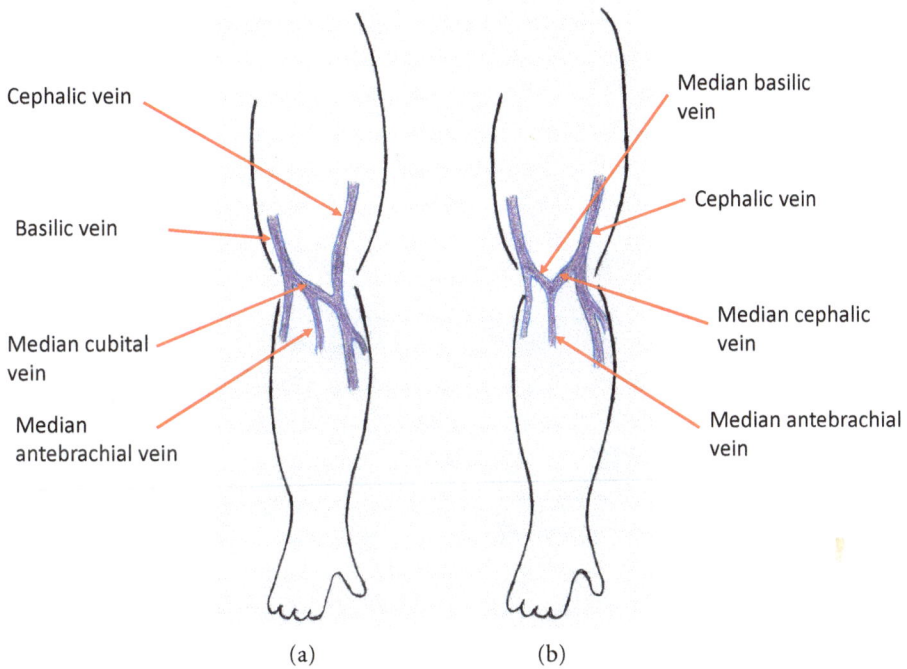

(a)   (b)

Fig. 16.   Common variations of antecubital veins anatomy (a) H-shape and (b) M shape

CHAPTER 4

Challenges in Tunnelled Catheter Insertion

Anil Gopinathan

Introduction

National Kidney Foundation Kidney Disease Outcomes Quality Initiative Guidelines (NKF DOQI) recommends patients with chronic kidney disease (CKD) to have a functioning arteriovenous fistula (AVF) 6 months prior to their expected date of first hemodiaysis (HD). Meanwhile, those already initiated to HD should preferably have a functional fistula within 90 days of the first HD.[1] Yet, according to 2012 annual report of US renal data system, 75% of patients begin their HD career with a central venous catheter (CVC).[2] For all practical purpose, almost every ESRD patient on HD will at some point in his/her lifetime undergo dialysis through a CVC. If HD through a CVC is required for more than 3 weeks, inserting a tunnelled HD catheter is the universal standard practice. This reflects the scope of tunnelled HD catheters in the management of ESRD.

Most often, a tunnelled HD catheter can be elegantly and safely inserted by any reasonably trained operator within 15–20 minutes. However, it is not unusual for this seemingly straight forward procedure to pose challenges that can make even the most experienced operators break into sweat. Every step involved in the catheter insertion (from site selection to post-procedural hemostasis) has the potential to be challenging. Henceforth, this chapter will visit the tunnelled HD catheter insertion procedure in a step-wise manner with emphasis on the commonly encountered difficulties at each stage and provide tips and tricks on circumventing or overcoming them.

Site Selection

The most preferred site for insertion of a tunnelled HD catheter is the right internal jugular vein (IJV) since it provides the most direct route to the superior vena cava (SVC) and right atrium (RA) and is associated with better patency and fewer complications than other sites. In the past, the right subclavian vein (SCV) used to be the most popular site for these catheters. However, subclavian cannulation is associated with almost 50% incidence of central venous stenosis; this may increase to 70% if the catheter gets infected.[3,4] Hence, SCV

4

should not be the first HD catheterization access site for a patient. The order of preference of veins for catherization, as stated in NKF DOQI guidelines is as such: right internal jugular vein, left internal jugular vein, right external jugular vein (EJV), and left external jugular vein. However, if the right IJV option is exhausted, the author prefers to cannulate the right EJV, before exercising the left sided option. The left IJV route is more prone to thrombosis than the right because of the two turns needed to reach the right atrium. Besides, it threatens the potential for a future fistula formation in the left upper limb.

If the EJV is reasonable in size and the anatomy of its union with the central veins is favorable, it is a good alternative site to IJV for vascular access.[5] It is punctured under ultrasound guidance and if required, patient may be asked to do a Valsalva maneuver to make it prominent. Unless the EJV has a very straight course in the neck, a venogram is recommended to delineate the angle at which it is joining the central veins (Fig. 1). An unfavorable orientation of the vein or stenosis at its central segment, are relative contraindications of using this access site. If the angle of union with central veins is very acute or if it shows reverse orientation (Fig. 2), it would be difficult to push down the large sized rigid dilators or peel away sheaths without causing venous rupture/ perforation. Besides, the chances of catheter kinking is high. Over the wire insertion of HD catheter would be safer for EJV access, but may be obstructed by the prominent valve at EJV-SCV junction making the use of a peel away sheath often essential.

Direct puncture of the right brachiocephalic vein (Fig. 3) is another technique, preferred by this author if other right sided veins are not available and there is a need for preserving the left sided veins. This is a reasonably simple and safe technique. However, it is feasible only when the brachiocephalic vein is patent and visible at ultrasound. USG transducer is placed in the right supraclavicular region with beam directed medially,

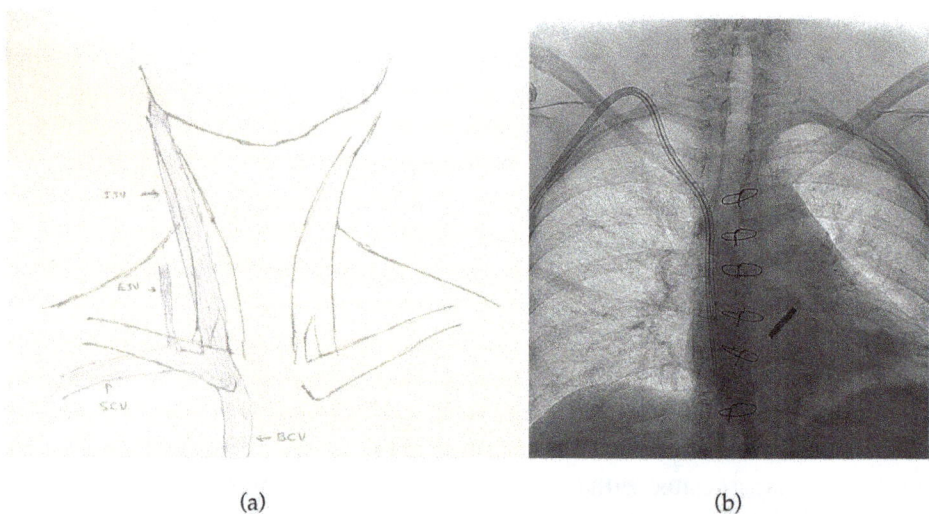

(a)   (b)

Fig. 1. In the line diagram (a), the EJV has a reasonably straight course in the neck and opens into the SCV at an obtuse angle. This is a favorable orientation for using this vein for tunnelled CVC insertion. The adjoining radiograph (b) shows a catheter inserted through such a vein. Note the smooth angulation as the vein transits through the EJV into the central veins.

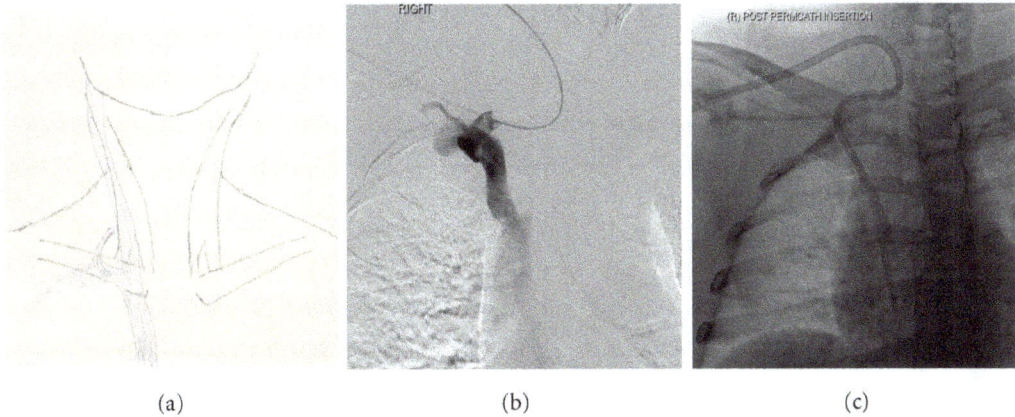

(a) (b) (c)

Fig. 2. The line diagram (a) shows an EJV with reverse orientation. If we access through such a vein, the angle at which the EJV joins SCV will make it essential to change the orientation at an acute angle, immediately after the venepuncture. This increases the risk of perforation during dilatation. If safely inserted, the catheter is likely to kink and lead to early obstruction. The angiogram (b) shows such a reverse orientation of the EJV. As this patient had no other suitable vein in the neck or chest, a catheter was inserted through this EJV with unfavorable anatomy. As expected, the catheter kinked at two points as noted in the post insertion radiograph (c). This catheter could not be used beyond two HD sessions; hence the patient later underwent a femoral catheter insertion (not shown here).

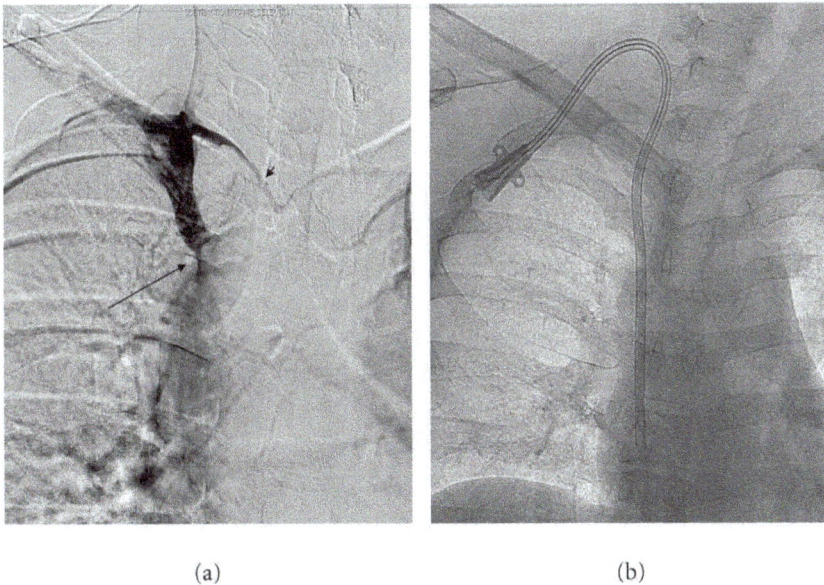

(a) (b)

Fig. 3. This patient had a thrombosed right arm AVG, with history of multiple previous right sided tunnelled HD catheters. He was due for a new AVF creation in the left arm. On table, ultrasound showed chronically occluded right IJV and EJV. Hence under USG guidance, the right brachiocephalic vein was directly punctured from the supraclavicular fossa with a sheathed needle and angiogram done (a). It showed moderate stenosis at the junction of the brachiocephalic veins (arrow) with reflux opacification of subclavian vein and mediastinal collaterals (arrow head). The stenosis was easily crossed with a guidewire and tunnelled HD catheter inserted in the usual fashion with post insertion radiograph showing good catheter position (b).

posteriorly and caudally would usually demonstrate the right brachiocephalic vein. A firm pressure on the transducer is required to visualize this vessel. The supraclavicular hollow may often interfere with good contact with the transducer surface; this may be overcome by creating a subcutaneous mound by infiltrating dilute lignocaine. Once the vein is visualized, it can be safely accessed with a micropuncture set. This approach has greater potential risk to cause pneumothorax compared to an IJV puncture. A somewhat similar but more aggressive approach has been described by Wellons *et al.* to directly puncture the SVC.[6] Through a transfemoral route, they do a venogram to demonstrate the SVC and then along the supraclavicular route puncture the vein under fluoroscopy guidance. The risk of hemorrhagic complications would be one of the major deterring factors for this approach to be used, unless left with no other alternative.

SCV can be punctured from a supraclavicular or infra-clavicular approach depending from where it is best seen and most accessible. Subclavian artery lies slightly superior and behind the course of the SCV, hence supraclavicular puncture should not be attempted without ultrasound. Traditionally, most of the subclavian vein canulations are done with blind punctures based on anatomical landmarks. A 22–23 G needle is inserted at a point 1 cm infero-lateral to the junction of the mid and medial third of the clavicle, into the deltapectoral groove, with the needle tip directed toward the opposite shoulder/suprasternal notch and needle trajectory as flat as possible to the chest wall, scraping the inferior surface of clavicle, with continuous suction, until venous blood is aspirated. The broad principles of SCV puncture for a tunnelled HD catheter insertion are the same, but the author prefers to perform this exclusively under real-time ultrasound guidance. An infraclavicular, right SCV puncture is easier with the operator (for right handed) standing on the head side of the patient, facing towards the foot; meanwhile left sided puncture is easier with the operator facing in the reverse direction. Slight Tredelenberg position may be given for its questionable benefit in making the vein prominent as well as to decrease risk of air embolism. Bracing the shoulders with a small pillow between the scapulae, turning the head towards the opposite and keeping the ipsilateral shoulder pulled downwards with the arm by the side is the standard position in which I target the SCV. Such positioning brings the medial segment of the subclavian vein (this is the segment we are aiming for in SCV canulations) closer to clavicle, slightly separates it from the accompanying artery, and tends to make it more engorged and hence easy to puncture (a flaccid vein may slip away from the penetrating needle).

If the neck veins and SCVs are not available, femoral veins may be used for insertion of HD catheter, notwithstanding the increased incidence of infection, and the risk of ilio-caval thrombosis interfering with possible future renal transplantation. If there are prominent neck or upper chest wall collaterals, it would be worthwhile to interrogate them with CT or venography to assess their suitability for catherization prior to puncturing the femoral vein (Fig. 4).

In a small subgroup of unfortunate patients, where all the aforementioned access sites are exhausted, more innovative and unconventional solutions for CVC insertion may need

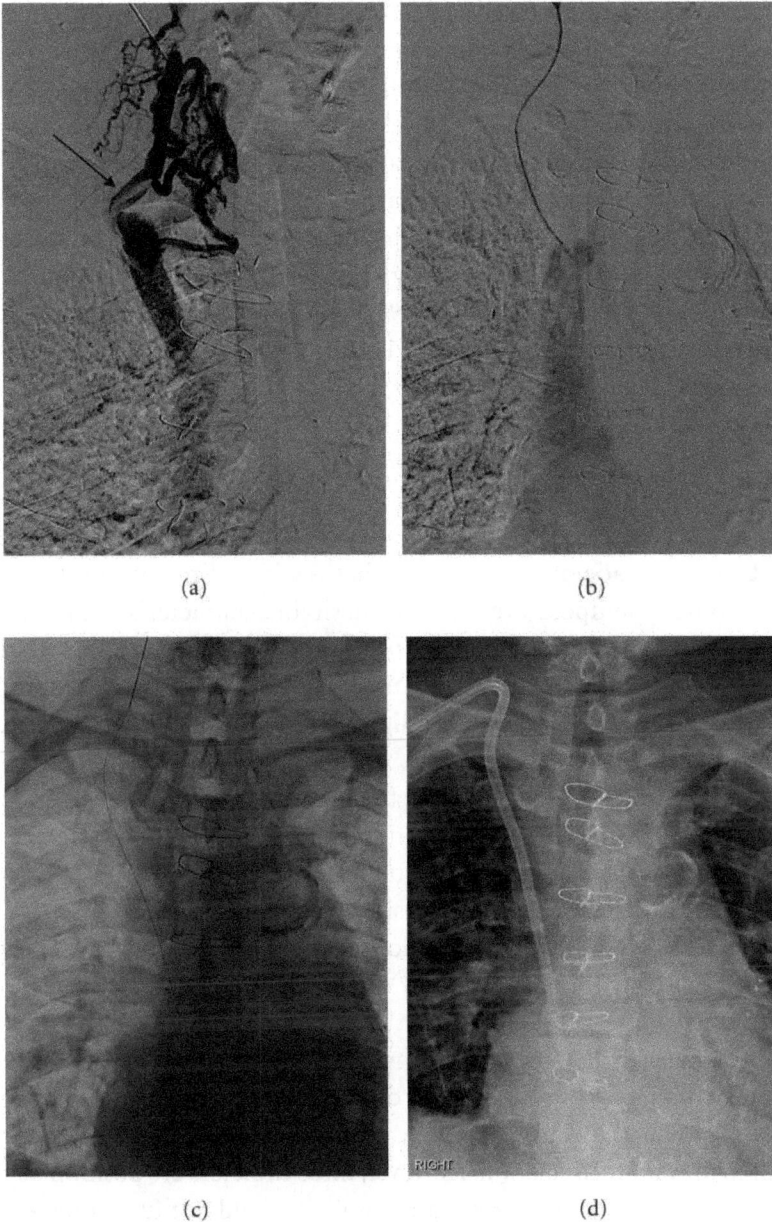

(a)

(b)

(c)

(d)

Fig. 4. This was another HD patient with bilateral neck, subclavian and femoral vein occlusions. Hence, we decided to interrogate for any other potential site for tunnelled CVC insertion. As he had visible prominent collateral veins in the neck, one of them in the upper to mid neck was punctured with a 23 G scalp vein needle and venogram was done (a). This venogram showed multiple collaterals that eventually drain in to the right brachiocephalic vein. One of these collaterals was prominent (arrow in Fig. 4a) with a less tortuous course into the central veins. Hence it was punctured with a micropuncture set under ultrasound guidance (b & c). As the course of the collateral was tortuous it was straightened with an Amplatz wire before dilatation. Subsequently, an over-the-wire tunnelled CVC was inserted as the collateral was deemed not of adequate caliber to insert a large sized peelaway sheath. The post-insertion radiograph (d) reflects the course of the collateral vein, but no significant kink is seen in the catheter.

to be adopted. Some centres use external iliac vein (EIV) as an alternative to femoral veins. It is associated with lower infection rates compared to latter as the catheter exit site will be above the groin. Besides, as it does not cross the hip joint, chances of catheter kinking and stretching are less, thereby decreasing chances of thrombosis compared to femoral catheters. A translumbar direct insertion of the catheter into the IVC has been described with risk of thrombosis and infection rates similar to traditional site.[7] Retroperitoneal hematoma is a potential major complication with this approach. A more drastic approach is transhepatic insertion of HD catheter through the hepatic veins, up through the IVC into the right atrium. Smith *et al.* reported 29% procedural complication rate, 31% early catheter dislodgment and 19% experienced early catheter failure (<7 days) with the transhepatic tunnelled CVC insertion.[8]

Selecting the Catheter

Several different catheters developed by different manufactures are available in the market. Most of them are made of polyurethane that is stiff at time of insertion but softens when exposed to body temperature. This is a desirable characteristic that decreases the vascular tissue injury when the catheter is *in situ* for prolonged periods. Carbothane is a polycarbonate-polyurethane copolymer used in the manufacture of tunnelled HD catheters. It allows the catheter walls to be slightly thinner while offering greater strength than polyurethane catheters.

Tunnelled CVC have a Dacron cuff close to their hub. This cuffed segment of the catheter should be placed in the subcutaneous tunnel, at least 3 cm away from the vascular entry site. The cuff incorporates into the adjoining subcutaneous tissue with fibrosis. This helps in keeping the catheter anchored in position as well as provides a mechanical barrier that prevents bacterial colonization. To reduce the risk of infection, catheters with special coatings (such as heparin, antibiotic or silver) are available; however their real value is still not fully ascertained.

To prevent catheter thrombosis and to ensure effective hemodialysis by reducing recirculation, manufacturers have come up with different tip designs (Fig. 5). In the staggered tip (single body) design, the outflow tip is located at least 25–30mm distal to the inflow tip with an intent to prevent recirculation. However, these catheters are position dependant i.e. if the arterial end abuts the SVC or right atrial wall, it would not function well. This position dependency is overcome in the split tip catheter geometry. Besides, the motion of the split ends of the catheter also helps in slowing fibrin sheath formation. Both the designs can come with or without side holes. During dialysis, if the catheter tip abuts the vessel wall, the consequent vacuum effect can cause the walls of the soft catheter to collapse on itself and thus impede the dialysis. This vacuum phenomenon can be overcome by having side holes at the tip. However, due to irregularity of their cut surfaces, side holes increase the incidence of thrombus formation around the catheter tip, thereby increasing potential for infection. In the relatively newer, symmetrical tip design, both the arterial and venous

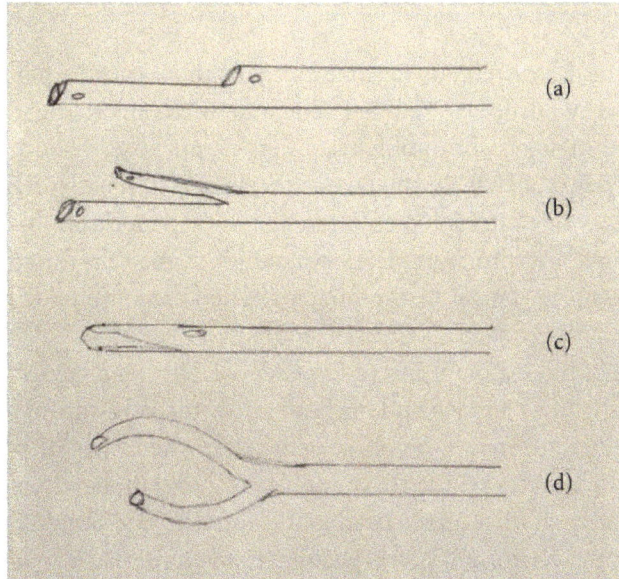

Fig. 5. Diagrammatic representation of few of the different tip designs of tunnelled HD catheters that are commercially available. (a) staggered tip with side holes; (b) split tip with side holes; (c) symmetrical tip; (d) self centering split tip.

lumina have the same length. Here the inflow occurs through the side slot and the most proximal portion of the end-hole, outflow occurs as a jet directed away from the catheter tip. In animal experiments, this design is found to decrease recirculation. In patients on long-term HD, it is quite common for the arterial end of a catheter to get blocked by clots or fibrin sheath or by abutting the vessel wall. In such situations, switching the lumina will clear the clots and push the lumen away from vessel wall. However such flow reversal tends to cause unacceptable levels of recirculation with split and staggered tip designs, while symmetric tip design catheters are equally efficient in both the standard and reverse configurations. Newer self centering split tip catheters have a unique design. The distal split ends are curved, so that the lumen always stays away from the vessel wall. Some early evidence on its superiority over the symmetric tip design has been found.[9]

Although most commercially available tunnelled HD catheters need antegrade tunneling from skin to venepuncture site, some of them can be tunnelled in the reverse direction. Some of them allow an over-the-wire insertion as well. In certain critical situations, over-the-wire insertion may be preferable and safer as it ensures continuous wire access throughout the procedure and reduces the chances of air-embolism.

To summarize, there is no clear winner among the various commercially available catheters, but it is essential for an operator to be aware of the usual hardware available as well as their major pros and cons. This will enable a wise selection of the catheter in a challenging situation where picking the right device is of utmost importance to ensure procedural success.

Venepuncture

Blind puncture of the access vein, using anatomical landmarks or by palpating a neighboring artery should be avoided. USG-guided venepuncture is mandatory. A high frequency probe (usually 7–10 MHz) is commonly used for this purpose. Transducers with smaller footprint are preferred, especially in small sized patients or thin patients. A lower frequency probe (3–5 MHz) may be required when greater acoustic penetration is required i.e. when the vein is deep seated. Pre-puncture, USG evaluation of the vein for approximately 5 cm (or as much permissible) central to the puncture site is recommended. This will enable detection of any underlying stenosis or occlusion.

Good exposure of the region of interest is essential. During a jugular venous puncture, the neck should be slightly extended and the head gently turned away from the side of puncture. This is especially important with obese patients or when the patient has a short neck. In such cases, the author places a folded drape or blanket beneath the scapula. Hyperextension of the neck or too much of turning the head to the opposite side should be avoided as it can cause the vein to change its normal lateral position to the carotid artery and bring it anterior to the artery. Thus, this increases the risk of arterial puncture. The IJV should be punctured low in the neck, between the two heads of the sternomastoid, approximately 1 cm above the medial end of the clavicle. A high puncture is more likely to cause the catheter to kink. The vein should be punctured along its anterolateral wall with the USG transducer placed perpendicular to the vein and needle travelling parallel to the long axis of the transducer (in-plane puncture) (Fig. 6). This approach would entirely preclude inadvertent arterial puncture since the needle is always under vision and even when you make a through and through puncture, the needle will escape the carotid artery. Although a single wall puncture to enter the vein is ideal, in patients with thick/edematous skin or in those with extensive scarring from multiple previous catheterization, a double wall puncture of the vein is often required. With thick overlying skin, in spite of a generous and deep incision the needle may encounter severe resistance and the vein tends to move away from the needle tip. In such cases a sharp, forceful stab with the needle is required, which often tends to puncture both walls. Then the needle is gradually withdrawn until the tip is within the lumen. The author regularly attaches a 10 cc slip-tip syringe filled with 3 ml of normal saline to aspirate blood and confirm adequate position of the needle tip within the lumen. A luer lock syringe should be avoided as one always runs the risk of losing the access while trying to unscrew the syringe. Meanwhile, 5 cc slip tip syringes do not snugly fix to the puncture needle.

In very thin patients or those with prominent supraclavicular hollow, the vein may not be well visualized as the anatomy restrains good contact between the ultrasound transducer and the skin surface. This can be overcome by using a thick layer of jelly that removes all the air from the interface between the skin and transducer surface. A superior option would be to efface the hollows on the body surface by injecting diluted lignocaine.

In most situations, the venepuncture can be done with an 18 G puncture needle. In challenging situations when the operator is not confident on what he/she may encounter

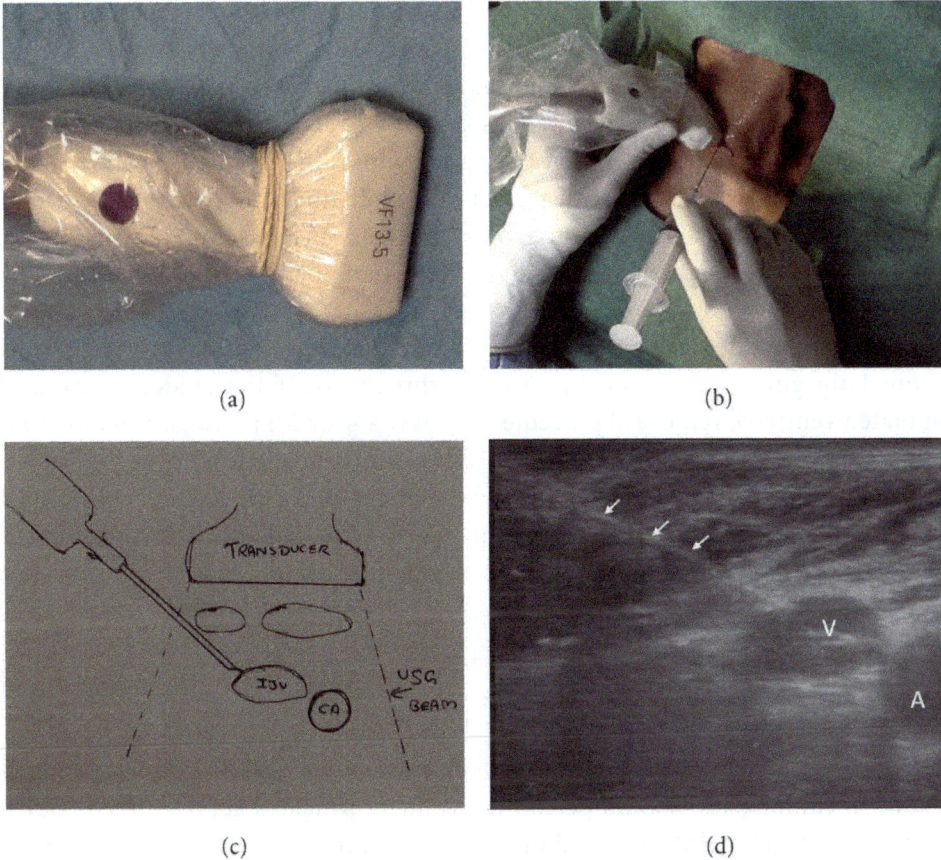

Fig. 6. A high resolution ultrasound transducer, draped in sterile cover is demonstrated in (a). An ultrasound guided left IJV puncture is shown in (b). Note the position of the needle tip in relation to transducer surface. Such an in-plane entry will ensure that the entire course of the needle is always under vision as depicted diagrammatically in (c) and on the ultrasound image (d). The echogenic linear structure (as shown by arrows) shows the needle from skin surface to its entry into the IJV (V) (A: internal carotid artery).

further along the course of the procedure or if the vein is of small caliber and sitting in a precarious location, a 21 or 22 G needle from a micropuncture set is the preferred puncture device. Once venepuncture is done with a 21 G needle, a 0.018" guidewire is passed through it into the central veins. If there is marked tortousity of the veins at the thoracic inlet or if there is a central stenosis, the wire would not go down freely. Gentle manipulation may be tried, but too much of fiddling when a guidewire is going through a metallic needle should be avoided or it may damage the guidewire. If the guidewire kinks or it is difficult to remove the guidewire through the needle, instead of trying to force it, the whole, needle-guidewire assembly should be removed out of the vessel together. If one manages to get the stiff metallic portion of a microwire, beyond the needle tip, well into the vessel, the needle can be removed and the 3 Fr inner dilator from the micropuncture set may be inserted over the wire into the vessel (this should not be attempted if only the floppy distal end of the guidewire is within the vessels as it is not stiff enough to support the catheter). Subsequently, venogram can be done through this dilator to assess the vascular anatomy. This dilator can also serve as a support catheter for safe guide wire manipulation. Once the guidewire is in a secured position, the 3 Fr dilator is removed, reassembled with the 4 Fr outer dilator from the set and the whole coaxial dilator assembly is re-inserted over the wire and used in the manner as a micropuncture set is designed to be used. All these manipulations should be done under fluoroscopy guidance.

If the venepuncture has been made with an 18 G needle, preferably a 0.035" moveable core J tip guidewire should be inserted through it. If the guidewire cannot be manipulated centrally, remove the needle and pass a 4 or 5 Fr vascular dilator (please note that these are different from relatively blunt soft tissue dilators) over this wire into the vessel. This dilator can be used in the same manner as described earlier with the 3 Fr dilator from the micorpuncture set. The tendency to pass a glidewire through a metallic needle should be resisted, as it can strip the hydrophilic coating that would get retained in the vessel wall.

Subcutaneous Tunneling

Subcutaneous track that leads from the skin exit site to the venous entry site should be at least 8–10 cm long and should permit the catheter to take a gently curved course to prevent it from kinking. Infiltrate 3–5 ml of 1% lignocaine at the proposed skin exit site and make a 1 cm long incision deep enough to expose the subcutaneous fat. Please note that a superficial incision which only reaches up to the dermis is inadequate. A 22G spinal needle is now inserted, parallel to the skin surface, through this incision, along the proposed course of the track towards the venous entry site. Diluted lignocaine (1 in 3 dilution of 1% lignocaine is routinely used) in copious amounts (usually up to 20 ml) is injected with this spinal needle along the entire track. Besides anesthetizing the track, it hydrodissects the tissue that makes the tunneling much easier. This maneuver of hydrodissection is especially useful in patients with thick chest wall skin or in those with extensive fibrosis (as in patients with multiple previous tunnelled CVC insertions). Subsequently, the trocar from HD catheter set is used to make the subcutaneous tunnel. In thick patients and those with excessive subcutaneous scarring, the trocar insertion may need some assistance in the form of pre-dissection with a curved artery forceps. Metallic trocars are often more effective and easier to use if you give it a gentle curve before inserting it. Such a manually curved trocar should be inserted with its concavity facing ventrally. In difficult cases, keeping the skin stretched by the other hand (or by an assistant) is often helpful. Once the trocar has emerged from the other end of the tunnel, the catheter is attached to it and pulled through the tunnel. It is best to pull the cuff of the catheter 5–6 cm into the tunnel. It is much easier to pull the catheter back later than advance it forward. The author avoids attaching the catheter to the trocar before tunneling through. A catheter dangling from the tail end of the trocar is often an impediment to elegant usage of the device. It is also more likely to come in contact with surfaces that may compromise its sterility.

Dilatation of the Entry Track

The dilatation of the entry track is done over the guidewire inserted through the venepuncture under fluoroscopic vision. Before inserting the dilators, a long segment of the guidewire needs to be placed into a stable straight segment of the vein. If the jugular or subclavian veins are being punctured, it is mandatory to secure the guidewire down into the IVC and document it with fluoroscopic picture. This is the only way to confirm that the operator had indeed achieved a venous access. This is also essential for smooth and safe dilatation of the entry track. If the guidewire is freely floating in the right atrium, it cannot support the large caliber dilators and risk of perforation is high. Dilatation should always be done under fluoroscopic guidance. The dilators should be advanced in the same direction (in both the planes) as the guidewire. The moment the guidewire starts to show any kink, the dilatation should be stopped and dilators should be advanced in a different direction with very short increment at each step. This step is especially crucial in patient with short neck or obese patients or if there is extensive scarring around the neck. The same holds true for EJV, subclavian, left IJV and collateral vein punctures. In the above situations, this author routinely does dilatation over a stiffer wire (usually a 0.035" Amplatz wire) than those provided in the usual tunnelled HD catheter set (Fig. 7). This stiff wire straightens out most of the sharp angles and tortousity in the course of the vein as well as provides adequate support for the dilators without inadvertently getting kinked.

Insertion of the Catheter

Although some of the commercially available catheters can be delivered over a wire, most are inserted through a peelaway sheath. In spite of pre-dilatation with the provided dilators in the catheter set, sometimes it is difficult to push down the peelaway sheath-dilator assembly over the guidewire. In such cases, slight widening of the skin incision at the puncture site, dissection of the subjacent tissue with artery forceps and pre-dilatation with a larger (than the sheath) sized dilator would do the trick.

The dilator pre-fixed to the peelaway sheath, usually extends 2–3 cm beyond the distal end of the sheath. Hence, once the sheath-dilator assembly is well within the vessel, one may just slide down the sheath over the dilator without pushing the latter all the way into the caudal SVC or right atrium. This is especially important during insertion of left sided catheters. With a left sided access, once the sheath has reached the brachiocephalic vein, no

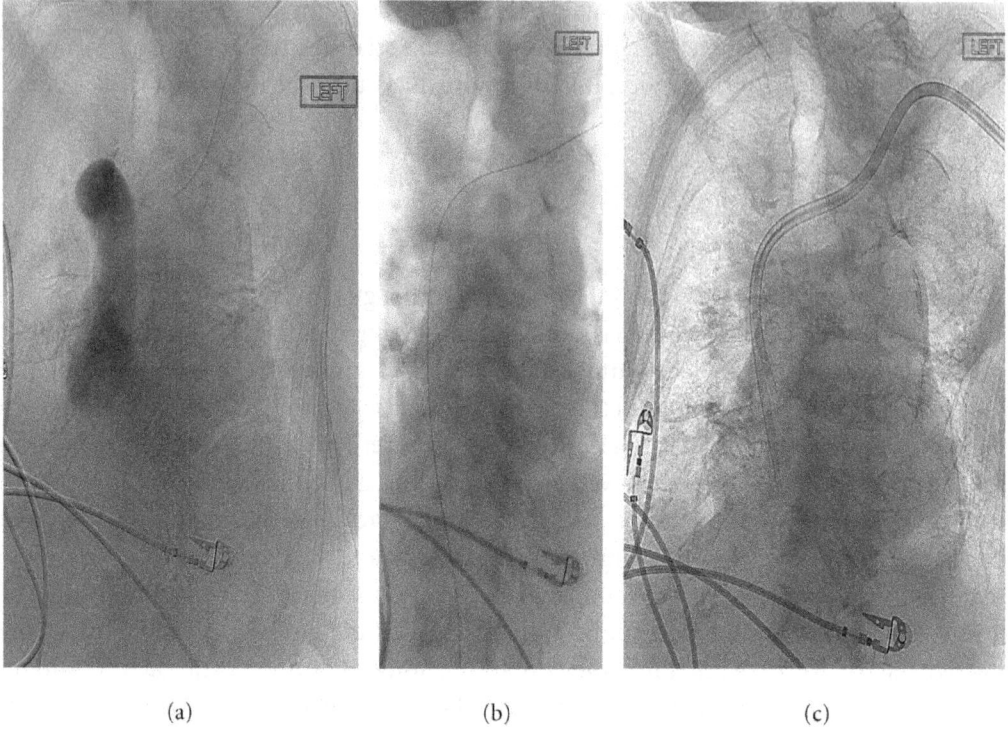

(a)   (b)   (c)

Fig. 7.   This patient with a difficult CVC access had multiple congenital chest wall deformities. The right sided veins were occluded from previous catheterizations. The left IJV was patent, but was tortuous and deep to the artery due to the skeletal deformity. Hence, the left IJV in the supraclavicular region was punctured along its left lateral wall (as the artery was anterior to it) with a micropuncture set. Angiogram was done through the dilator of the micropuncture set. It showed marked tortuosity of the central veins with left brachiocephalic vein almost forming a complete loop before draining into the SVC (a). The loop was crossed using a 4 Fr diagnostic catheter and hydrophilic Glidewire and catheter advanced all the way into the IVC. Next the Glidewire was removed and a stiff guidewire (0.035" Amplatz Superstiff) was inserted through the catheter. The stiff guidewire straightened out the curls in the redundant/ tortuous venous system (b). Subsequently, the tunnelled HD catheter was inserted in the usual manner (c).

attempts should be made to push the dilator across the second bend into the SVC. Most catastrophic bleeds happen during this final step of sheath delivery due to venous perforation by the sharp dilators. Hence, it is essential for the patient to be absolutely still during this step. Once within the vessel, there should be no resistance to further movement of the sheath. The operator should stop at the first encounter of any kind of resistance and make subtle changes in the direction of advancement, as it is often due to the sheath hitting against the vessel wall.

Once the peelaway sheath is in good position, the dilator and wire is removed. The sheath usually has a valve at its hub to prevent air from getting sucked after removal of the dilator. For this valve to function optimally, it is essential to withdraw the guidewire tip into the dilator and then remove them together. Some operators make the patient do maneuvers that keep the intra-thoracic pressure elevated to prevent air embolism. In my experience, most patients are not in a mental or physical frame to adequately comply with such instructions. Hence, it is better to allow patient to continue with his/her regular breathing and make best use of the valve at the hub of the sheath. As a second safety measure, you may place a thumb over the hub immediately after the dilator is removed.

Sometimes there may be resistance to inserting the catheter through the peelaway sheath. If this is at the level of the skin entry, some dissection of the tissue locally would help. More often, it is seen at places where the vessel is changing direction. In both scenarios, this may be addressed with a 0.035" stiff-guidewire with floppy tip and hydrophilic coating inserted through one of the ports of the HD catheter to render it some additional strength and pushability.

Position of the Catheter Tip

The best position of the tip of a tunnelled HD catheter is a contentious topic with different (somewhat conflicting) recommendations from different expert groups such as US FDA and NKF DOQI. NKF DOQI itself changed its target from SVC to right atrium in 2006. The most recent NKF iteration as well as most expert reviews, recommend the catheter tip to be in the right atrium with its arterial lumen facing the mediastinum. However these recommendations are like a shot-gun approach as they do not distinguish between the different tip designs of the currently available dialysis catheters and the dynamic interplay between catheter function, side-hole design, and resulting complications.[10] To overcome this problem, Tal *et al.* have suggested the concept of a 'functional tip.' The functional tip is the part of the catheter from the most proximal side hole to the catheter tip. They recommend the entire functional tip to be in the right atrium provided the tip does not touch the floor of the right atrium. Another issue towards which all the recommendations turn a blind eye is the exact location of the right atrium. On fluoroscopy or radiograph, it is impossible to accurately decide where the SVC ends and right atrium begins. The vena cavae open along the posterior wall of the atrium and not exactly at their cranial or caudal end to recognize it on frontal radiographic views. Besides, in the supine position with intra-abdominal contents pushing the diaphragm up, the mediastinal structures are often a few cm higher in their location as compared to where they would be if the patient is in an erect position. This effect is protracted in patients with large habitus. For that matter, even on cross sectional CT images, experienced thoracic radiologists would often find it difficult to tell where exactly the SVC ends unless the CT study is cardiac gated. In summary, despite the various expert guidelines, the tip of catheter is always a rough estimate in reality.

Another issue to be considered is the drag on the catheter by the chest wall. Once the patient assumes an erect posture, there is a constant drag on the catheter by the anterior chest wall. This tends to pull the catheter back into the SVC or even into the brachioce-phalic vein. This effect is profound in obese patients and in women with big breasts. Hence, in such patients, I prefer using slightly longer catheters that can reach all the way into the lower atrium.

For catheters inserted from femoral approach, the ideal location of the tip is even more unclear. Some reviews consider lower IVC as a good site for placing the catheter tip. If the common iliac vein is of reasonable size and patient is not due for a renal transplant soon, we generally place the catheter tip in the common iliac veins. Stenosis and thrombus devel-oping in the common iliac vein are easier to treat through an endovascular approach. IVC is used as site of parking the catheter tip only after the common iliac vein as an option is exhausted.

Bleeding from the Track

Some amount of venous ooze from the subcutaneous tunnel following a HD catheter insertion is common. They usually respond to manual compression. Rarely a more pro-longed compression of 15–20 min may be required. A few ml of diluted (1:10,000) adrena-line injected into the track followed by few minutes of compression is often helpful to stop this bleeding. Some operators have found injection of small amounts of thrombin into the track is helpful to stop this ooze. The author prefers to take a few hemostatic purse string stitches with 3-0 or 2-0 silk around the catheter exit site and further along the track if required. These stitches should be removed in 24 hours time to reduce the risk of infection. Caution has to be exercised not to catch the catheter with the needle.

Catheterization in Pre-Existing Venous Stenosis

It is not unusual to find the central veins to be stenosed on the angiogram done after the initial venepuncture. A mild to moderate stenosis does not need any additional treatment and it is often sufficient to place the catheter across the stenosis (Fig. 8). However, if it is tight enough to obstruct a smooth insertion of the catheter, it should be pre-dilated with balloon angioplasty. In some situations, one may need to even stent the vein to place the catheter through (Fig. 9) into the right atrium. Compared to stenosis, a venous occlusion may be more difficult to cross; but having crossed it, it may be dealt with in the same manner as venous stenosis.

Sometimes, with tight occlusions, the guidewire may be able to cross the lesion but a catheter over it may not follow suit. Without getting a catheter down, an angiogram to confirm successful traversal of the occlusion cannot be done. In such situations, it is recom-mended to come from a femoral access, snare out the wire traversing the occlusion from

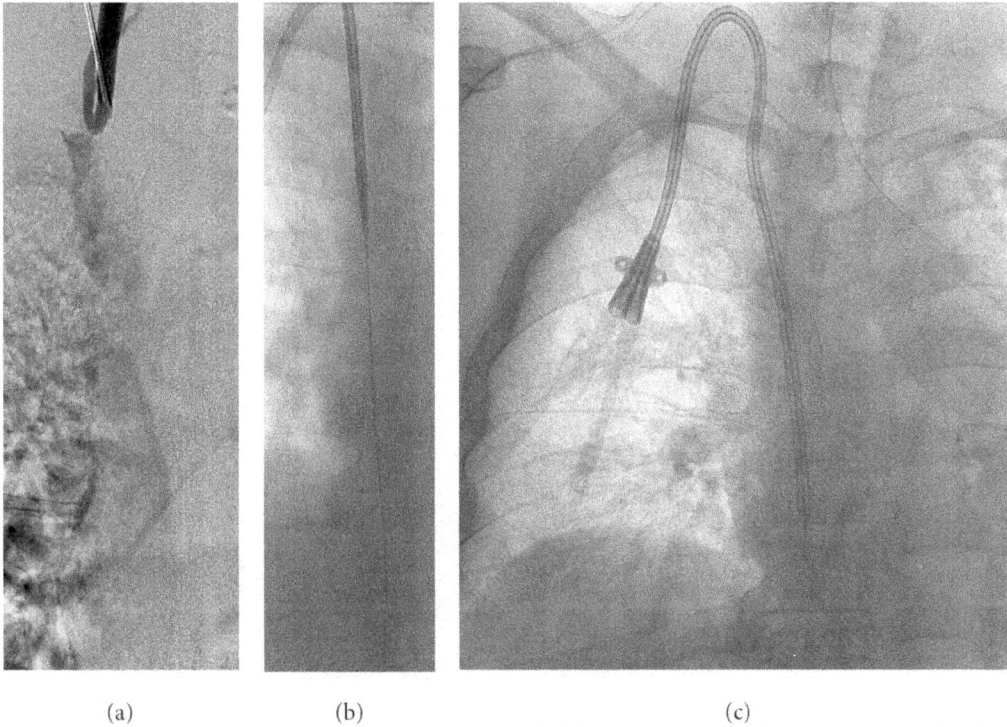

(a)   (b)   (c)

Fig. 8. This patient with history of previous right sided catheterization arrived for tunnelled HD catheter insertion. The right IJV in the neck was patent but small in caliber. It was punctured in the lower neck with an 18 G needle. As the guidewire could not slide down smoothly, angiogram was done through the puncture needle. It showed a focal tight stenosis in the upper brachiocephalic vein (a). With the cause of obstruction identified, it was negotiated with a guidewire. This stenosis was serially dilated (dotterisation) over the wire with the dilators from the HD catheter set (b). Subsequently the tunnelled HD catheter was inserted through this access with the catheter tip placed in the mid-lower atrium (c). As the stenosis could be easily crossed with the dilators and it was far from the right atrium where the catheter tip had to be parked, we did not attempt angioplasty for the venous stenosis.

the IVC or iliac veins, bring it out through the groin puncture and establish body floss. This confirms that the wire is within the vessel and has not perforated. A body floss wire also provides extra support to push down balloon dilatation catheters, as it can be pulled from both ends and held taut. After predilatation, such obstinate occlusions often need a stent to keep them adequately open.

When other options are exhausted, sharp recanalization of a central venous occlusion for insertion of a tunnelled HD catheter can be done. This is especially feasible and safe for short segment occlusions with a relatively straight segment of vein proximal and distal to the occlusion. Different techniques for sharp recanalization have been described in literature, some using simple hardware like a Chiba needle or TIPS needle while some have used more sophisticated apparatus like Outback vascular re-entry device (Cordis, NV, US).[11,12]

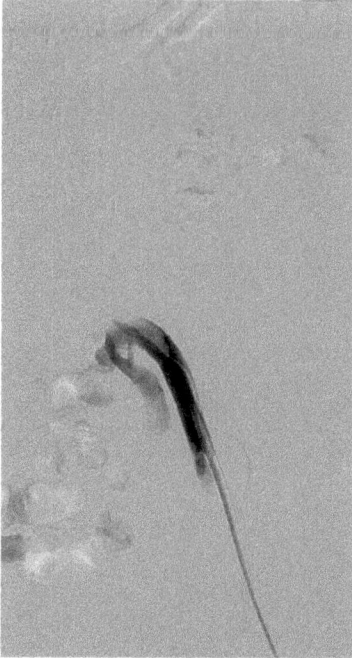

(a)

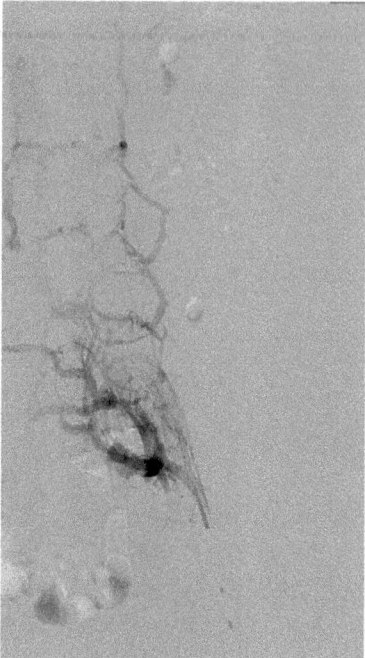

(b)

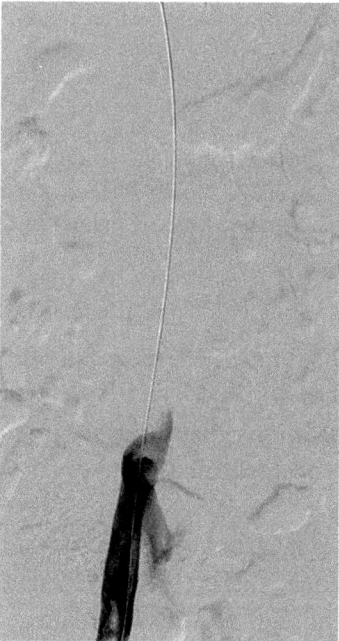

(c)

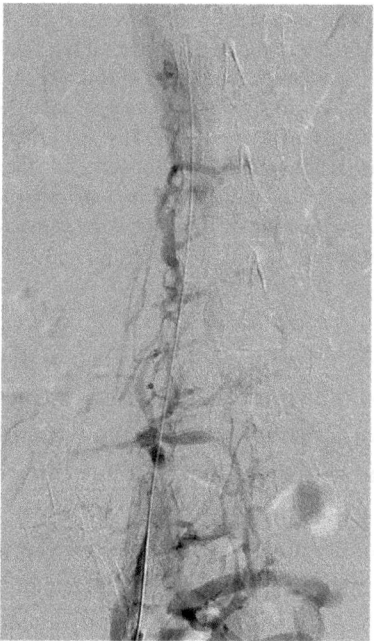

(d)

Fig. 9. (*Continued*)

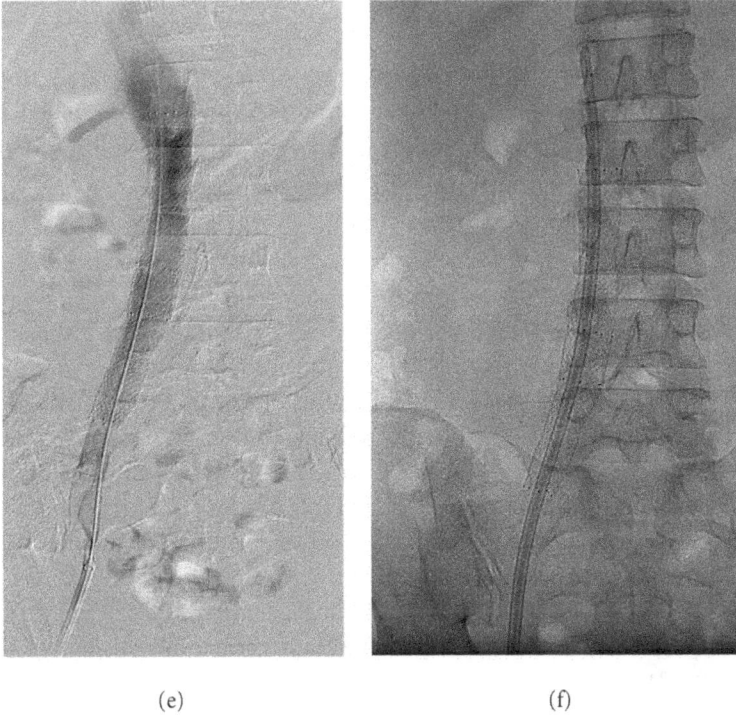

(e)   (f)

Fig. 9.   This elderly patient with multiple comorbities besides ESRD, had chronically occluded central veins with no suitable site for HD access from the neck or upper extremities. He was known to have chronic right iliac venous occlusion from an earlier CT scan. He arrived with a non-functioning left femoral non-tunnelled HD catheter. A venogram done through this catheter showed left common iliac vein thrombosis (a) with reflux into the retroperitoneal veins (b). As there was a reasonable thrombus load in this common iliac vein, we did not attempt recanalization on this side keeping in view the risk of embolization of the clots, in the background of his poor health status. As the right iliac vein occlusion was long standing, we decided to attempt recanalization on that side. Right femoral venogram showed the known right iliac venous occlusion. The occlusion was crossed with a guidewire (c & d). The occlusion extended into the lower IVC. The chronic right iliac and IVC occlusions were initially balloon dilated. However, they were refractory to balloon angioplasty. Hence, lower IVC and right common iliac vein stenting was done with good outcome (e). Subsequently, a right femoral tunnelled HD catheter was inserted through the stents with its tip placed in a normal segment of the IVC (f).

Conclusion

Good endovascular practices enable an operator to safely and successfully overcome most of the challenging situations that can be encountered during a tunnelled HD catheter insertion. In difficult situations, some minor improvizations are often required to succeed in a procedure and this can be learnt only through experience.

References

1. K/DOQI Clinical practice guidelines for vascular access. *Am J Kidney Dis.* 2006; **48**: (suppl 1): S183–S247.
2. US Renal Data System, USRDS 2012 Annual Data Report: Atlas of Chronic Kidney Disease and End-Stage Renal Disease in the United States, National Institutes of Health, National Institute of Diabetes and Digestive and Kidney Diseases, Bethesda, MD, 2012. http://www.usrds.org/adr.aspx
3. Schillinger F, Schillinger D, Montagnac R, *et al.* Post catheterization vein stenosis in haemodialysis: Comparative angiographic study of 50 subclavian and 50 internal jugular accesses. *Nephrol Dial Transplant.* 1991; **6**: 722–724.
4. Trerotola SO, Kuh-fulton J, Johnson MS, *et al.* Tunnelled infusion catheters: increased incidence of symptomatic venous thrombosis after subclavian versus internal jugular venous access. *Radiology.* 2000; **217**: 89–93.
5. Vats HS, Bellingham J, Pinchot JW, Young HN, Chan MR, Yevzlin AS. A comparison between blood flow outcomes of tunnelled external jugular and internal jugular hemodialysis catheters. *J Vasc Access.* 2012; **13**(1): 51–54.
6. Wellons, ED, *et al.* Transthoracic cuffed hemodialysis catheters: A method for difficult hemodialysis access. *J Vasc Surg*, 2005; **42**(2): 286–289.
7. Lund GB, Trerotola SO, Scheel PJ Jr. Percutaneous translumbar inferior vena cava cannulation for hemodialysis. *Am J Kidney Dis.* 1995; **25**(5): 732–737.
8. Smith TP, Ryan JM, Reddan DN. Transhepatic catheter access for hemodialysis. *Radiology.* 2004; **232**(1): 246–251.
9. Balamuthusamy S. Self-centering, split-tip catheter has better patency than symmetric-tip tunnelled hemodialysis catheter: single-center retrospective analysis. *Semin Dial.* 2014; **27**: 522–528. doi: 10.1111/sdi.
10. Athreya S, Scott P, Annamalai G, Edwards R, Moss J, Robertson I. Sharp recanalization of central venous occlusions: a useful technique for haemodialysis line insertion. *Br J Radiol.* 2009; **82**(974): 105–108.
11. Anil G, Taneja M. Revascularization of an occluded brachiocephalic vein using Outback-LTD re-entry catheter. *J Vasc Surgery.* 2010; **4**: 1038–1040.

CHAPTER 5

Assessment of Vascular Access Maturity and Long Term Performance

Jackie P. Ho

Access Maturation and Successfully Used for Homodialysis

There are various definitions[1–3] for "failure to mature" (FTM) for a vascular access. The most fundamental purpose for vascular access creation is to:

(1) Allow repeated successful cannulation for hemodialysis;
(2) Sustain enough blood flow to provide adequate clearance;
(3) Perform in a timely manner.

To be easily cannulated by trained dialysis nurses, there are some basic physical criteria to fulfil:

• A luminal size about 6 mm or above;
• Depth from skin surface <6 mm;
• A relatively straight part at least about 10 cm or two relatively straight parts of 4 cm each.

Individual skilful staff may be able to cannulate access that does not reach all of the requirements. The criteria of repeated cannulation success can be two needles in six consecutive hemodialysis sessions or in more than 2/3 of hemodialysis session within one month.

Amount of flow in the vascular access determines the amount of blood going through the dialysis circuit to provide clearance of body waste. This is usually less of a problem for AVG as thrombosis will occur if the flow is too low. In AVF, the flow may still present but unable to provide adequate clearance. For Caucasian patients, usually the lowest HD machine set flow rate Qb threshold of 300 ml/min or 350 ml/min is required. However in Asians or patients with small body size, a Qb of 200 ml/min could also be sufficient though higher Qb is preferred. AVF able to sustain threshold or higher Qb is considered good for hemodialysis. Another method to evaluate flow is the access flow, the threshold value is 600 ml/min.

Some clinicians use three months after creation as a cut off (early dialysis suitability failure) and some use six months (late dialysis suitablility failure) for diagnosis of FTM. Claude *et al.*[3] showed in an Asian cohort, the FTM rate dropped from 60% to 33% if we push the cut off time from three to six months period. This signifies that the maturation is a progressive process. Duration of maturation varies widely and depends on the original vessel size, location of the artery and vein connected, surgical skill as well as patient's demographics and co-morbidities.[2]

For AVF, KDOQI recommends cannulation beyond four weeks of AVFs creation to minimize access bleeding problem. This recommendation is applicable in non-arterialized vein and not for secondary AVF constructed by arterialized outflow vein in a failing or failed access (type 1). Cannulation can be done within 1–2 days for secondary AVF creation with arterialized vein and provided that the size is good. One session of heparin-free hemodialysis may be required after the operation to avoid wound bleeding. For AVG, the recommendation is two weeks after creation. This recommendation applies to most ePTFE graft to allow time for the surrounding tissue to incorporate onto the graft and facilitate hemostasis. There are newer synethetic grafts (rapid access graft) with special engineering features that allow early cannulation (within 1–3 days).[4,5]

The first clinic follow-up assessment after vascular access creation should happen 1–2 weeks post-operation to check:

- Wound condition (any wound infection, cellulitis or wound dehiscence);
- Any swelling of the limb due to venous hypertension and the severity (if severe, or persistent need to rule out central vein obstruction);
- Size and thrill of an AVF (a tourniquet can be applied over the upper arm to check the size and obviousness of the fistula for easy cannulation), thrill or pulsation of an AVG Auscultate for bruit if thrill or pulsation is not obvious;
- Any symptom or sign of steal syndrome.

For AVGs, if there is no feature of infection or steal syndrome and not much soft tissue edema, cannulation of the graft usually can be started after the first clinic visit.

For AVFs, usually more than one clinic visit is needed to assess the maturity before they are ready for cannulation. Quick ultrasound study in clinic provides valuable information about the readiness of AVF for cannulation and cause of poor maturation (if any).

When an AVF or AVG is considered ready for cannulation, it is better for the clinician to mark out on the patient's limb the course of the dialysis access — preferred sites for "A" and "V" cannulation using permanent marker. A photograph record can also be made using the patient's own mobile phone for the dialysis nurses to look at. A memo explaining how the artery and vein (and also graft in AVG) are connected would aid the nurses to understand the vascular access better.

After starting cannulation of the created vascular accesses, patients will be followed up again 2–3 weeks later to evaluate success of cannulation and to detect any difficulty encountered.

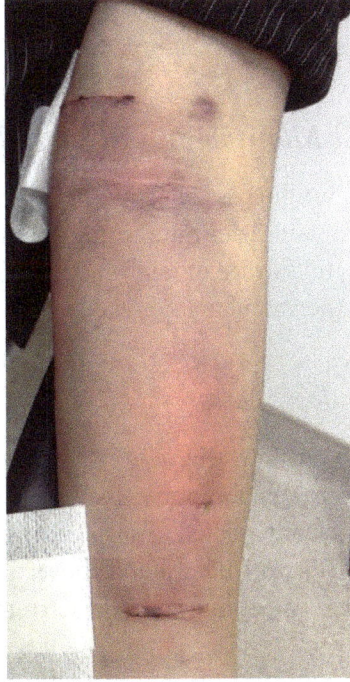

Fig. 1. Post-operative inflammatory response of the patient towards the newly implanted loop BB AVG. White cell count and procalcitonin levels were normal. Her inflammatory response subsided with two weeks of COX 2 inhibitor treatment.

If persistent soft tissue swelling is detected during the first post-operative visit, light pressure bandage (e.g. Tubigrip) can be prescribed for mild condition. Investigation for central vein obstruction has to be considered if moderate to severe persistent swelling is noticed. Excessive swelling of the limb is dangerous as it may lead to wound dehiscence and increases the risk of cellulitis.

Investigation is recommended if an AVF is not considered ready for cannulation by 6–8 weeks.[6] A simple ultrasound scan to assess the condition of AV anastomosis, size of the fistula and depth of the fistula from skin level will help clinicians to understand the underlying issues of poor maturation. For pre-emptive AVFs, longer observation time can be given for maturation process. In incident ESRF patients already using tunnelled CVC for hemodialysis, it is reasonable to suggest intervention if the maturation is not satisfactory by 8–12 weeks. Strategies to manage poor maturation will be discussed in Chapter 7.

Sometimes, difficulty in cannulation of either AVG or AVF is due to the misunderstanding of the access condition between the dialysis nurses and the clinicians. In other situations, the difficulties are related to the uneven size, depth, tortuosity of the fistula or graft. Clinicians marking out the preferred sites with or without ultrasound guidance will facilitate subsequent cannulation. Direct communication and maximal utilization of written and photographic images will help in many occasions. In certain conditions, additional interventional or surgical procedure is needed to enable easy cannulation of the accesses.

The flow chart of Fig. 2 summarizes the strategy to follow up after access creation and review their maturation and readiness for cannulation.

Long-term Vascular Access Monitoring and Surveillance

While it is still controversial about the cost-effectiveness of duplex surveillance of vascular accesses,[7,8] clinical, hemodialysis parameter monitoring and flow evaluation of the vascular accesses enable early detection and salvage of failing accesses, which are certainly helpful in maintaining longer patency of the vascular accesses.[9]

Mode of Failure of a Vascular Access

(1) Outflow obstruction due to outflow vein stenosis or central vein obstruction;
(2) Reduction of inflow due to arterial anastomosis stenosis or inflow arterial stenosis;
(3) Fistula or in-graft stenosis usually related to surgical or needling trauma;
(4) A combination of any of the factors mentioned above;
(5) Other complications include infection, pseudoaneurysm, skin erosion and bleeding etc. (see Chapter 13).

Basic assessment methods include clinical, hemodialysis parameters and flow surveillance. It is more accurate when the data from various categories are analysed together.

Clinical Assessment

(1) Appearance of the fistula

It is important to assess the skin condition over the vascular access. Some patients may be allergic towards cleansing reagent (Fig. 3) or adhesive tapes during dialysis, resulting in rashes. Bad skin condition may lead to infection of the fistula or graft. Occasionally, one may notice hematoma formation (bruising or hardening of subcutaneous tissue) (Fig. 4) around the fistula due to difficult cannulation. Further needling over that region should be withheld for a few weeks to avoid causing irreversible damage to the access. Cluster of needling over the same site of an access may result in aneurysmal change (Fig. 5) or pseudoaneurysm of the accesses. Erosion of skin may develop over the aneurysmal area and may eventually turn into an ulcer. Skin ulceration may lead to infection of the vascular access or torrential bleeding. In slim patients with prominent brachial artery, inadvertent needling of the brachial artery resulting in pseudoaneurysm of the brachial artery had been observed. The dialysis center should be instructed not to further needle that area; sometimes endovascular or surgical repair might be indicated. Difficult needle site hemostasis after hemodialysis might also reflect underlying venous outflow obstruction and high venous pressure.

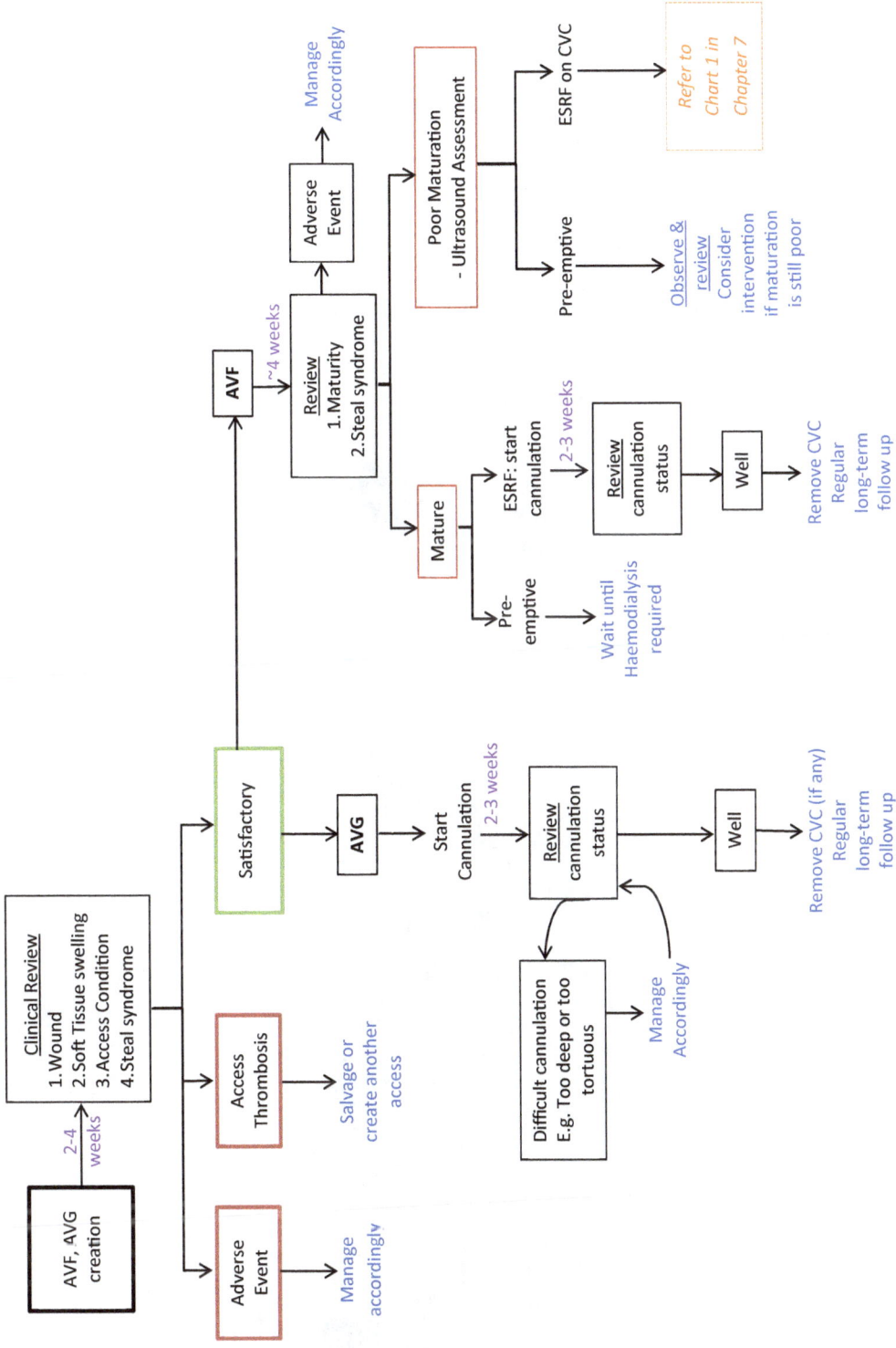

Fig. 2. Flow chart on the strategy to follow-up and review the maturation and readiness for cannulation of vascular accesses after creation.

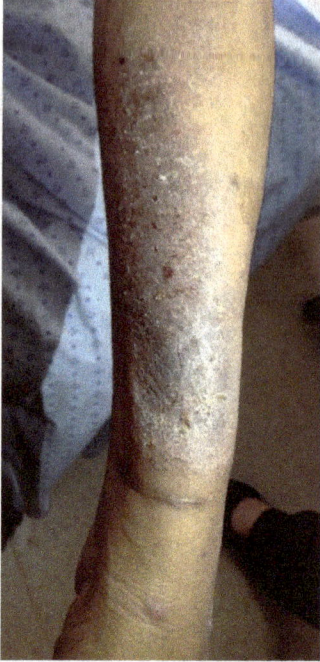

Fig. 3.   Clinical photo showing patient's left forearm skin allergy to antiseptic solution used for cleansing before cannulation.

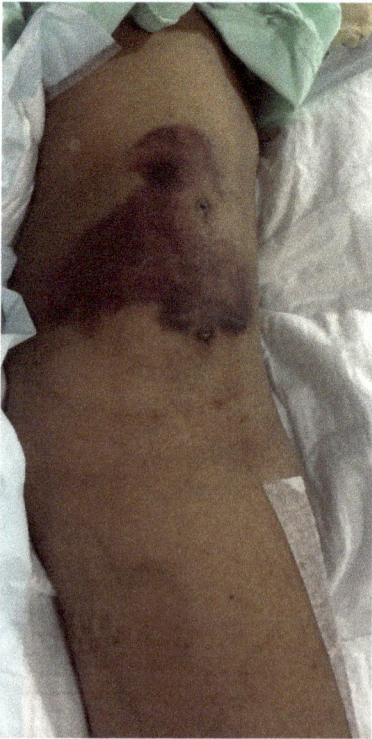

Fig. 4.   Bruising and hematoma formation around the left BC AVF.

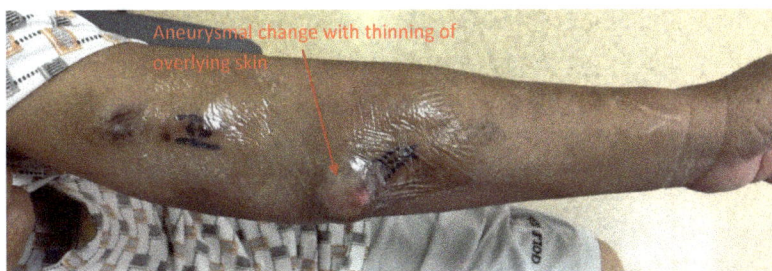

Fig. 5. Cluster needling resulting in aneurysmal change and thinning of the skin overlying the aneurysmal region.

(2) Thrill of AVF and AVG

Thrill can be felt prominently along the pathway of AVF from the AV anastomosis to more proximal region. Thrill is also commonly palpable in AVG near the arterial anastomosis or over the outflow vein. It may not be obvious over the graft, especially for ring supported graft. Auscultation with a stethoscope for bruit may help to determine presence of flow in situations of ringed AVG, deep seeded AVF or presence of edema or hematoma. Reduction or loss of thrill of an AVF or AVG may imply inflow reduction. If a pulsation can be felt over an AVF, it is likely due to a stenosis present downstream along the fistula, outflow vein or central vein obstruction. Such change in AVG is more subtle as the graft wall is thicker and less compliant than native vein wall.

(3) Localized hardening of a segment of venous fistula or vein next to the vein-graft anastomosis

This may signify the presence of a segment of venous stenosis along the fistula or anastomotic region of a graft. It is not uncommon to see presence of venous fistula stenosis over the anastomosis, juxta-anastomotic region and needle clustering sites. Whereas for AVG, stenosis is commonly seen over vein-graft anastomosis or outflow vein nearby.

(4) Generalized swelling of the vascular access limb

If the swelling involves the whole upper limb, it is most likely due to central vein obstruction. There may be presence of facial swelling or dilated subcutaneous vein around the chest or shoulder region in the presence of central vein obstruction. Deep vein stenosis may develop if the vascular access is connected to the deep vein (e.g. BA AVG, forearm loop brachial artery to brachial vein AVG). Rarely, it could be related to sudden onset deep vein thrombosis of the upper limb. One needs to rule out cellulitis and low grade infection if the swelling is only localized around an AVG or AVF.

(5) Although the majority of steal syndrome occurs soon after the vascular access creation, delay onset of steal syndrome is not uncommon. This is usually due to progressive dilatation of outflow vein or development of arterial occlusive disease over the limb with time.

Clinicians have to look out for various symptoms and signs of steal syndrome during follow-up assessment. The clinical features include pain, coldness, numbness, weakness or area of gangrene of the hand or fingers of the ipsilateral limb. Simple assessment of the arterial supply of the hand and finger can be performed by palpation of the radial and ulnar pulses and measuring the individual finger tip's oxygen saturation using pulse oximeter.

Hemodialysis Parameter Monitoring (Fig. 6) and Access Flow Surveillance

(1) Flow: Qa and Qb

Qa, access flow, measures the amount of blood flow through the hemodialysis circuit and is a surveillance tool for vascular access. There are various methods to measure access flow (e.g. duplex velocity measurement with bolus saline dilution, thermal dilution, needle reversal etc.), which will not be discussed in detail here. The normal value of a functional vascular access may vary from 600 ml/min to 2000 ml/min. Thus both actual Qa value and its trend reflect the performance of an access. NKF KDOQI guidelines recommend referral of the patient for fistulogram if Qa <600 ml/min or more than 25% reduction of Qa over a four-month period. In Singapore, we do see some small sized patients with functional AVF maintaining long period adequate hemodialysis at access flow range 500 ml/min to 600 ml/min. It is advisable for clinicians to review clinical, dialysis parameters together with flow value for decision of necessary intervention or treatment.

Table 1.   Comparison of Qa and Qb measurement.

	Access flow Qa	Blood flow rate Qb
Method to measure	Require special machine	A parameter of the dialysis machine
Modality	Surveillance	Monitoring
Availability	Available in some centers	All dialysis centers are able to provide
Sensitivity to reflect failing vascular access	A low actual value (<600 ml/min) or decreasing trend is significant	Qb may remain unaffected until the flow is significantly reduced. Less sensitive to reflect presence of stenosis.
Pitfall	May not drop significantly in the early phase of venous outflow obstruction. It will decrease when more significant stenosis set in. Inaccurate value if the machine is not calibrated or the timing of measurement is wrong. May vary with fluctuation of cardiac output and blood pressure.	Unlikely to drop if the failing mode is mainly venous outflow obstruction. It will reflect only when very severe venous obstruction is present.

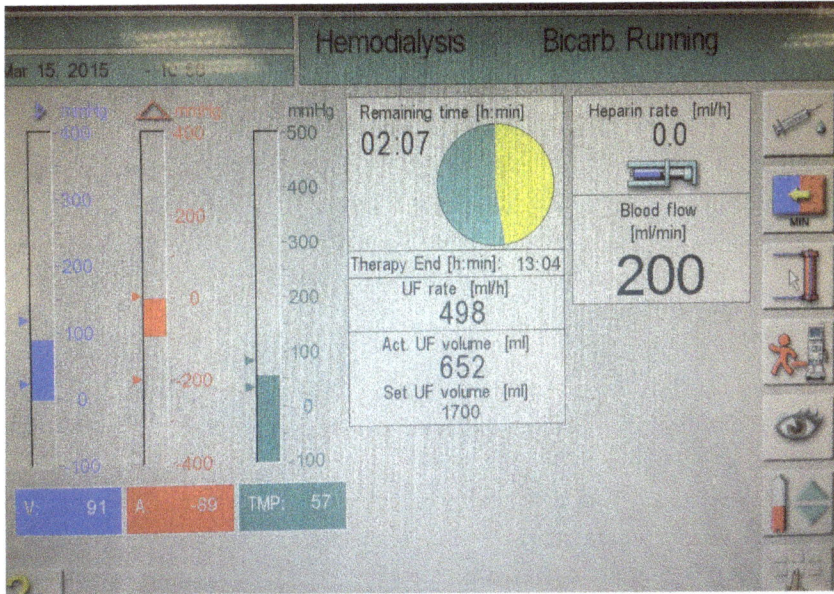

Fig. 6. Dialysis parameters shown on the hemodialysis machine panel.

Remarks: Access flow measurement should be performed in the first 1.5 hours of hemodialysis in order to obtain a more reliable value. The accuracy of access flow measurement could be affected by improper calibration of machine, fluctuation in cardiac output and blood pressure.

Qb, blood flow rate, is the blood flow rate set by the pump of the dialysis machine. The usual value range from 200 ml/min to 350 ml/min. This is the minimal flow rate driven by the machine to run the circuit. Alert signal might be given out due to excessively negative arterial pressure or high venous pressure with the set Qb value in a failing vascular access.

(2) Arterial pressure AP and in sucking

When there is a stenosis over the inflow, in-sucking will be noted. The arterial pressure will be excessively negative. Normally the arterial pressure is less than −140 mmHg with the normal range of Qb. It may vary according to needle gauge, direction and location of "A" needle and sometimes angulation of the needle.

(3) Venous pressure VP

Venous pressure measures the resistance of blood return from the dialysis machine back to the venous system of patients. One can measure either static or dynamic venous pressure. Usually the dynamic venous pressure parameter is more available as it is automatically recorded by the dialysis machine. The dynamic venous pressure recorded at the beginning of hemodialysis (Qb set at 200 ml/min) is more reliable. The threshold of venous pressure to consider presence of venous outflow obstruction varies with machine manufacturers, tubing and needle size. For simplicity, we usually take 150 mmHg as the threshold. Trend of venous pressure change is also important. Progressively increasing VP or persistent VP >150 mmHg likely signifies venous outflow obstruction or central vein obstruction (Fig. 7). However, in vascular access with tortuous course, close proximity of needle tip with a vein wall or graft wall or the presence of a large sized hematoma around the needling area may give a false positive result.

(4) Recirculation

Recirculation can be measured by access flow measuring machine. Increase in percentage of recirculation indicates outflow obstruction. Usually 20% is the threshold.

Hemodialysis parameters are available in all dialysis centers. Provided the staff in dialysis centers recorded the parameters properly, clinicians can easily get hold of the data for access evaluation. Machine to measure access flow is not essentially available in all centers. Those centers with the machine will usually perform the measurement about once every month. In Singapore, usually the dialysis center will either provide the hemodialysis chart with Qb, VA and VP, or they will provide the access flow Qa value.

Duplex Ultrasound Surveillance

KDOQI(2006) recommended duplex surveillance study for AVG as well as AVF.[9] However, duplex ultrasound is also costly and not available in the dialysis center. Patients have to

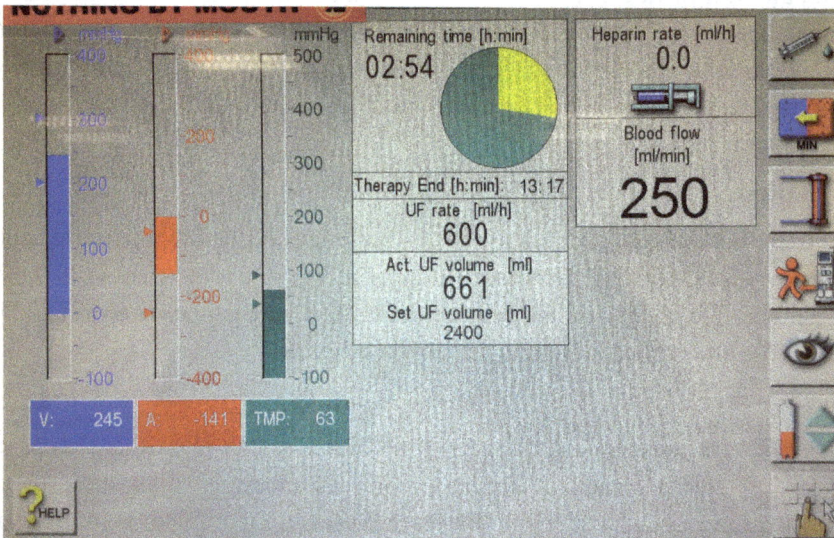

Fig. 7. Dialysis parameters of a failing fistula with elevated arterial and venous pressure.

take extra time off for the surveillance study in either a hospital or a clinic. Most clinicians would recommend duplex surveillance for high risk vascular access which are namely, vascular access with previous stenosis or thrombosis, precious vascular access, patients with a track record of frequent stenosis in previous vascular accesses. Clinical and dialysis parameter monitoring bear a high specificity of access stenosis but rather low sensitivity, as poor as 35.8%, which is demonstrated in the study by Malik *et al.*[10] When there is clinical or dialysis parameter indicating failing vascular access, diagnostic duplex study of the vascular access is not necessary. It will delay the salvage procedure for the failing vascular accesses as well as increase the medical cost involved. Duplex ultrasound surveillance aims to pick up stenosis that is not clinically obvious and enables early intervention to prevent sudden thrombosis. Although the findings of the randomized controlled studies are not congruent, many clinicians believe there is a value of duplex surveillance to prolong the patency of stenosis-prone high risk vascular access, or at least reduce hospital admission[11] due to acute thrombosis. Nonetheless, the healthcare reimbursement policy and the economic resources for dialysis patients also greatly determine the practicability of a duplex surveillance program.

Duplex ultrasound study of a vascular access provides both structural as well as flow information. It is able to detect hemodynamically significant stenosis (by measuring the velocity ratio proximal and distal to a stenotic area and diameter reduction on B mode), measure the flow of the vascular access (usually measure the flow in the inflow artery), pick up lesions outside the access that cause mechanical compression (e.g. hematoma compressing the fistula) and identify other structural problems like pseudoaneurysm, thrombus dissection etc. In terms of the velocity criteria for duplex diagnosis of stenosis, our previous study[12] found a good congruence between duplex study velocity ratio criteria for artery stenosis (Figs. 8a, b) and the angiographic stenosis percentage. However, some researchers suggest that the criteria of velocity ratio >2 representing >50% stenosis in arterial disease may over-estimate the percentage of stenosis in a vascular access.

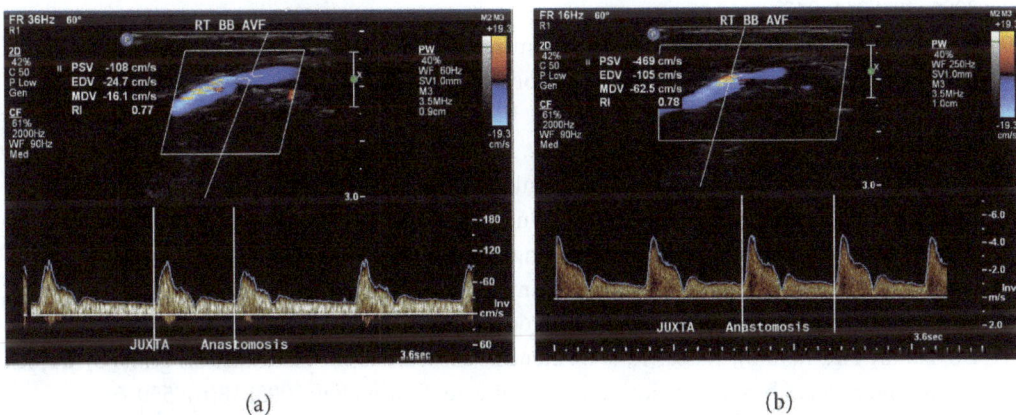

(a)　　　　　　　　　　　　(b)

Fig. 8.　Duplex assessment showing high grade juxta-anastomosis stenosis of a BB AVF with significant velocity increase over the stenotic site (b) compared to pre-stenotic segment (a).

Simple Ultrasound Study

The author found that quick ultrasound study performed by clinician on vascular accesses with any unfavorable feature in the clinic facilitates clinical decision of necessary intervention or surgical treatment for failing accesses. Although the quick scan may not be highly accurate, it can easily pick up segments of stenosis along the accesses, hematoma, thrombus and pseudoaneurysm.

Angiogram of the vascular access (fistulogram) is recommended if the clinical features, hemodialysis parameters, flow surveillance or duplex ultrasound result suggest inflow or outflow problem. We will discuss the approach of fistulogram in the next chapter. Intervention should be performed in the same angiographic session if significant lesion(s) being identified.

References

1. Beathard GA, Arnold P, Jackson J, *et al.* Aggressive treatment of early fistula failure. *Kidney Int.* 2003; **64**: 1487–1494.
2. Lok CE, Michael Allon M, Moist L, *et al.* Risk equation determining unsuccessful cannulation events and failure to maturation in arteriovenous fistulas (REDUCE FTM I). *J Am Soc Nephrol.* 2006; **17**: 3204–3212.
3. Renaud CJ, Ho P, FRCS, Lee EJ, *et al.* Comparative outcomes of primary autogenous fistulas in elderly, multiethnic Asian hemodialysis patients. *J Vasc Surg.* 2012; **56**: 433–439.
4. Schild AF, Schuman ES, Noicely K, *et al.* Early cannulation prosthetic graft (Flixene™) for arteriovenous access. *J Vasc Access.* 2011; **12**(3): 248–252.
5. Aitken EL, Jackson AJ, Kingsmore DB. Early cannulation prosthetic graft (Acuseal™) for arteriovenous access: A useful option to provide a personal vascular access solution. *J Vasc Access.* 2014; **15**(6): 481–485.
6. National Kidney Foundation. Kidney disease outcomes quality initiative clinical practice guidelines for vascular access: 2006 updates — treatment of fistula complications. *Am J Kidney Dis.* 2006; **48**: S234–242.
7. Robbin ML, Oser RF, Lee JY, *et al.* Randomized comparison of ultrasound surveillance and clinical monitoring on arteriovenous graft outcomes. *Kidney Int.* 2006; **69**(4): 730–735.
8. Malik J, Slavikova M, Svobodova J, *et al.* Regular ultrasonographic screening significantly prolongs patency of PTFE grafts. *Kidney Int.* 2005; **67**(4): 1554–1558.
9. KDOQI 2006 Updates Clinical Practice Guidelines and Recommendations. Clinical Practice Recommendations for Vascular Access. Clinical Practice Recommendations for Guideline 4: Detection of Access Dysfunction: Monitoring, Surveillance, and Diagnostic Testing.
10. Malik J, Slavikova M, Malikova H, *et al.* Many clinically silent access stenoses can be identified by ultrasonography. *J Nephrol.* 2002; **15**(6): 661–665.
11. Dossabhoy NR, Ram SJ, Nassar R, *et al.* Stenosis surveillance of hemodialysis grafts by duplex ultrasound reduces hospitalizations and cost of care. *Semin Dial.* 2005; **18**(6): 550–557.
12. Raju AV, May KK, Zaw MH, *et al.* Reliability of ultrasound duplex for detection of hemodynamically significant stenosis in hemodialysis Access. *Ann Vas Dis.* 2013; **6**(1): 57–61.

CHAPTER 6

CO_2 Angiography: Application and Caution

Kyung J. Cho

In the 1950s and early 1960s, carbon dioxide (CO_2) gas was used as an intravenous contrast agent to delineate the right heart for evaluation of suspected pericardial effusion.[1,2] This imaging developed from animal and clinical studies, which demonstrated that CO_2 use was safe and well tolerated with peripheral venous injections. With the advent of digital subtraction angiography (DSA) in 1980, CO_2 became a useful contrast agent for vascular imaging.[3] Now with the availability of high-resolution DSA, stacking software, and reliable gas delivery system, CO_2 angiography has become widely used for vascular diagnosis and endovascular procedures.

The development of each of these clinical applications demonstrates the unique benefits of the use of CO_2 in vascular imaging. Among these benefits, CO_2 is non-allergenic and non-nephrotoxic, allowing its use in patients with renal failure and iodinated contrast allergy for the arterial and venous imaging except for arterial imaging above the diaphragm.[4] Unlimited total volumes of CO_2 can be injected for diagnostic angiography and complex endovascular procedures when CO_2 injections are separated by two to three minutes. Finally, when used as a contrast agent with small amounts of added iodinated contrast medium, CO_2 can reduce the risk of nephrotoxicity of iodinated contrast.[5,6] This chapter discusses the principles, techniques, and practices of CO_2 angiography with emphasis on the use of CO_2 in patients with renal failure and hemodialysis.

Unique Physical Properties of CO_2

A thorough understanding of the physical properties of CO_2 combined with the development of a facile catheterization and imaging techniques are essential in obtaining a successful CO_2 angiogram. CO_2 is a colorless, odorless, and naturally occurring gas, making up 0.035% of the atmosphere. CO_2 is a buoyant and compressible gas with low viscosity (1/400 times iodinated contrast). The low viscosity of CO_2 allows delivery of diagnostic quantities of the gas through end-hole catheter (eliminating the need for a pigtail catheter), micro-catheters, skinny needles (22 gauge to 25 gauge), between the guidewire and the catheter, and through the side port of the sheath of the stent delivery system.

When CO_2 is injected into a vein, it rapidly dissolves to form carbonic acid and is transported to the lungs as bicarbonate or gas bubbles in the blood stream into the pulmonary arteries, where the gas is removed by the lungs. CO_2 is 20 times more soluble than oxygen. The solubility of CO_2 in blood can be demonstrated by using a DSA technique. When 5 cc of CO_2 is trapped in the right atrium in the left lateral decubitus position (right side up position), the CO_2 will be dissolved within 30 seconds in swine. When CO_2 bubble is seen in the pulmonary outflow tract following gas injection into the hepatic vein or IVC, the gas should dissolve within 15 to 20 seconds.

CO_2 is a negative contrast agent with molecular weight of 42 (molecular wt. of O_2 and N_2 are 32 and 28, respectively), and therefore, CO_2 imaging requires the DSA technique with good contrast resolution. CO_2 is visible under fluoroscopy during the procedure, allowing its use for a test injection to confirm the position of the catheters, balloons and stents. Peripheral intravenous CO_2 injections will visualize the veins for targeting the basilic or cephalic vein for a PICC placement. When CO_2 is injected in a vessel, the gas will not mix with the blood, allowing its use as a flushing agent (injection of 5–10 cc of CO_2 every 2 minutes) to maintain catheter patency.

Digital subtraction imaging of CO_2 is obtained from the negative gas density produced by displacing the blood within the vessel being studied. The buoyancy and compressibility of CO_2, may lead to incomplete displacement of blood, and thus result in incomplete visualization of the posterior branches of the aorta (Fig. 1). However, the advantage of the CO_2 buoyancy is filling of the anterior branches including visceral arteries as well as renal transplant artery. When the renal arteries originate from the posterolateral aspect of the aorta, the renal artery can be better visualized by turning the patient to position the kidney anterior to the injection site.

CO_2 can be compressed within the syringe and catheter during injections and will expand upon exiting the catheter (explosive delivery). Purging the catheter with 5 cc of CO_2 prior to the injection of a diagnostic volume of CO_2 decreases gas compression and minimizes explosive delivery.

Properties of CO_2

- Endogenous
- Invisible (colorless and odorless)
- Non-nephrotoxic
- Non-allergenic
- 20 times more soluble than O_2 in the blood
- 400 times less viscous than iodinated contrast
- Buoyant
- Compressible
- Immiscible with blood
- Effective removal of CO_2 by the lungs

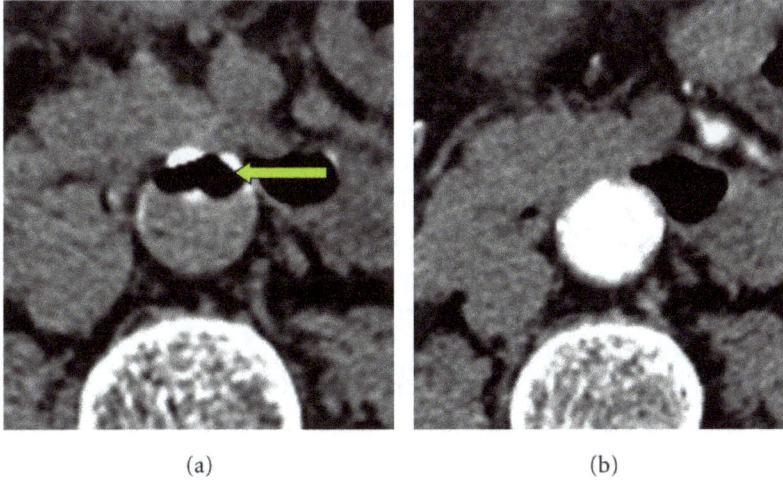

(a)   (b)

Fig. 1.   Axial CT scan at the level of abdominal aorta with the intra-aortic injection of CO$_2$ (a) and iodinated contrast (b). The buoyant CO$_2$ can be seen in the anterior (nondependent) part of the aorta (arrow), whereas the contrast medium has been mixed with the blood filling the entire lumen.

- Injection of unlimited total volumes of CO$_2$
- Shorter examination times
- Eliminates or decreases contrast medium volumes

CO$_2$ Delivery Methods

Many CO$_2$ delivery systems have been used, including the hand-held syringe method, computer-controlled CO$_2$ injectors, the plastic bag system,[7,8] power injectors, and CO$_2$mmander/AngiAssist. Once a syringe has been filled with CO$_2$ from a highly-pressurized CO$_2$ cylinder, its pressure should be decreased to the atmospheric pressure by quickly opening and closing the stopcock. This CO$_2$-filled syringe should not be left on the procedure table with the opened stopcock before injection. Preferably, the syringe should be filled with CO$_2$ just before injection. Once the CO$_2$-filled syringe is opened, the CO$_2$ in the syringe is rapidly replaced by room air due to the difference of CO$_2$ partial pressure between the syringe and atmosphere. This syringe system is inconvenient as the syringe should be re-filled before each of the multiple injections.

The plastic bag system (AngioDynamics, Queensbury, NY, US) has been discontinued due to a serious complication resulting from incorrect use of the device. A similar plastic bag system is available from Merit Medical Systems, Inc. (South Jordan, Utah, US) (Fig. 2). Once the plastic bag system is opened, all connection of the delivery system should be checked for an air-tight connection. The bag should be filled and emptied with CO$_2$ three times before final filling to remove air from the bag. Once the CO$_2$-filled bag is connected to the delivery system, the stopcock attached to the bag is closed, and a 30 cc or 60 cc

Luer-lock syringe is connected to the injection port. An air-tight connection of the system should be checked for any air leaks by aspirating the syringe. The delivery system should be flushed with CO_2 into the room via the three-way stopcock to remove air from the delivery system.

Another recently introduced, FDA-approved portable CO_2 delivery system (Fig. 3) is the CO_2mmander with AngioAssist (PMDA, LLC, Ft. Myers, FL, US). It allows for the delivery of CO_2 at a low pressure to any reservoir such as the plastic bag or a large syringe with the Luer-lock fitting. The AngiAssist is a unique stopcock system with valves that control direction of gas flow, allowing gas delivery in a non-explosive fashion using a 60 cc reservoir syringe and a 30 cc injection syringe.

CO_2 may be delivered without filtration from the CO_2 cylinder for venography or arteriography. However, the use of a filter (0.2 μm) can effectively remove particulate contamination and bacteria (0.5–5.0 μm). At the University of Michigan, gas filtration is used for CO_2 delivery with a new filter for each procedure. We currently use a FDA-approved Filter, Syringe Pharmassure 0.2 micron w/HT Tuffryn Membrane Pall Corporation #HP1002 (Ann Arbor, MI, US).

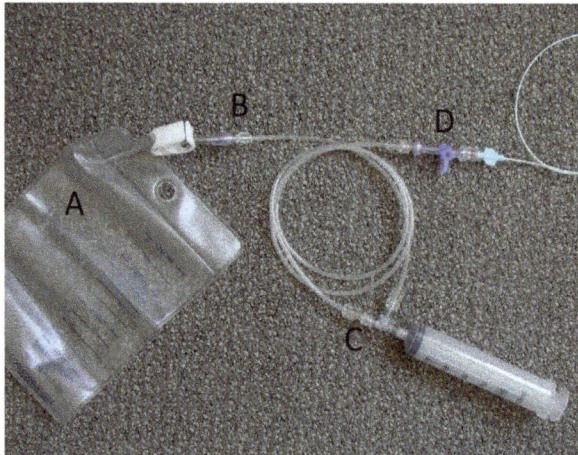

Fig. 2. The Plastic Bag System (Merit Medical Systems, Inc. South Jordan, UT, US) with multiple check flow valves, eliminating the need for stopcock manipulation. The plastic bag (A) contains 1200 cc of CO_2. Once the bag is filled with CO_2 and connected to the delivery system (B), the check flow valves (C) allow aspiration of CO_2 from the bag and injection into the catheter through the 3-way stopcock (D).

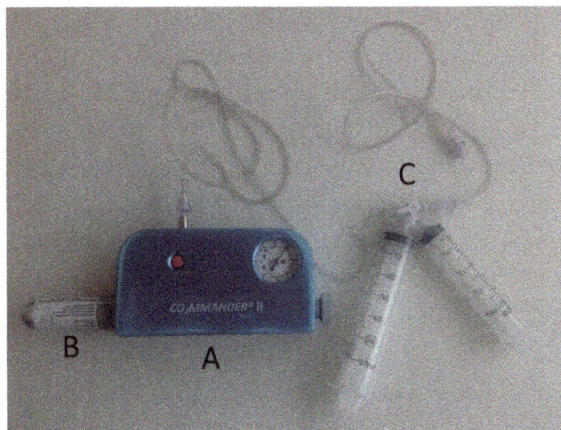

Fig. 3. CO$_2$mmander (PMDA, LLC, N. Ft. Myers, FL, US) (A) is a FDA-approved portable medical CO$_2$ delivery system. It can deliver CO$_2$ at a low pressure to any reservoir such as the plastic bag or a syringe with the standard Luer-lock fitting. The CO$_2$mmander utilizes 25 g cartridge of medical grade CO$_2$ (B), USP, containing 10,000 ml of CO$_2$. The K-valve of the AngiAssist (C) system located between the tubing from the CO$_2$mmander and the tubing to the catheter controls gas flow from the CO$_2$mmander to the 60 cc syringe, to the 30 cc syringe, and then into the catheter. The 30 cc syringe is used for CO$_2$ delivery into the catheter.

CO$_2$ Injection and Imaging

The injection rate and volume of CO$_2$ usually depend on the diameter and length of the vessel being studied. In the majority of cases, the duration of CO$_2$ injection is 1–2 s. The injection rates are: 30–40 cc for the IVC and aorta and 10–20 cc for selective injection into the renal, iliac and femoral arteries. Care must be taken not to reflux CO$_2$ during upper extremity fistulography or brachial arteriography to avoid potential neurologic complications (cerebral embolism).[4] The amount of CO$_2$ injected for fistulography should be less than 15 cc/1.5 seconds. The injection of CO$_2$ into a small diameter vessel such as brachial artery or an injection into an AVF with venous outflow obstruction may increase central gas reflux.

The basic components of the digital subtraction system should include a high-output X-ray generator, a high quality image intensifier, appropriate television imaging, and computer image processor including stacking software for integrating a series of images to solve problems associated with the breakup of CO$_2$ bubbles following injection. As motion degrades CO$_2$ images, additional mask images and rapid exposure (4 to 6 frames /sec) are often essential for post processing of the images.

Monitoring Patients Undergoing CO$_2$ Angiography

The current practice in angiography suite includes continuous monitoring of vital signs, including pulse oximetry (oxygenation), ECG/pulse (circulation), blood pressure (circulation) and respirations (ventilation). Adding end tidal CO$_2$ (ETCO$_2$) can improve patient safety during CO$_2$ angiogram. Capnography is the most reliable monitor for air contamination or pulmonary vapor lock since it can monitor ventilatory and hemodynamic function of the lungs. Bradycardia or arrhythmia during CO_2 angiography may occur due to gas embolism of the coronary artery from inadvertent CO_2 injection into the thoracic aorta or due to paradoxical embolism. Pulmonary artery pressure increases mildly lasting less than 5 minutes with the intravenous injection of a diagnostic volume of CO_2. Inadvertent injection of excessive volume of CO_2 or room air will cause a decrease in systolic blood pressure and $ETCO_2$ but may cause little change in SpO_2 initially. When capnography is not available, blood pressure should be monitored one minute after CO_2 administration. The maximum hemodynamic effect of inadvertent administration of a large volume of CO_2 will occur at one minute after the injection. If an air embolism is suspected, the patient should be placed in the left lateral decubitus position (right side position) and the Trendelenburg position to trap the gas in the right atrium to allow blood to flow under the trapped gas.

Vital Signs Monitoring

- Systemic blood pressure (1 minute after CO_2 injection)
- ECG
- SpO_2 (Pulse Oximetry)
- Respiratory rate
- End-tidal CO_2 (Capnography)

Advantages, Disadvantages and Limitation of CO_2

The most important advantage of CO_2 is the lack of nephrotoxicity and allergic reaction that allows its use in unlimited amounts during diagnostic and endovascular procedures in patients with impaired renal function or iodinated contrast allergy. Of additional imaging benefit, CO_2 reflux allows visualization of central vessels that may not be visualized with iodinated contrast.

Disadvantages associated with the use of CO_2 as a contrast agent are related to its unique physical properties. It requires a special delivery system to avoid air contamination and delivery of excessive volume resulting from a direct connection between the catheter and CO_2 cylinder. Additionally, CO_2 cannot be used for imaging of the thoracic aorta, coronary artery and cerebral circulation although not required in hemodialysis access management.

Potential degradation of CO_2 imaging is related to its property as a negative contrast. Respiratory motion, peristalsis and any other motion can degrade CO_2 imaging. As mentioned before, the buoyancy of CO_2 causes incomplete filling of the aorta, only filling the nondependent portion of its lumen. For renal arteries originating posterior to the injection site, they will not fill with CO_2.[5]

CO_2 should be used with caution in patients under nitrous oxide general anesthesia as tissue nitrogen can diffuse into the CO_2 bubble resulting in expansion of the gas. Another caution with CO_2 use is in patients with COPD and patent foramen ovale. While the author has not seen any significant complications from CO$_2$ use in patients with COPD or paradoxical gas embolism from an intravenous injection of gas, caution is recommended. If there is any concern, minimal amounts of CO$_2$ should be injected with the patient turned to his or her right side up position while separating each gas injection by 3–5 min.

Use of CO$_2$ in Patients with Renal Failure and Hemodialysis Access

When dialysis access is not functioning properly, physical examination is initially performed to evaluate the problem, followed by an ultrasound to obtain greater details of the status of the access. Diagnostic angiography is generally only requested when necessary information about the dialysis access and its inflow and outflow status cannot be obtained by ultrasound. Interventional procedures are usually performed to correct the malfunctioning access at the time of diagnostic angiography.

CO$_2$ fistulography and central venography can provide much of the diagnostic information that can be obtained with a contrast medium in patients with failing hemodialysis access.[6] The use of CO$_2$ decreases the amount of iodinated contrast medium required, helps preserve residual renal function and eliminates the need for steroid preparation in patients with iodinated contrast allergy. CO$_2$ venography can be performed in an outpatient facility and requires no pre-procedure lab testing. No or minimal conscious sedation is required. CO$_2$ can be used as a contrast agent in various interventional procedures for failing dialysis access including balloon angioplasty and stenting for arterial or venous stenosis, catheter-directed thrombolysis, and mechanical thrombectomy for clotted dialysis access.

Upper extremity venography

Ultrasound imaging is the first line imaging modality for preoperative vascular mapping of the upper limb arterial and venous systems. However ultrasound has limitation to assess central vein or proximal vein in very obese patients. In patients with renal insufficiency who require upper extremity proximal or central vein evaluation and recanalization, CT venography cannot be used because of the risk of contrast-induced nephropathy (CIN). Gd-enhanced MRA should not be utilized for a roadmap because of the risk of nephrogenic systemic fibrosis. Due to its physical properties, CO$_2$ can be used as an alternative contrast agent for evaluation of the entire upper extremity venous system prior to placement of a dialysis access in patients with residual renal function or contrast allergy (Fig. 4).[9,10]

The patient is placed supine on the fluoroscopic table and the arm being studied is placed in the anatomical position on an arm board. The site for peripheral venous access for CO$_2$ injection depends on the veins to be visualized. For visualization of the subclavian and innominate veins, venous access can be any vein of the arm or hand. The low viscosity of CO$_2$ allows its injection into any vein using a 23 gauge angiocath. Since CO$_2$ tends to flow through the vein being injected, the forearm cephalic vein is likely to fill with CO$_2$ injection into a vein on the radial side of the dorsum of the hand. Once the forearm cephalic vein is filled, CO_2 will continue to flow through the axillary and basilic veins in the upper arm into the axillary and subclavian veins. When the axillary vein fails to fill, a rubber tourniquet is applied to better fill the previously unfilled veins (Fig. 5). Multiple injections of CO_2 are often necessary for imaging the entire arm veins. Each CO_2 injection should be separated by 2 min to allow absorption of CO_2 injected.

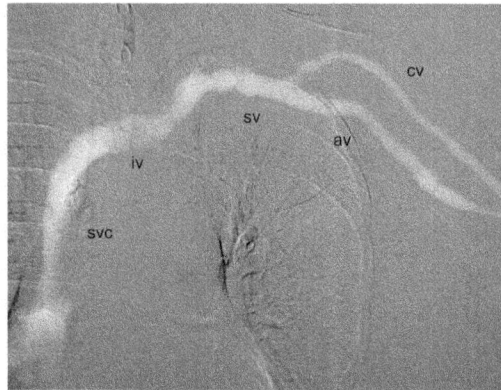

(a)

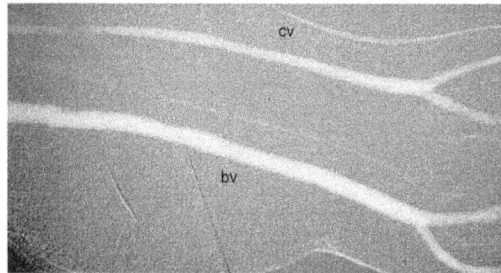

(b)

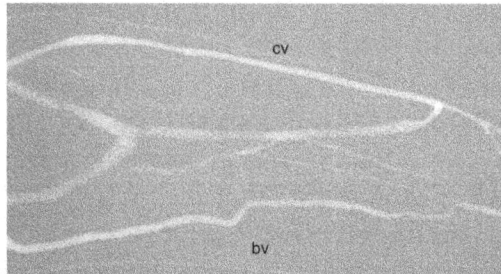

(c)

Fig. 4. CO_2 left upper extremity venogram prior to dialysis access surgery. CO_2 was injected into an intravenous line placed on the dorsum of the hand. (a) Central venogram shows patent axillary (av), cephalic (cv), subclavian (sv), and innominate (iv) or brachiocephalic veins and superior vena cava (svc). (b) Venogram of the upper arm showing patent cephalic (cv) and basilic (bv) veins. (c) Venogram of the forearm showing patent cephalic (cv) and basilic (bv) veins.

The plastic bag delivery system is useful as it prevents air contamination when used correctly and also allows multiple injections. Unlike iodinated contrast medium, inadvertent CO_2 extravasations during the injection will unlikely cause tissue injury. CO_2 injection into a small peripheral vein may cause pain at the site of injection due to explosion of the compressed gas. Prior small amount intravenous injection of lidocaine (40–60 mg) and purging the delivery tubing with CO_2 will help reduce pain from gas injection.

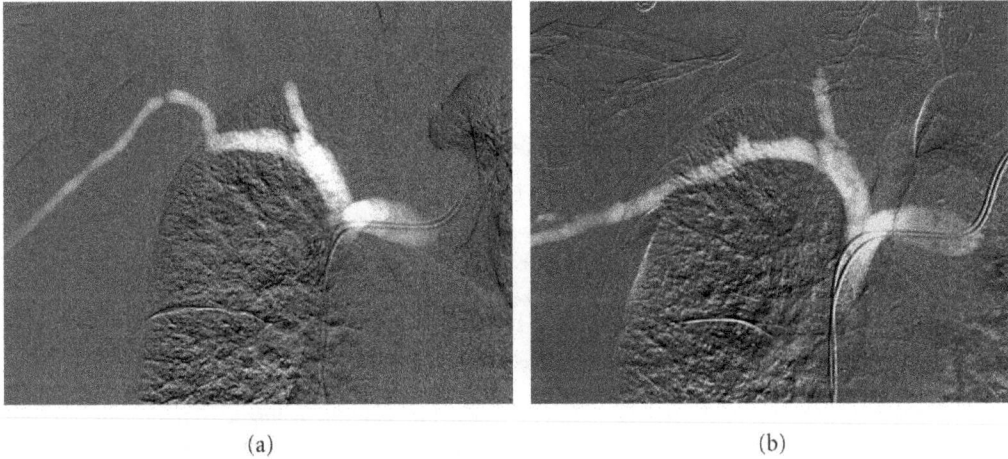

(a)   (b)

Fig. 5. CO_2 right subclavian venograms with or without use of a tourniquet. (a) Venogram with the injection of CO_2 into a peripheral venous access showing filling of the cephalic, subclavian and brachiocephalic veins. There is no filling of the axillary vein. (b) Repeat CO_2 venogram with application of a rubber tourniquet above the elbow. The axillary vein was filled due to compression of the cephalic vein.

Fistulography

CO_2 can be used as a contrast agent in the angiographic imaging of malfunctioning hemodialysis fistula and graft. When a stenosis or occlusion is present, balloon angioplasty or stenting can be performed using CO_2 as a contrast agent. CO_2 should be used in patients with residual renal function, high risk factor for CIN (e.g. failing transplant kidney and the fistula is failing, pre-emptive fistula failed to mature) or hypersensitivity to iodinated contrast.

The techniques used for CO_2 imaging of hemodialysis graft and fistula is similar to that used with contrast medium. Regardless of the type and location of AVF and AVG, access is made to the venous outflow vein toward the arterial anastomosis. Initially 15 to 20 cc of CO_2 is injected into the venous outflow to evaluate the venous anastomosis and central veins. The injection rate of CO_2 should be adjusted to avoid gas reflux through the arterial anastomosis into the brachial artery. CO_2 should not be injected directly into the brachial artery because gas may reflux into the subclavian artery and cerebral circulation (Fig. 6). A CO_2 injection into the cerebral circulation, the coronary artery and even the mesenteric circulation can result in serious complications.[11–15]

For evaluation of a RC AVF, access is made to the outflow vein below the elbow toward the arterial anastomosis using the micropuncture technique. Once the vein is punctured using a 21 gauge needle under ultrasound guidance, a 0.018" Torq-Flex wire is advanced toward the arterial anastomosis. A 3 Fr dilator is then inserted over the wire, and CO_2 DSA is performed with 15–20 cc of CO_2 to image the forearm, upper arm and central veins. If a stenosis is present, the coaxial catheter pair is inserted over the wire and the inner 3 Fr dilator is removed. Over a 0.035" Terumo wire advanced into the proximal radial artery, a 6 Fr sheath (5 cm) is inserted, and 10–15 cc of CO_2 is injected to re-demonstrate the stenosis before balloon angioplasty. A completion venogram is performed with the injection of 15 cc of CO_2 into the outflow vein near the anastomosis.

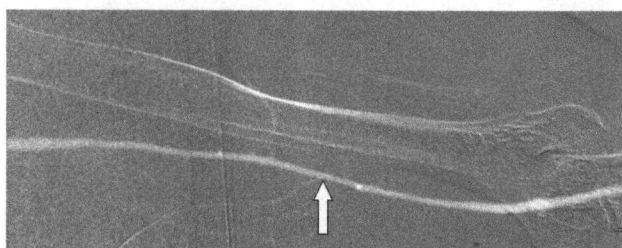

Fig. 6.   Preoperative CO_2 brachial arteriogram in a 51-year-old woman with failing renal transplant and failed bilateral upper extremity dialysis fistula. After accessing the brachial artery, CO_2 arteriogram was performed with CO_2 injection into the distal brachial artery. Note central reflux of the gas toward the axillary artery. The patient had transient lightheadedness, bradycardia and hypotension.

Malfunctioning hemodialysis fistulas and grafts

Percutaneous interventions are usually successful in restoring blood flow in malfunctioning hemodialysis access and have largely replaced surgical revision as the preferred treatment for failing or clotted accesses. Percutaneous interventions have increased the duration of access site patency and saved the venous system for future access creation, and thus reduced the need for temporary hemodialysis catheters.

Angiographic evaluation is usually indicated if clinical or ultrasound assessment suggests either clotted or failing access. The indications for percutaneous catheter interventions include: (1) evaluation for the early failure or immature AVFs; (2) inflow arterial stenosis and anastomosis stenosis; (3) venous outflow stenosis (Figs. 7 and 8): (4) central vein stenosis or occlusion (Fig. 9); (5) access thrombosis; (6) graft aneurysms; and (7) arterial steal with hand ischemia.

The usual entry sites for percutaneous intervention are the same as for diagnostic fistulography and central venography. The access approach is determined on the basis of

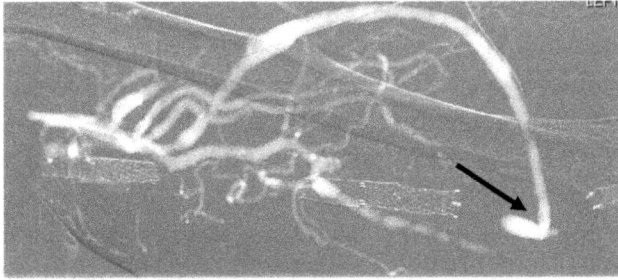

(a)

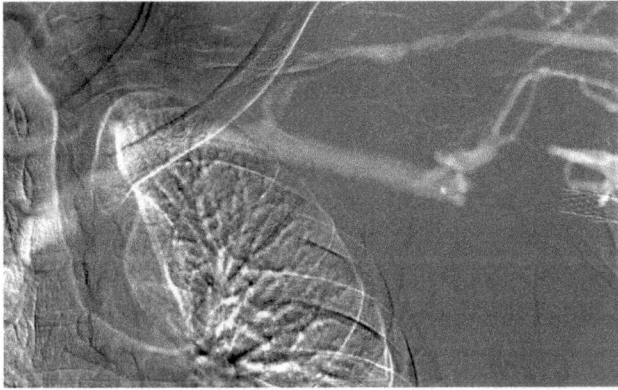

(b)

Fig. 7. CO$_2$ fistulogram in a patient with failing arteriovenous graft and iodinated contrast allergy. (a) After accessing the graft, CO$_2$ venogram shows partially patent graft with patent brachial artery anastomosis (arrow). Stenoses are present at the venous anastomosis. There are multiple stents previously placed for failing brachio-basilic arteriovenous fistula. (b) Central venogram shows occlusion of the proximal axillary vein, and patent subclavian and innominate veins.

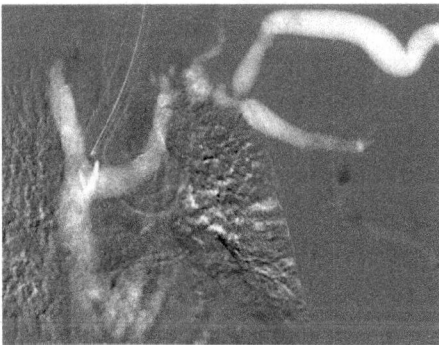 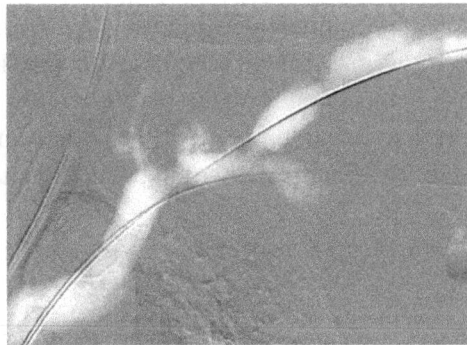

(a)   (b)

Fig. 8. A 77-year-old female with end-stage renal disease and a left brachiocephalic fistula presented with left arm swelling. (a) Venogram with CO$_2$ injection into the fistula showing stenosis at the cephalic vein to axillary vein junction and subclavian vein. (b) After a 10 mm balloon dilatation of the anastomosis and subclavian vein, CO$_2$ venogram showed improvement of the stenoses.

clinical problem, fistula examination and ultrasound evaluation. Percutaneous transfemoral approach may be required for recanalization of a central vein occlusion and stenting with large balloon catheters or stents. CO_2 can be used as a contrast agent in all percutaneous intervention with small amounts of contrast as needed.

The procedure for a forearm loop graft (BC or BBAVG) declotting (Fig. 10) is begun by placing a 6 Fr sheath (5 cm) within the inflow portion of the access toward the venous outflow. A 5 Fr Kumpe catheter (Cook, Inc. IN, USA) is advanced through the venous anastomosis into the outflow vein, and CO_2 venogram is performed to evaluate venous outflow and central veins. Small amounts of contrast or CO_2 are injected into the clotted access to avoid displacing clots into the arterial inflow. Venoplasty of the venous anastomosis is performed using a 6 or 7 mm diameter angioplasty balloon on the assumption of the presence of stenosis at the venous anastomosis. For declotting thrombosed AV graft, mechanical thrombofragmentation and aspiration could be performed using a fragmentation device such as Trerotola PTD (Arrow International, Reading, Pennsylvania, US), the Cleaner (Argon Medical Devices, Inc. Plano, TX, US) or AngioJet (Boston Scientific, Natick, MA, US). A second 6 Fr sheath (5 cm) is inserted into the venous portion of the graft toward the arterial anastomosis in the criss-cross fashion. Pulse spray pharmacomechanical thrombolysis (PSPMT) using tPA or urokinase may be performed as an alternative intervention using two infusion catheters such as Unifuse or SpeedLyser PRO Infusion System (AngioDynamics, Queensbury, NY, US) placed in the criss-cross fashion without prior sheath placement. After completion of thrombofragmentation of the arterial side of the graft, a 4 Fr Fogarty embolectomy catheter (Edwards Lifesciences, Irvin, CA, US) is used to remove the arterial plug from the arterial anastomosis. Once graft blood flow has been established, a fistulogram is performed with CO_2. Recanalization and angioplasty of central vein stenosis or occlusion is performed from the arm and/or femoral vein approach. Stents are used for significant recoil following balloon dilation or vein perforation. The use of self-expandable nitinol covered stents such as FLARE stent (BARD Peripheral Vascular, Tampe, AZ, US) for stenosis at the AV graft venous anastomotic stenosis has achieved longer patency of the access. Furthermore, CO_2 can be used to guide dialysis access catheters such as percutaneous transhepatic hemodialysis catheter placement. CO_2 can also be used safely as a contrast agent in inferior venacavography for various indications including recanalization of occluded IVC and vena cava filter placement in patients with renal failure and at risk for contrast-induced nephropathy.[16,17]

CO_2 angiogram may also be used to guide tunnelled CVC insertion (Fig. 11).

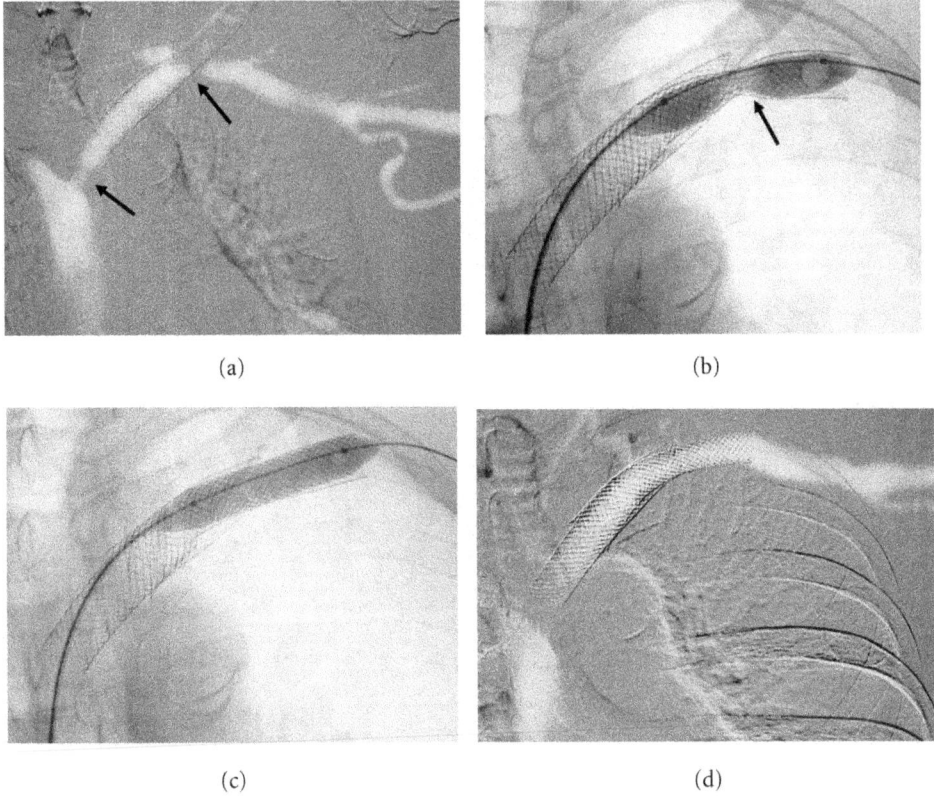

(a)

(b)

(c)

(d)

Fig. 9. Subclavian and brachiocephalic vein stenosis were treated with balloon angioplasty and stent placement in a patient who had previously undergone stent placement in the brachiocephalic vein. (a) CO_2 venogram shows stenosis at the proximal and distal ends of the stent (arrows). (b) After placement of a 12 mm diameter Wallstent, the angioplasty balloon shows waist at the stenosis (arrow). (c) Upon further inflation, the balloon was fully inflated. (d) Completion venogram shows successful dilation of the lesion. The proximal stenosis was subsequently successfully treated with balloon angioplasty (not shown).

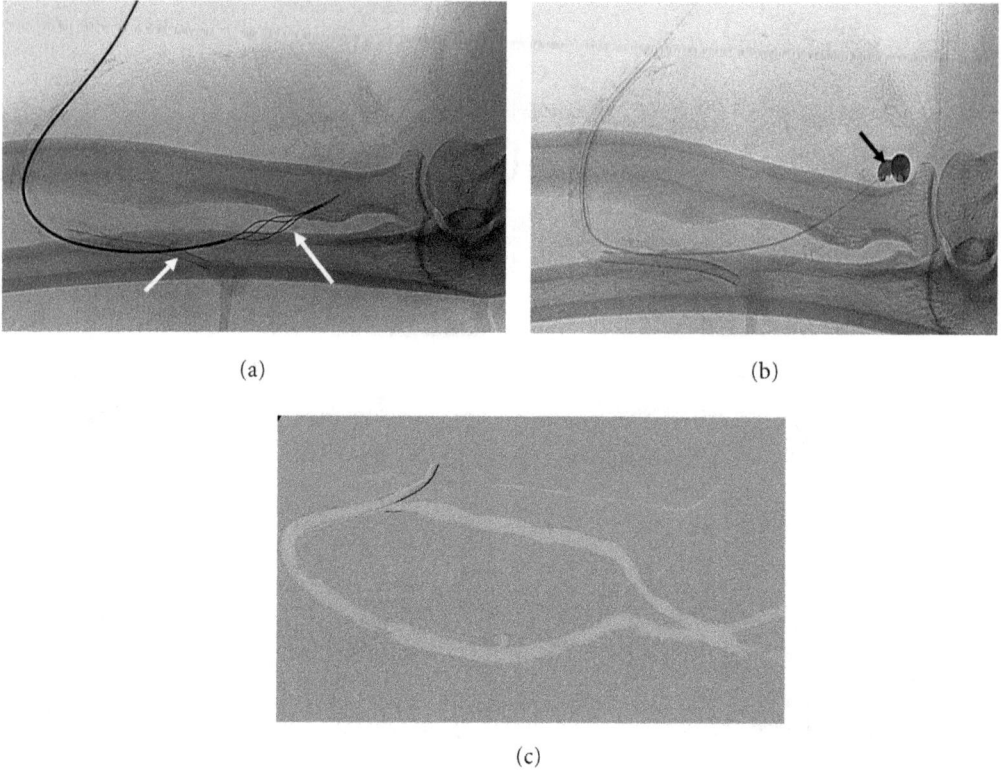

(a)

(b)

(c)

Fig. 10. CO_2-guided declotting of a thrombosed forearm loop graft in a 55-year-old woman with iodinated contrast allergy. The graft was punctured initially on the arterial side toward the venous anastomosis and the venous anastomosis was dilated to 6 mm, followed by thrombofragmentation using Trerotola PTD. (a) From 2nd access to the venous side, thrombofragmentation was done along the arterial side of the graft using Trerotola PTD (longer arrow). The initial access sheath is in the arterial side of the graft (shorter arrow). (b) The arterial plug was removed with a 4 Fr Fogarty Embolectomy Catheter (arrow) (Edwards Life sciences, Irvine, CA, US). (c) A completion venogram with CO_2 shows patent graft with a good blood flow.

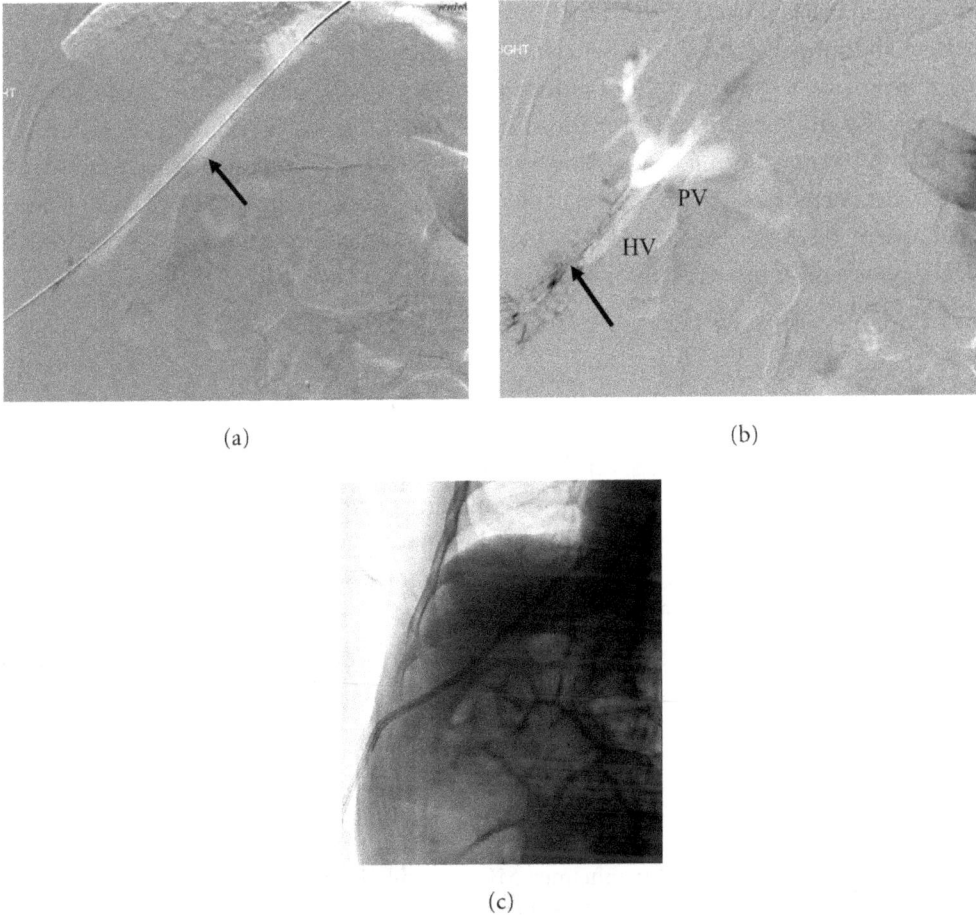

(a)

(b)

(c)

Fig. 11. Percutaneous transhepatic dialysis access placement in a 15-year-old man with Prune-Belly syndrome, failed renal transplant thrice, and occluded superior and inferior vena cavas. (a) After accessing the right hepatic vein using a 22 gauge needle, CO_2 hepatic venogram using a 5 Fr sheath showing patent right hepatic vein (arrow). (b) Tractogram with the injection of CO_2 into the transhepatic tract showing transgression of a peripheral portal vein branch (arrow). Both the hepatic (HV) and portal (PV) veins filled with CO_2. Because a peripheral portal vein was transgressed, the decision was made to place the dialysis catheter using the initial access. (c) After dilation of the transhepatic tract to 10 Fr, a 10 Fr × 19 cm cuff to tip pediatric dialysis catheter (Medical Components, Inc. Harleysville, PA, USA) was placed. Two weeks later, the dialysis catheter was exchanged for a 15 Fr long-term hemodialysis catheter. The patient tolerated the procedure and underwent uncomplicated hemodialysis.

Potential Complications and Management

Complications associated with CO_2 use are uncommon and usually related to operator errors. These include adverse reactions caused by air contamination or injection of gases other than CO_2, inadvertent administration of excessive amounts of CO_2, vapor lock and cardiac arrhythmia or neurologic symptoms associated with central CO_2 reflux from brachial artery injection.[10–20] Nausea or vomiting associated with CO_2 injection into proximal abdominal aorta is transient and requires no specific treatment. Body position change from side-to-side may help reduce pain. CO_2 wedge hepatic injection may cause capsular perforation with resultant bleeding in the presence of right heart failure and coagulopathy. Prior purging of the delivery catheter with CO_2 and avoidance of guidewire injury of the peripheral hepatic vein during cannulation may help avoid this complication. If the patient develops bradycardia or hypotension following initial CO_2 injection, the delivery system must be checked for any source of air contamination before additional CO_2 injections.

Summary

CO_2 is a safe, useful contrast agent for diagnostic and interventional procedures for various hemodialysis access problems including preoperative proximal to central vein evalution vein mapping, central vein visualization, evaluation of failing hemodialysis access,[22] for guiding dialysis catheter placement, and intervention for the treatment of stenosis and thrombosis in hemodialysis fistulae and grafts. Since CO_2 is non-allergenic and non-nephrotoxic, not inducing nephrogenic systemic fibrosis,[21] it can serve as a safe and effective alternative intravascular contrast agent in patients with renal impairment or an iodinated contrast allergy.[22–27]

References

1. Stauffer HM, Durant TM, Oppenheimer MJ. Gas embolism: roentgenologic considerations, including the experimental use of carbon dioxide as an intracardiac contrast material. *Radiology.* 1956; **66**: 686–692.
2. Scatliff JH, Kummer AJ, Janzen AH. The diagnosis of pericardial effusion with intracardiac carbon dioxide. *Radiology.* 1959; **73**: 871–883.
3. Hawkins IF. Carbon dioxide digital subtraction arteriography. *AJR.* 1982; **139**: 19–24.
4. Shifrin EG, Plich MB, Verstandig AG, Gomori M. Cerebral angiography with gaseous carbon dioxide CO_2. *J Cardiovasc Surg (Torino).* 1990; **31**: 603–606.
5. Hawkins IF Jr, Wilcox CS, Kerns SR, Sabatelli FW. CO2 digital angiography: a safer contrast agent for renal vascular imaging? *Am J Kidney Dis.* 1994; **24**(4): 685–694.
6. Ehrman KO, Taber TE, Gaylord GM, Brown PB, Hage JP. Comparison of diagnostic accuracy with carbon dioxide versus iodinated contrast material in the imaging of hemodialysis access fistulas. *J Vasc Interv Radiol.* 1994; **5**: 771–775.
7. Hawkins IF Jr., Caridi JG, Kerns SR. Plastic bag delivery system for hand injection of carbon dioxide. *AJR Am J Roentgenol.* 1995; **165**(6): 1487–1489.
8. Hawkins IF, Caridi JG, Klioze SD, Mladnich CRJ. Modified plastic bag system with o-ring fitting connection for carbon dioxide angiography. *AJR.* 2001; **176**: 229–232.
9. Sullivan KL, Bonn J, Shapiro MJ, Gardiner GA. Venography with carbon dioxide as a contrast agent. *Cardiov Interv Rad.* 1995; **18**: 141–145.

10. Hahn ST, Pfammatter T, Cho KJ. Carbon dioxide gas as a venous contrast agent to guide Upper-arm insertion of central venous catheters. *Cardiovasc Intervent Radiol.* 1995; **18**: 146–149.

11. Lambert CR, DeMarchena EJ, Bikkina M, *et al.* Intracoronary carbon dioxide on left ventricular function in swine. *Clin Cardiol.* 1996; **19**: 461–465.

12. Rundback JH, Shah PM, Wong J, *et al.* Livedo reticularis, rhabdomyolysis, massive intestinal infarction, and death after carbon dioxide arteriography. *J Vasc Surg.* 1997; **26**: 337–340.

13. Linstedt U, Link J, Grabener M, Kloess W. Effects of selective angiography of the carotid artery with carbon dioxide on electroencephalogram somatosensory evoked potentials and histopathologic findings. A pilot study in pigs. *Invest Radiol.* 1997; **32**: 507–510.

14. Caridi JG, Hawkins IF Jr. CO_2 digital subtraction angiography: potential complications and their prevention. *J Vasc Interv Radiol.* 1997; **8**: 383–391.

15. Dimakakos PB, Stefanopoulos T, Doufas AG, *et al.* The cerebral effects of carbon dioxide during digital subtraction angiography in the aortic arch and its branches in rabbits. *AJNR Am J Neuroradiol.* 1998; **19**: 261–266.

16. Boyd-Kranis R, Sullivan KL, Eschelman DJ, Bonn J, Gardiner GA. Accuracy and safety of carbon dioxide inferior vena cavography. *J Vasc Interv Radiol.* 1999; **10**(9): 1183–1189.

17. Waybill MM, Waybill PN. Contrast media-induced nephrotoxicity: Identification of patients at risk and algorithms for prevention. *J Vasc Interv Radiol.* 2001; **12**: 3–9.

18. Cho DR, Cho KJ, Hawkins IF Jr. Potential air contamination during CO_2 angiography using a hand-held syringe: Theoretical considerations and gas chromatography. *Cardiovasc Intervent Radiol.* 2006; **29**(4): 637–641.

19. Cho KJ, Hawkins IF. *Carbon Dioxide Angiography: Principles, Techniques, and Practices.* Informa Health Care, New York, London. 2007.

20. Criado E, Kabbani L, Cho K. Catheter-less angiography for endovascular aortic aneurysm repair: A new application of carbon dioxide as a contrast agent. *J Vasc Surg.* 2008; **48**(3): 527–534.

21. Perez-Rofrigues J, Lai S, ehst BF, Fine DM, Bluemke DA. Nephrogenic systemic fibrosis: incidence, associations, and effect of risk factor assessment report of 33 cases. *Radiology.* 2009; **250**: 371–377.

22. Kariya S, Tanigawa N, Kojima H, Komemushi A, *et al.* Efficacy of carbon dioxide for diagnosis and intervention in patients with failing hemodialysis access. *Acta Radio.* 2010; **51**: 994–1001.

23. Hawkins Ir, Cho, KJ, Caridi JG. Carbon dioxide in angiography to reduce the risk of contrast-induced nephropathy. *Radiol Clin N Am.* 2009; **47**: 813–825.

24. Heye S, Fourneau I, Maleux G, Claes K, Oyen KR. Preoperative mapping for hemodialysis access surgery with CO_2 venography of the upper limb. *Eur J Vascular Endovasc Surg.* 2010; **39**: 340–345.

25. Cho KJ, Hawkins IF Jr. Discontinuation of the plastic bag delivery system for carbon dioxide angiography will increase radiocontrast nephropathy and life-threatening complications. *AJR.* 2011; **197**: W940–W941.

26. Moos JM, Ham SW, Han SM, *et al.* Safety of carbon dioxide digital subtraction angiography. *Arch Surg.* 2011; **146**: 1428–1432.

27. Fujihara M, Kawasaki D, Shintani Y, *et al.* Endovascular therapy by CO_2 angiography to prevent contrast-induced nephropathy in patients with chronic kidney disease: A prospective multi-center trial of CO_2 angiography registry. *Catheter Cardiovasc Interv.* 2014 Nov 7. doi: 10.1002/ccd.25722.

CHAPTER 7

Strategies to Assist Fistula Maturation and Successful Use for Hemodialysis

Jackie P. Ho

Maturation of AVF is a continuous process. Usually, AVFs are expected to mature between 8–12 weeks. AVFs are considered as "early fistula failure" if they cannot be cannulated and used for hemodialysis three months after creation.[1] The Society of Vascular Surgery/ American Association of Vascular Surgery and the North American Vascular Access Consortium further define[2]:

Early dialysis suitability failure: an AVF for which, despite interventions (radiologic or surgical), was not possible to use the AVF successfully for hemodialysis by the third month following its creation.

Late dialysis suitability failure: an AVF for which, despite interventions (radiologic or surgical), was not possible to use the AVF successfully for hemodialysis by the sixth month following its creation.

Fistula Used Successfully for Hemodialysis (FUSH): an AVF is considered functionally successful for hemodialysis use. The criteria can be percentage of successful two needle cannulation during HD sessions, volume of dialysate flow achieved, or target clearance achieved.

After AVFs creation, all patients should be encouraged to do isometric hand exercise to promote fistula growth. When an AVF shows poor maturation on follow-up, clinicians need to decide whether to continue longer observation period, intervene to assist its maturation or abandon the current fistula and create another vascular access. In general, it is advisable to try preserve and assist the maturation process of the created AVF as every AVF is precious to a renal failure patient and performance of AVFs are generally superior to AVGs.

Nonetheless, because of wide variation of individual patient's medical condition, clinical situation and fistula status, the benefit of preserving a native vein fistula has to be balanced against the disadvantage of prolonged tunnelled CVC usage and the risk of injury to the fistula induced by intervention for each non-mature case.

Several factors have to be considered to help clinical decision on intervene or not, when to intervene and what to do for "failure to mature FTM" fistula:

(1) **Urgency of the vascular access**

Is the AVF created for pre-emptive purpose, secondary fistula for failing existing access, or incident ESRF in patients carrying tunnelled CVC? One could afford to wait for a longer time in pre-emptive condition but earlier action would be needed in patients already on CVC hemodialysis to avoid CVC-related complications.

(2) **The cause of failure to mature**

Failure to cannulate the AVF successfully could be due to venous fistula failure to grow, or the venous fistula did grow in size but being either too deep or too tortuous for easy cannulation. Failure of venous fistula to grow again could be due to small or stenosed arterial inflow, stenosis of AV anastomosis stenosis of juxta-anastomotic region or general poor quality of the vein. Failure to mature due to different causes requires different strategies to tackle. Some may be easier to overcome, whereas some may require more sophisticated intervention.

(3) **Patient's medical condition and willingness to accept further intervention**

Does the patient have reasonable life expectancy to justify procedure in assisting maturation of the fistula, especially if repeated procedures are expected? What is the risk to patient if a regional or general anesthesia is required for a maturation procedure? Even more importantly, is the patient willing to undergo one or more additional procedures to make the fistula functional?

Ultrasound and Duplex study is an essential part when AVFs are found to have poor maturation or cannulation difficulties on follow-up. The causes of unsuccessful use for hemodialysis can be classified into three main types and could be a combination of those.

(1) Venous fistula of reasonable size but with tortuous course;
(2) Venous fistula fail to grow;
(3) Venous fistula of reasonable size but deep in the subcutaneous tissue.

Tortuous Venous Fistula

Venous fistula could be tortuous and also uneven in depth, making cannulation difficult. Hematoma may easily form and further complicates the situation. Resting time to let hematoma resolve is required. For tortuous fistula, clinician may mark out the course of the fistula and indicate suitable location superficial enough for cannulation to HD nurses.[3] If there is outpatient hemodialysis center in the hospital, the patient can be brought into the hospital dialysis center for the first few sessions of cannulation so that an established track can be followed afterwards. Ultrasound-guided cannulation of the fistula could assist localization of a good spot for needling at the beginning (Fig. 1). Buttonhole cannulation may help develop a constant site

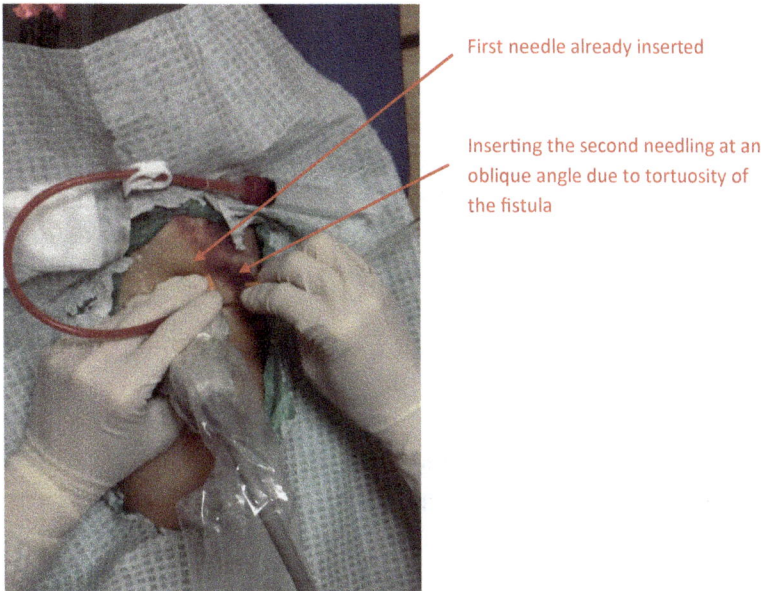

First needle already inserted

Inserting the second needling at an oblique angle due to tortuosity of the fistula

Fig. 1.   Ultrasound-guided needling of left BC AVF for an obese patient with tortuous course in the mid and distal arm. Only a short segment of the fistula is superficial and the rest is deep in subcutaneous tissue

for needling. Good communication with the staff in community dialysis center using diagrams or pictures is important for subsequent smooth transition of care back to the community.

Venous Fistula Fail to Grow

This could be due to AV anastomosis stenosis, inflow small sized artery or arterial stenosis, presence of large sized or too many outflow branches in the fistula inflow segment, juxta-anastomotic stenosis, or diffuse stenosis or small size of the venous fistula. Ultrasound study of the AVF is able to pick up most of the aforementioned situations. Clinically, it is not difficult to identify single big sized branch in the inflow segment taking away too much flow. Thrill might be palpable over the branch and digital compression of the branch will augment the thrill over the main venous fistula. Ultrasound study will reveal the size of the venous fistula becomes smaller beyond the branch site. A simple stab incision and ligation of the branch 1–2 cm away from the main fistula under local anesthesia would be good to divert flow back into the main fistula track. USG may pick up obvious stenosis along the AVF pathway but more accurate diagnosis require Duplex study.

For other causes, if contrast use is not contraindicated, a diagnostic angiogram to assess inflow artery, AV anastomosis and the venous fistula would provide helpful information for intervention to be carried out at the same session. Several endovascular strategies can be used to tackle various conditions impeding fistula maturation.

AV anastomosis stenosis and juxta-anastomostic segment stenosis are common causes for poor maturation. Occasionally, inflow arterial stenosis existed *de novo* or due to clamp injury

(Fig. 2) may account for poor maturation. Usually simple balloon angioplasty for the stenotic lesion(s) will facilitate maturation of the fistula.[1] High pressure balloon angioplasty might
be needed occasionally. For AV anastomosis, the minimum size should be 4 mm. For juxta-
anastomotic stenosis, the size of angioplasty balloon has to be chosen based on the adjacent
relatively normal venous fistula, usually size 4–6 mm. Repeated procedure might be needed.

Personally, I would consider surgical revision of the anastomosis if there is significant
recoil of anastomotic stenosis after high pressure or cutting balloon angioplasty despite
several attempts or the fistula shows no progress 3–4 weeks after the angioplasty (and no
other causes apart from anastomosis re-stenosis can be identified).

For long segment stenosis or diffusely small sized venous fistula, method of balloon
angioplasty maturation (BAM) with several sessions to progressively and aggressively dilate
the usable segment of the venous fistula with or without controlled vein injury can be
adopted. Usable segment of AVF is a relatively straight segment of the venous fistula that can
be cannulated for both "A" and "V" needles. In RC AVF, the usable segment is usually the
forearm cephalic vein and maybe median antebrachial vein. For BC AVF, the usable segment
would be the cephalic vein in the distal or mid arm. BAM can be performed antegradely
through arterial puncturing or using retrograde approach puncturing the venous fistula
proximally.[4] If the venous access is still small in size, arterial access is preferred. In partially
dilated venous fistula, the venous access can be achieved using ultrasound guidance. The
balloon angioplasty procedure can be performed using either ultrasound or fluoroscopy
guidance.[4] Usually long balloon is preferred. The balloon angioplasty can be performed every
2–3 weeks and sequentially stepping up the balloon size 2 mm each time. The target size is at
least 6–8 mm for forearm cephalic vein fistula and at least 8–10 mm for arm fistula.

In resistant venous fistula stenosis, high pressure cutting balloon or scoring balloon angioplasty are required. Some centers advocate controlled vein injury method. The basic concept
is to cause a mild rupture of the vein with a control of inflow pressure. Thus, mild haemor-
rhage around the fistula wall will cause remodelling and fibrosis of the fistula resulting in a
fibrous tube for cannulation. This can be performed by digital compression over the AV anastomosis to reduce the inflow pressure during the whole process of balloon inflation and defla-
tion for high pressure BAM.[4] Completion angiogram is performed to ensure no frank venous
rupture before releasing the pressure over AV anastomosis. Regional anesthesia or local anesthesia over the high pressure angioplasty site could reduce pain experienced by the patient.

For BAM with or without controlled vein injury, the small branches along the usable
fistula segment will either be damaged or shrink in size due to flow competition. This will
result in re-routing of flow into the main fistula track.

Adopting this aggressive endovascular strategy, specialized centers reported[4,5] high
successful maturation rate (>90%) for fistula of small size with a low complication rate
(anastomotic or fistula rupture). Repeated endovascular interventions are usually required
to maintain the patency of these assisted AVFs.[6]

Furthermore, if major branch(es) are noted over the inflow segment of venous fistula
on fistulogram, surgical ligation or endovascular coil embolization can be performed to
divert flow back into the main track.

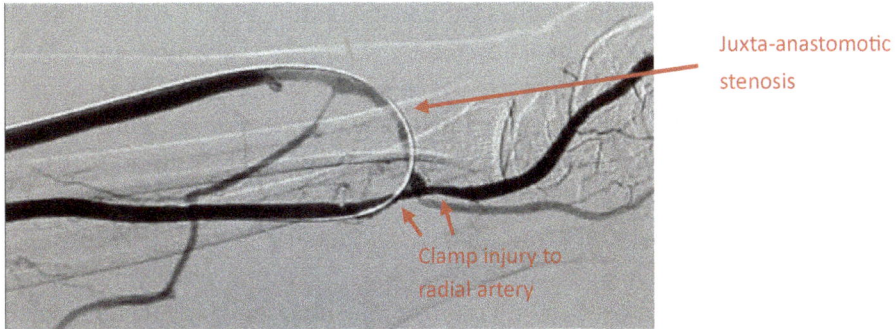

Fig. 2. Stenosis of the radial artery proximal and distal to the AV anastomosis likely due to clamp injury. There is also juxta-anastomotic stenosis present in this FTM RC AVF.

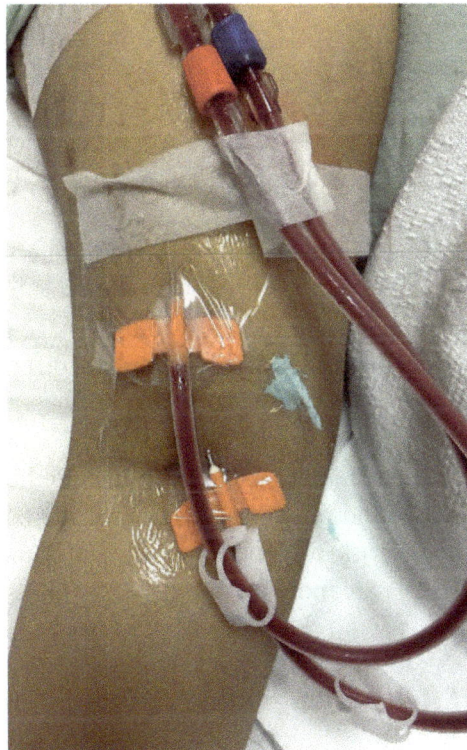

Fig. 3. Pointing both "A" and "V" needle towards the venous outflow in a patient with only short segment superficially situated arm BC AVF.

Venous Fistula Deep in the Subcutaneous Tissue

Under this category, one needs to determine whether the whole usable track of the AVF is deep or part of it is deep. If just part of the track is deep, and the superficial part is marginally adequate for two needles insertion. Pointing both needles towards the venous end (Fig. 3) may accommodate two needles cannulation with a relatively short segment of AVF.

In general, there are two approaches to manage matured AVFs with whole or most of its usable track deep in subcutaneous tissue: (1) superficialization and (2) balloon angioplasty maturation. The choice to use which approach depends on the actual size of the venous fistula and the thickness of subcutaneous tissue. For marginal sized venous fistula (e.g. 5–6 mm diameter) and subcutaneous tissue only slightly thick (7–10 mm), balloon angioplasty maturation may assist further dilatation of the venous fistula and gradually become more superficial as it grow. Miller *et al.*[4] reported a high clinical success rate using aggressive endovascular approach to treat fistula 6–8 mm diameter and >6 mm deep. Although the primary patency at 6 month is only 17%, the secondary patency at 12 months was able to maintain at 72%.

For good sized fistula (>6 mm) but situated deep in the subcutaneous tissue (>10 mm), superficialization of the fistula is a better option. The methods to achieve superficialization include:

(1) Lipectomy — several transverse incision made to excise the adipose tissue superficial to the fistula.
(2) Incision along the fistula and raised it above the subcutaneous fat — not preferred because the future cannulation will be on the surgical scar. The risk of wound breakdown is high.
(3) Transposition — dissect the fistula away from surrounding tissue and transpose to a more superficial location.
(4) Ultrasound-guided liposuction — under ultrasound image guidance and tumescent solution infusion to subcutaneous tissue, the subcutaneous tissue above the fistula is removed using liposuction technique.[7]

The potential complications of superficialization include hematoma or seroma formation, injury to fistula and wound problems. As most of the patients requiring superficialization are morbidly obese, risk of general anesthesia is usually substantial. Local, regional or tumescent anesthesia are more preferred.

Potential Downside of Balloon Angioplasty Maturation

There is a dilemma between too early aggressive angioplasty to the non-matured fistula which may induce more intimal hyperplasia,[6] and sitting on a FTM fistula which will lead to complications of prolonged CVC and even fistula thrombosis. Aggressive balloon angioplasty of the non-matured fistula will inevitably inflict trauma to the endothelium and underlying smooth muscle cell and, in turn, induce intimal hyperplasia. Based on various literatures, frequently multiple procedures are required to successfully dilate a venous fistula to be ready for use. Fistulae matured with angioplasty are also more likely to require further angioplasty treatment to maintain the secondary patency. The frequent interventions also imply a significant economic and healthcare resource demand to the patients and the society. On the other hand, the advantage of a native vein fistula over AVG and CVC is also obvious. Therefore clinicians should evaluate FTM fistula patients thoroughly in terms of their urgency of requiring the fistula, physical condition of the fistula, medical comorbidities, psycho-social status and life expectancy before making a decision to intervene.

Other Means to Facilitate AVF Maturation

Non-invasive methods are available to promote fistula maturation. A small scale study comparing patients receiving structured isometric hand exercise on the fistula limb after operation to control group showed increased maturation rate and more favorable ultrasonic parameters including vein diameter, wall thickness and blood flow rate.[8] Lin *et al.* applied external far infrared therapy to the fistula region twice weekly and showed in a randomized controlled trial that far infrared therapy group had significantly higher physiological and clinic maturation rate[9] than control group at 3 months and 12 months assessment.

A clinical strategy flowchart as in Chart 1 is suggested.

Case 1

A 38-year-old gentleman with glomerulonephritis and ESRF underwent left RC AVF creation in another country six weeks ago. Tunnelled CVC was in for two months already. Attempted cannulation of the fistula resulted in large hematoma and he was admitted to the hospital. Clinically the thrill over the cephalic vein fistula is weak. Bedside ultrasound identified localized stenosis over the AV anastomosis. The average size of the cephalic vein fistula was good (~4.5 mm).

Angiogram was performed via left brachial artery access under local anesthesia (Figs. 4 to 6) using a 4-Fr sheath.

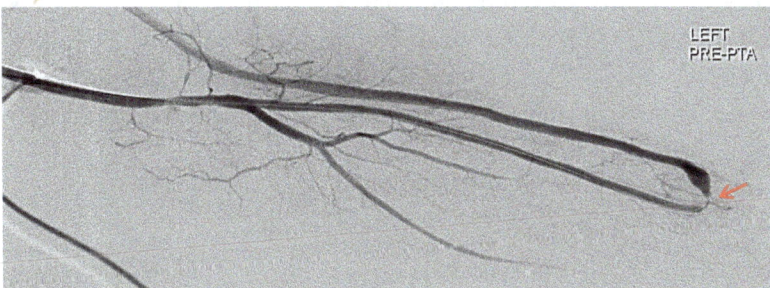

Fig. 4. Angiogram via left brachial artery access showed localized AV anastomotic stenosis. The radial artery and the outflow cephalic vein fistula are patent.

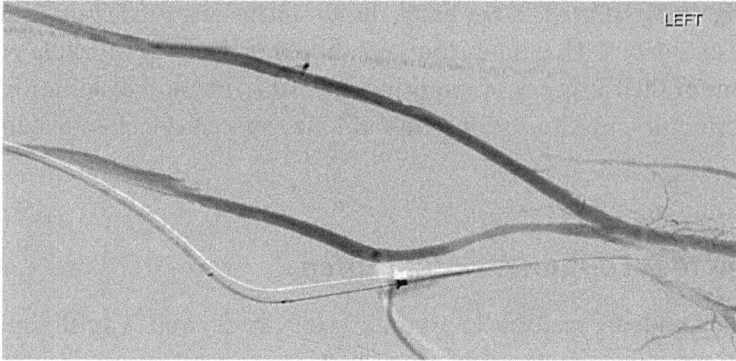

Fig. 5.   Angiogram showing patent cephalic vein and basilic vein over the left arm.

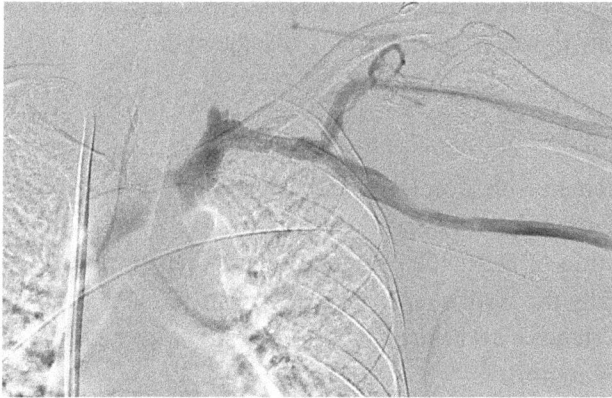

Fig. 6.   Angiogram showing patent central vein and right IJV tuennlled CVC *in situ.*

Angioplasty of the left radiocephalic anastomosis performed using Advance LP18, 4 mm/40 mm balloon (Cook Medical Inc, Bloomingdale, USA) (Fig. 7).

Thrill of left RC AVF improved after the intervention. Cannulation of his left forearm fistula commenced one week after the procedure. His CVC was removed three weeks after the intervention.

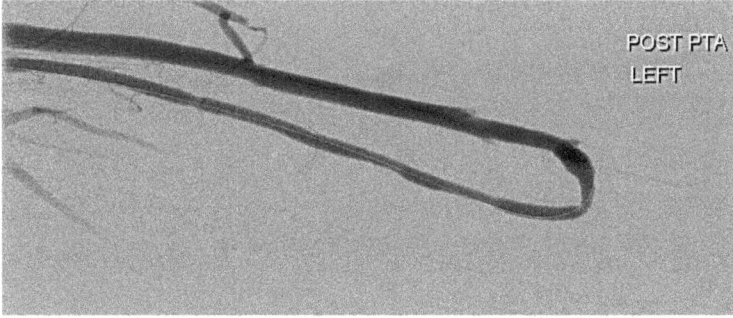

Fig. 7. Post-angioplasty angiogram.

Case 2

A 76-year-old gentleman with diabetic nephropathy developed ESRF. Tunnelled CVC via right IJV for hemodialysis had been started six months ago. Left RC AVF was created in another hospital around the time of CVC insertion. Cephalic vein size was about 5 mm, when used for HD, in-sucking over the "A" side was noted. Fistulogram was performed by a colleague showed AV anastomotic stenosis and juxta-anastomotic stenosis (Fig. 8).

Fistuloplasty had to be stopped pre-maturely due to excessive pain experienced by the patient during the angioplasty. The procedure was then repeated under sedation and local anesthesia injection around the anastomotic and juxta-anastomotic region (Figs. 9 and 10).

Angioplasty of the anastomosis and juxta-anastomotic region was performed using Advance LP18 4 mm and 5 mm balloon (Cook Medical Inc, Bloomingdale, USA) respectively. After the fistuloplasty procedure, hemodialysis was able to be sustained using the RC AVF. However, the in-sucking condition recurred again one month later. Surgical revision of distal forearm RC AVF to mid-forearm was then performed. The fistula can be used immediately after the revision surgery. CVC was removed three weeks after the surgery.

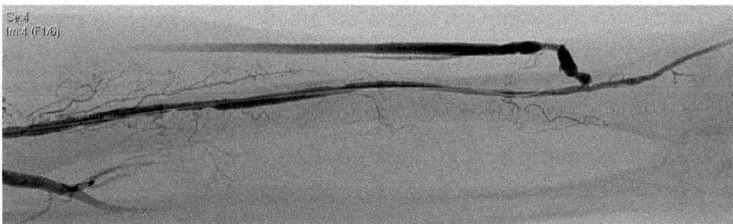

Fig. 8. Angiogram showed AV anastomotic and juxta-anastomotic stenosis of left RC AVF.

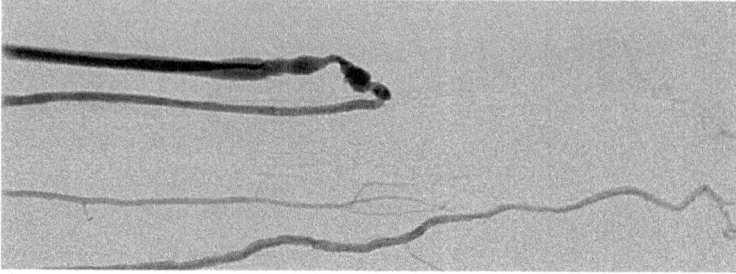

Fig. 9.   Repeat fistulogram showed same lesions as above.

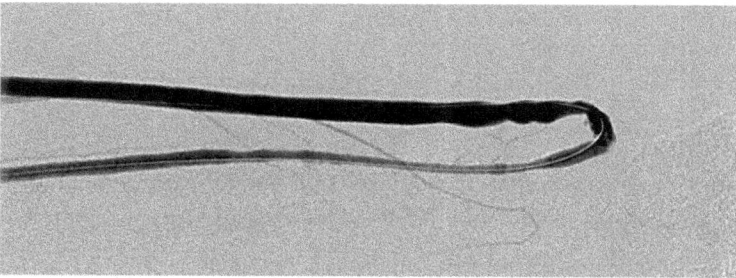

Fig. 10.   Angiogram after fistuloplasty of the AV anastomosis and juxta-anastomosis regions.

References

1. Beathard GA, Arnold P, Jackson J, *et al.* Aggressive treatment of early fistula failure. *Kidney Int.* 2003; **64**: 1487–1494.
2. Lee T, Mokrzycki M, Moist L, *et al.* Standardized definitions for hemodialysis vascular access. *Semin Dial.* 2011; **24**(5): 515–524.
3. WC Jennings, KE Taubman. Alternative autogenous arteriovenous hemodialysis access options. *Semin Vasc Surg.* 2011; **24**: 72–81.
4. Miller GA, Goel N, Khariton A, *et al.* Aggressive approach to salvage non-maturing arteriovenous fistulae: a retrospective study with follow-up. *J Vasc Access.* 2009; **10**: 183–191.
5. Chawla A, DiRaimo R, Panetta TF. Balloon angioplasty to facilitate autogenous arteriovenous access maturation: a new paradigm for upgrading small-caliber veins, improved function, and surveillance. *Semin Vasc Surg.* 2011; **24**: 82–88.
6. Chaudhury RR, Lee T, MD, Woodle B, *et al.* Balloon-assisted maturation (BAM) of the arteriovenous fistula: the good, the bad, and the ugly. *Semin Nephro.* 2012; **32**(6): 558–563.
7. Causey MW, Quan R, Hamawy A, *et al.* Superficialization of arteriovenous fistulae employing minimally invasive liposuction. *J Vasc Surg.* 2010; **52**: 1397–1400.
8. Salimi F, Majd Nassiri G, Moradi M, *et al.* Assessment of effects of upper extremity exercise with arm tourniquet on maturity of arteriovenous fistula in hemodialysis patients. *J Vasc Access.* 2013; **14**: 239–244.
9. Lin CC, Yang WC, Chen MC, *et al.* Effect of far infrared therapy on arteriovenous fistula maturation: an open-label randomized controlled trial. *Am J Kidney Dis.* 2013; **62**(2): 304–311.

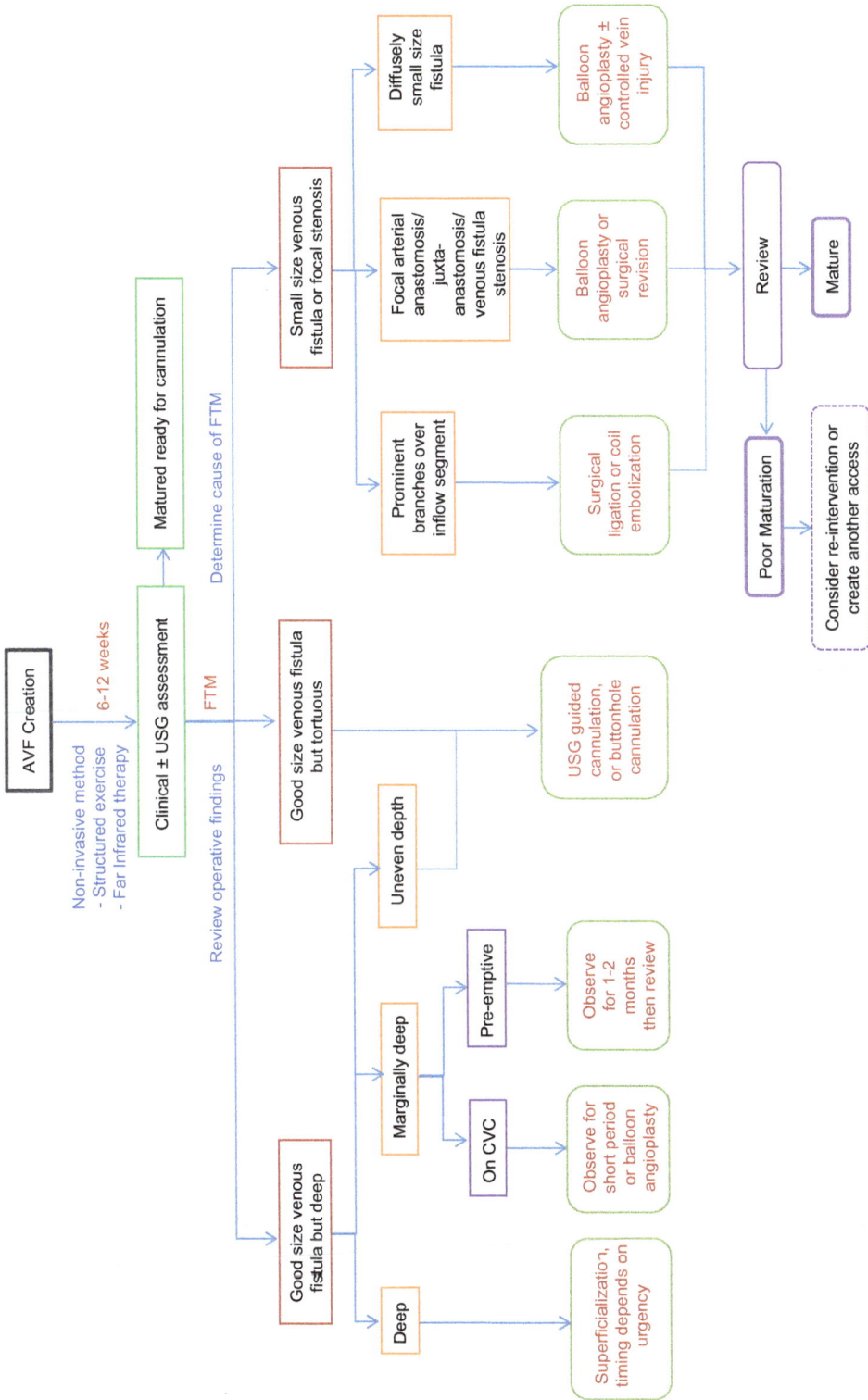

Chart 1. Strategies to manage failure to mature AVFs.

CHAPTER 8

Endovascular Salvage for Failing Hemodialysis Access

Jackie P. Ho

Hemodynamically significant stenosis may develop along the course of vascular access due to intimal hyperplasia or repeated cannulation trauma. Central vein obstruction may also develop due to prolonged central venous catheterization with tunnelled catheter or simply due to presence of vascular access on the ipsilateral limb.

It is far better to treat the hemodynamically significant stenosis of a vascular access before it blocks than to deal with it after thrombosis has happened. Endovascular salvage is minimally invasive and posts lower risk to patients. Thus, it is usually adopted as the first line of treatment for failing vascular access.

The basic principles of endovascular salvage for failing accesses are:

- **Angiographic interrogation of inflow, fistula/conduit and also outflow (all the way to SVC);**
- **Treat all lesions with minimal number of puncture(s).**

There are a variety of endovascular strategies and approaches to treat failing vascular accesses. Different clinicians may use different methods based on their training, past experience, patient characteristics, facility condition and service structure etc. There is no absolute right or wrong way. Nonetheless, there are several factors to consider when planning the procedure:

(1) *Where is/are the lesion(s)?*

Understanding the common sites of stenosis of different types of vascular accesses, the patient's clinical presentation and past access history, one can roughly deduce the likely location(s) of the stenosis.

For native vein AVF, the usual sites of stenosis are the AV anastomosis and juxta-anastomotic regions. Along the venous fistula, stenoses may also present, especially over cluster cannulation sites.[1] For AVG, the stenotic site is most frequently (nearly always) the vein-graft anastomosis,[2] which may also extend into the adjacent outflow vein. Another common site of stenosis is in-graft where cluster needling occurred. Stenosis of the artery-graft anastomosis is rare and seldom happens as the only site of stenosis. Occasionally, for

105

both AVF and AVG, the artery near the arterial anastomosis may develop stenosis that is probably related to clamp injury which occurred during access surgery. Central vein stenosis or occlusion is more common with previous tunnelled CVC placement, especially if the duration of CVC is long. It can also happen without CVC history, merely with a vascular access created on the ipsilateral limb for a long time.

As a general rule, the stenotic sites of a particular vascular access identified and treated in previous endovascular intervention(s) are likely to recur again.[3,4] Previous fistulogram information will be useful for the planning of current intervention.

Clinical features of failing vascular access like reduction in thrill, pulsation of the fistula, firm cord-like structure felt, upper limb swelling, dilated upper limb superficial veins have high positive predictive power of stenotic lesions along the access. However, the absence of these features has poor negative predictive power. That is, stenosis may present in the absence of any obvious clinical features. Hemodialysis parameters will also give hints on the site of stenosis. Elevated venous pressure points to venous outflow obstruction. The site of stenosis is likely beyond the "V" needling site. Reduction in access flow can be due to inflow or outflow problems (see Chapter 5).

(2) *Distance of puncture site and lesion*

It would be easier to tackle the stenotic lesion if the sheath insertion is directly upstream or downstream to the lesion. However in all endovascular interventions, one also has to leave a practical distance between the sheath entry site (puncture site) and the lesion to enable sheath parking securely in the vascular system. Usage of 7 cm paediatric sheaths are preferred in fistula interventions to reduce the required distance between the access and the lesion.

(3) *Flow direction*

A good and clear angiogram is easier to obtain by injecting contrast along the flow (antegrade, arterial side of vascular access or arterial end of the access) than injecting contrast against the flow (retrograde, venous side of vascular access). If one chooses to place the sheath entry over the venous side, reducing the inflow and obstructing the outflow (by manual compression or tourniquet application) during contrast injection through the sheath or a catheter parked into the arterial side are required. In native vein fistula, a competent valve inside the vein may pose a challenge for guidewires to enter in a retrograde manner.

(4) *Potential risk of puncture site*

The risk of bleeding, thrombosis, hematoma formation and pseudoaneurysm is slightly higher with arterial puncture than the graft, fistula or venous puncture. The consequence if any complication happen is also more serious with artery than with vein or graft. The risk of puncture site complication also varies depending on sheath size (determined by type of device, wire platform of device, size of target vessel), anticoagulant usage, location of vessel (femoral higher risk than brachial and radial artery), body habitus (obese patient with deep seeded vessel higher risk), size of access vessel, quality of vessel (calcified vessel higher risk). For brachial artery puncture, the sheath size is preferred to be limited to only 4 Fr or 5 Fr. Sheath size limitation is less for fistula or outflow vein puncture. For puncture of the fistula or outflow vein, vasospasm may occur. One may not be able to evaluate the segment of vein/fistula where the sheath parked due to vasospasm. Ultrasound guided puncture for sheath insertion is recommended. Clinicians may choose a better location for puncture and minimize misfiring. Nonetheless, there is a learning curve for ultrasound guided vessel access.

(5) *Number of puncture site*

If the puncture site is over the arterial side, then inflow, fistula/conduit and outflow lesions can be tackled with one entry port (sheath). If the puncture site is over outflow vein retro-gradely, then only the inflow and fistula/conduit lesions, not the central outflow lesions, can be tackled with a single entry port. Similarly, if the puncture site is antegradely over the fistula/conduit near the inflow, only fistula/conduit and outflow lesions, not the inflow lesions, can be tackled. The difference may not be much between one or two puncture sites, though less puncture sites are more preferable. From the patient's point of view, one or two puncture site(s) over the upper limb probably makes no big difference. However, a groin puncture in addition to an upper limb puncture may become a concern. It would be better to prime the patient first if a groin puncture is planned.

(6) *Availability of pre-intervention or peri-intervention ultrasound study*

Depending on the healthcare resources, expertise availability and clinician preference, a detailed duplex study of the vascular access to locate the site of stenosis might be available before an endovascular intervention. However, this practice will certainly result in additional healthcare cost. Nowadays, many intervention suites, cardiac catheterization laboratories and operation theatres are equipped with simple ultrasound devices. A quick ultrasound scan of the vascular access just before the intervention procedure will provide helpful information to the proceduralist in planning the puncture site and approaches. One needs to be aware that the diagnostic criteria of a hemodynamic significant stenosis is mainly by velocity change together with diameter reduction. Therefore, there is limitation on the accuracy of a quick scan. Furthermore, duplex and ultrasound studies are not able to evaluate the central vein.

For example, a patient with left BC AVF is noted to have reduction in access flow from 1200 ml/min to 700 ml/min. Clinical examination showed matured BC AVF with thrill present but not very strong. The needling marks were clustered towards two regions of the left arm fistula. As discussed before, the most common sites of stenosis in native vein AVF

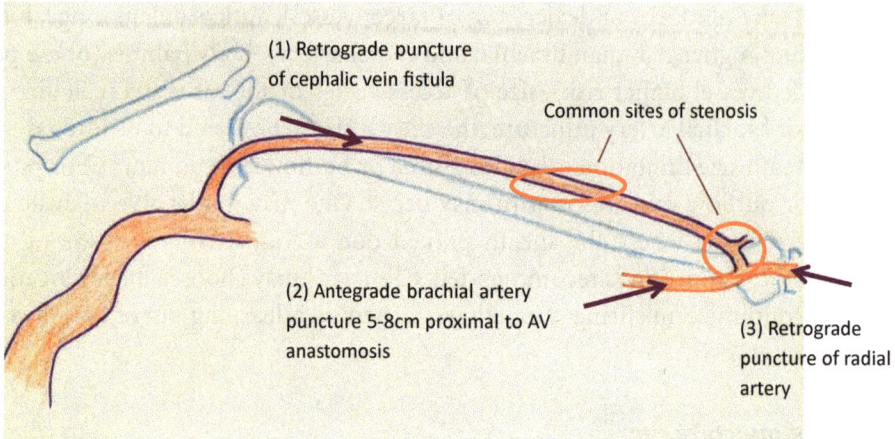

Fig. 1.   Schematic diagram of different approaches for endovascular intervention for a failing BC AVF.

are AV anastomosis, juxta-anastomotic region and fistula cannulation sites. Probability of central vein obstruction depends on any previous history of tunnelled CVC placement and the duration of the catheter *in situ.*

There are a few options of endovascular approaches:

(1) Retrograde cephalic vein fistula puncture over upper arm region. Due to the direction of blood flow, angiogram of the AV anastomosis, juxta-anastomotic region and cephalic vein fistula caudal to the sheath require compression over the inflow brachial artery ± the outflow cephalic vein beyond the sheath. Alternatively, one can park a catheter beyond the AV anastomosis into the inflow artery and inject contrast via the catheter. Any stenotic lesion over AV anastomosis, juxta-anastomotic region and cephalic vein fistula caudal to the sheath can be tackled with a direct path. Angiogram of the outflow cephalic vein and central vein can be performed as usual but lesion detected cannot be treated by the same sheath introduced. If there is any central vein lesion, an additional antegrade fistula puncture and sheath insertion would be needed. This approach may not be feasible if the cephalic vein fistula stenosis is very extensive, involving a long segment of cephalic vein in the arm. The advantages of this approach are the relatively easy hemostasis for venous fistula after the procedure and the direct path in tackling AV anastomotic and juxta-anastomotic lesions. The disadvantages are more troublesome contrast injection, inferior angiographic quality and inability to tackle any proximal outflow vein or central venous lesion. The proceduralist will be in close proximity to the fluoroscopy machine during the treatment.

(2) Antegrade puncture of brachial artery enables screening and treatment of lesions along the whole course of the AVF using a single entry port (including AV anastomosis to central vein). Good quality angiogram can be done easily. No effort is required to

compress the inflow or the fistula for good angiographic images. However, puncture of the brachial artery around mid-arm region can be challenging as the brachial artery is deeper underneath the soft tissue proximal to the cubital fossa. Also because of this reason, the distance between the puncture site and the AV anastomosis may only be of a short distance. Length of sheath that is able to park inside the brachial artery becomes limited. As the brachial nerve is in close proximity to the brachial artery, the patient may suffer significant pain if one hits the brachial nerve. There are several ways to overcome this challenge:

(i) Hyper-extend the elbow by putting support underneath the elbow joint;
(ii) Ultrasound-guided puncture;
(iii) Infuse small amount of local anesthetic agent around the artery (not only subcutaneous);
(iv) Use 20G IV cannula or micro-puncture set to puncture the brachial artery which minimizes the risk of arterial dissection;
(v) Use a paediatric sheath (7 cm in length), usually only half of the sheath is inserted into the brachial artery.

After the procedure, dedicated effort is needed to secure hemostasis and prevent pseudoaneurysm formation. Personally I will observe three criteria to ensure the manual compression is effective:

(i) No bleeding around the puncture site;
(ii) No swelling around the puncture site;
(iii) The brachial pulse can be felt underneath the finger of compression.

This approach shares the same disadvantage of close proximity between the operator and the fluoroscopy machine as retrograde fistula approach.

(3) Retrograde radial artery approach has the advantage of one puncture tackling all lesions along the BC AVF pathway. The radial artery is rather superficial and easy for manual compression hemostasis after the procedure. The operator may also stand a bit farther from the fluoroscopy with this approach. The disadvantages are:

(i) Angle to pass the guidewire into the cephalic vein fistula could be difficult, which worsens with the presence of high grade AV anastomosis stenosis;
(ii) Because of the distance between the puncture and lesion, and the need to pass through several turns, manipulation of guidewire to cross difficult lesions (e.g. CTO lesion) of cephalic vein fistula and central vein could be challenging;
(iii) Angiogram of the fistula may require a catheter parked inside the brachial artery or into the fistula, making the procedure more clumsy. Alternatively a longer sheath can be used.

Each of the above approaches has advantages and disadvantages. There are also other approaches besides the three discussed. Clinicians have to take into consideration the patient

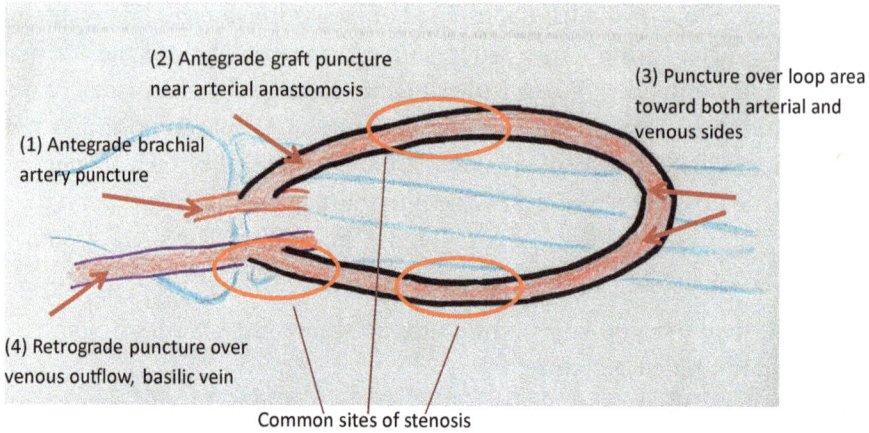

Fig. 2. Schematic diagram of different approaches for endovascular intervention for a failing forearm loop BB AVG.

factors, vascular access, facility conditions and personal experience in order to make a wise strategy.

> *Trick: Use a 20G IV cannula to make the puncture and advance a short 0.025" wire (available inside the paediatric sheath set) to track the IV cannula. A preliminary angiogram can then be performed using this 20G cannula. One can then commit if sheath insertion is needed and of what size.*

Another example, a 47-year-old gentleman with DM nephropathy and ESRF. He uses left forearm loop BB AVG for hemodialysis. Recently, his venous pressure raised to 180 mmHg for several dialysis sessions. Qb was normal 250 ml/min. No access flow measurement is available in his center. Pulsation of the AVG present.

Most common sites of stenosis would be vein-graft anastomosis and outflow vein. Second to that would be stenosis along the graft where frequent cluster needling took place. Rarely, there may be stenosis of artery-graft anastomosis or central vein lesions.

(1) Antegrade brachial artery puncture enables a single entry port to treat all possible lesions. The downsides are issues of brachial artery puncture, limitation on sheath size, and relatively indirect route to tackle vein-graft anastomosis and venous outflow lesions, especially if central vein obstruction present.

(2) Antegrade puncture of the graft near the arterial anastomosis enables treatment of lesions along the whole graft, vein-graft anastomosis as well as venous outflow. The route may also be indirect for venous outflow lesion similar to brachial artery puncture. The puncture and hemostasis would be easier than brachial artery approach. However, if the in-graft stenosis is close to the arterial anastomosis, this approach may miss out that lesion.

(3) Puncture over the loop region of the graft towards the venous limb (or both arterial and venous limbs) enables a direct route to tackle the most common vein-graft anastomosis stenosis and in-graft stenosis in both arterial and venous limbs. Treatment for central vein obstruction would also be more direct. The puncture and hemostasis of the graft over loop area is relatively simple. As the dialysis nurses nearly never cannulate the AVG loop region for HD, the loop region is mostly stenosis free. Personally, I will use a 20G IV cannula to puncture each side of the graft over loop region and perform fistulogram of both arterial and venous limb of the graft. The need for sheath insertion and its size will be guided by the fistulogram finding.

(4) Hemostasis for a retrograde puncture over the outflow vein (basilic vein) is simple and bears low risk. This approach allows treatment of the inflow, in-graft and vein-graft anastomosis lesion but not to the outflow vein beyond the puncture and central vein lesions. Contrast injection requires a catheter passing into the arterial limb of the graft or a strong compression over the inflow artery proximal to sheath insertion, which is a bit clumsy. I do not prefer this approach because the outflow vein is a precious structure for future vascular access. Any manipulation may lead to stenosis development subsequently. In this instance, when the loop AVG is finally not salvageable, a good sized and healthy basilic vein may be used for secondary AVF (single stage brachio-basilic AVF creation and transposition).

Besides the "Approach and strategy" of endovascular intervention, there are several areas the proceduralist needs to find out in order to determine the best way that works for him/her.

Position of the Patient on the Procedure Table

It depends on the relative position and range of movement of the operating table and the fluoroscopy unit. The goal is to have the fluoroscopy unit cover a screening area of the inflow, fistula/conduit and outflow (all the way to SVC) together with the puncture site of intervention. When remote puncture site is planned (e.g. femoral vein and upper limb fistula puncture for central vein occlusion), the coverage of fluoroscopy unit should be checked with caution.

Guidewires, Catheter and Angioplasty Balloon

Individual clinician may have his/her preference for difference guidewires as the workhorse. 0.035" or 0.018" guidewires are the basic tools for vascular access intervention. The choice between 0.035" and 0.018" platform is also related to the preferred angioplasty balloon. For those who prefer high pressure balloon[5] for vascular access treatment, 0.035" wires are used. Clinicians hoping to downsize the access sheath may consider 0.018" system. 0.014" guidewires may be required for intervention to small calibre or diseased radial and ulnar artery to improve the inflow or salvage steal syndrome. Usually the distance between vessel puncture site and lesion is relatively short. Therefore short guidewires (length between 150 cm to 190 cm), catheters (e.g. 65 cm), and angioplasty balloons (40–90 cm shaft length) are preferred. Exceptions occur when remote puncture site was planned for specific conditions. I prefer hydrophilic 0.035" or 0.018" guidewire with an angled tip as the default wires to facilitate selective cannulation into the vascular accesses.

Anti-coagulation Therapy during Intervention

ESRF patients usually have platelet dysfunction and the procedure time for vascular access intervention is usually shorter than the intervention for peripheral arterial disease. Anti-coagulation is not a "must" in vascular access interventions. However, in situations where the patient's vessel size is exceptionally small, existing severe arterial disease, poor flow over the vessel distal to angioplasty, or prolonged angioplasty time is planned, anti-coagulation therapy would be necessary.

Hemostasis After Intervention

The simplest way to achieve hemostasis for puncture site is manual digital compression. The compression time varies and depends on whether an artery/fistula/graft was punctured, size of the sheath, and any anti-coagulant given. In general, compression time to achieve hemostasis in ePTFE graft and artery is longer than vein or fistula with the same sheath size. Patients under stress with high blood pressure will make hemostasis more difficult. In conditions where a relatively large sized sheath (e.g. 6 Fr or 7 Fr) was used on fistula or graft, one could apply a figure of eight stitch (or even simple single stitch) over the skin and part of subcutaneous tissue of the entry site and then continue compression for a short period of time. This will reduce the compression time. Be careful not to take any part of the fistula or graft by the stitch. Clear instructions on when to remove the stitch should be given to the patient and the dialysis center.

Pain Control and Sedation

Angioplasty for AV anastomosis, juxta-anastomosis and venous fistula stenosis is often painful because frequently higher pressure (>10 atm) is required compared to peripheral arterial disease. Pain control is important as patients are likely to require repeated endovascular interventions within their dialysis years. A terrible experience may turn away patients from future necessary treatment. If expertise and resources are available, regional block (e.g. brachial plexus block) for patients undergoing fistuloplasty would be ideal. However, anesthetic support is not always readily available in many healthcare settings and this will also increase the procedure time as well as the cost. Alternatively, pain control can be achieved by mild sedation and intravenous analgesia (in my practice, Midazolam 1–2 mg and Fentanyl 20–50 mcg IVI depending on age, cardiopulmonary function and body weight) together with local anesthesia (1–3 ml 1% Lignocaine subcutaneous infusion) around the segment of fistula to be treated. Clinicians opt for sedation and analgesia need to pay attention to ensure appropriate credentialing, proper cardiorespiratory monitoring personnel and equipment, reversal medications, and resuscitation equipment are available in the facility where these endovascular interventions are being carried out.

Service Structures

Endovascular intervention to salvage failing vascular accesses can be organized as an in-patient service or a day procedure service depending on the social culture, healthcare facility structure etc. Nowadays, more and more hospitals and specialized centers prefer to provide day procedure service. Day procedure service minimizes the disruption of therapeutic procedure to patients' daily life. It also reduces manpower required to monitor and care for patients overnight. Nonetheless, safety of patients comes first. To enable safe and yet cost-effective day procedure service, modification of service structure in several aspects is needed:

- *Patient selection*: set up guidelines to sift out ultra-high risk and medically unstable patients from day procedure (e.g. patients with frequent hyperkalaemia and arrhythmia, patients with angina). Patients with mechanical heart valve replacement and requiring strict anti-coagulation therapy are also not suitable for day procedure.
- *Hemodialysis schedule*: For patients on three sessions per week diaysis regimen and relatively stable fluid and electrolytes balance, the endovascular intervention can be scheduled in between their hemodialysis days. If the timing is not allowed, the procedure can also be arranged a few hours after or before the hemodialysis session. More caution on patients' hemodynamic assessment and electrolytes monitoring is required. For patients with two sessions per week, it would be better to add on an extra session before or after the endovascular procedure. If the procedure is exceptionally difficult with large volume of contrast and fluid used, special arrangement of dialysis regimen with the dialysis center would be needed.
- *Procedural consideration*: The puncture sites for endovascular intervention are preferably over upper limb than femoral artery. Closure devices may help to secure femoral artery hemostasis and enable early ambulation but will increase the cost of the procedure. Standard practice of meticulous hemostasis on all puncture sites has to be achieved before patient returns to day ward. Special events that happened during intervention (e.g. small perforation of fistula during angioplasty, hypotension during procedure) have to be clearly and properly documented so that appropriate post-procedure monitoring and investigation can be carried out.

- *Post-procedural review:* This should be conducted by either the proceduralist who performed the procedure, or one of the assistants who participated in the procedure. A checklist for review will help to standardize the service. Nonetheless, special caution and monitoring are required for specific conditions of the patient or the procedure (e.g. longer monitoring time for excessive drowsiness after sedation, look out for fluid overload after exceptionally long procedure for difficult lesions, review limb and arterial condition if hematoma noted over puncture site).

Case 1

Madam O, 78 years old, DM, HT, IHD, ESRF, she has been using left BC AVF for hemodialysis for three years. She has defaulted follow-up for the past year. The access flow dropped gradually from 800 ml/min to 350 ml/min over a few months and the renal physician had convinced her to come back for further management.

Clinical examination revealed hardening of the mid-arm cephalic vein fistula. Pulsation rather than thrill was felt over distal arm fistula. No thrill can be felt in the upper arm region. Fistulogram and fistuloplasty was arranged right away.

An antegrade mid-arm brachial artery puncture under ultrasound guidance was performed using a 20G IV cannula and a 5 Fr paediatric sheath (7 cm) was inserted. The fistulogram (Fig. 3) showed a high grade juxta-anastomotic stenosis together with occlusion of the mid-arm cephalic vein fistula.

The cephalic vein fistula occlusion was crossed with a 0.035" angled Glidewire (Terumo Medical Co. Somerset, NJ, US) and supported by a 4 Fr 65 cm straight flush catheter (Cordis Co. Fremont, CA, US). The guidewire was parked just over the opening of the straight flush catheter so both together act like a rod (Fig. 4a). Gentle force was given to advance the two through the occluded cephalic vein fistula.

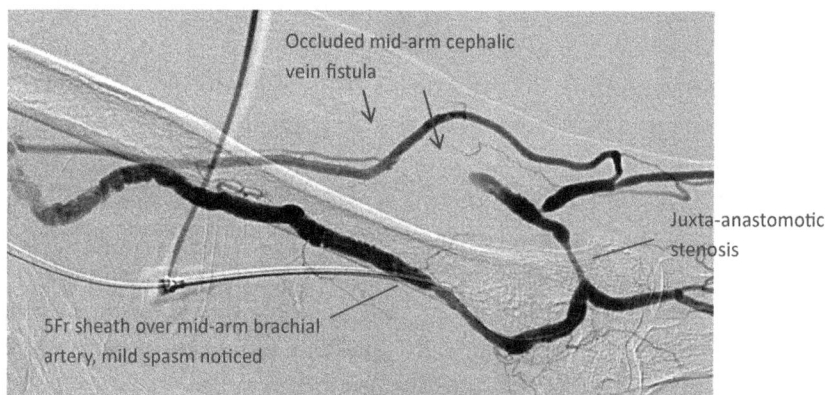

Fig. 3.   Fistulogram of the failing left BC AVF.

After advancing for a certain distance, the Glidewire was taken out. A good backflow of blood through the straight flush catheter indicates the catheter is in a patent segment of the fistula. Angiogram can then be done at this point to evaluate more proximal vein (Fig. 4b). Central venogram was performed without obstruction detected (not shown here). The Glidewire was then parked into the central vein. Balloon angioplasty of the occluded cephalic vein and stenosed juxta-anastomotic cephalic vein was performed using 6 mm and 5 mm Mustang balloon (Boston Scientific, MA, US) respectively. Completion fistulogram is shown in Figs. 5a–c. The thrill of the left BC AVF regained after the fistuloplasty.

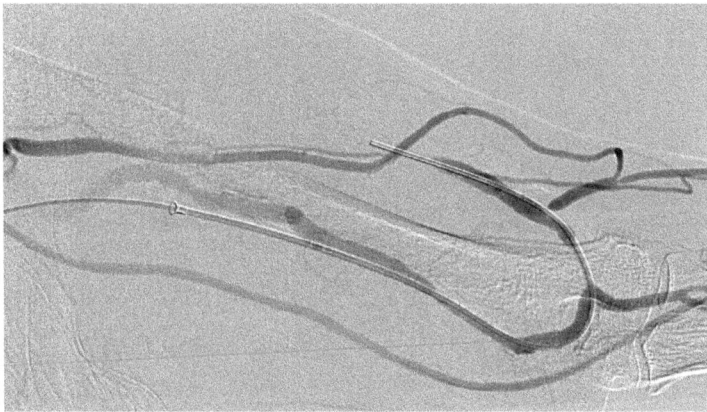

(a)

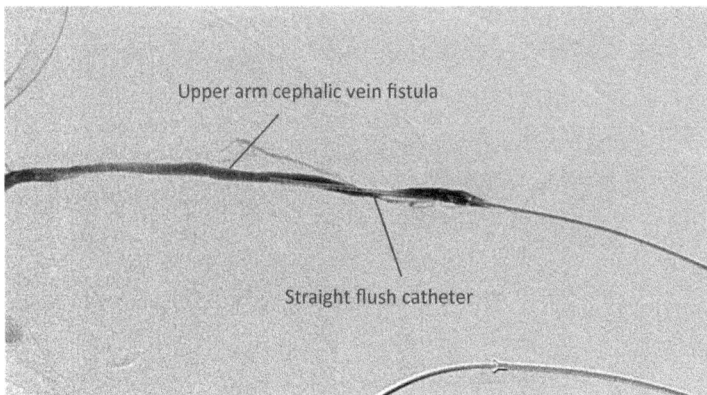

Upper arm cephalic vein fistula

Straight flush catheter

(b)

Fig. 4. Fluoroscopic angiographic images showing crossing of the cephalic fistula occlusion.

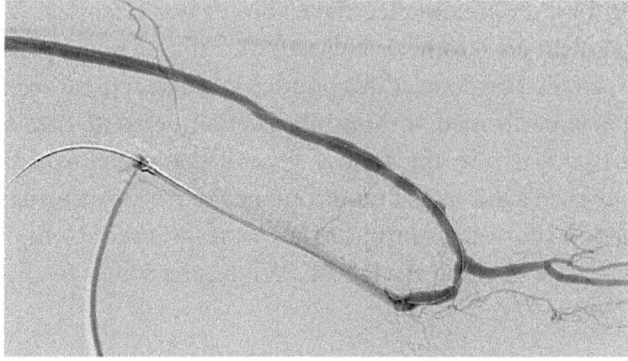

(a)

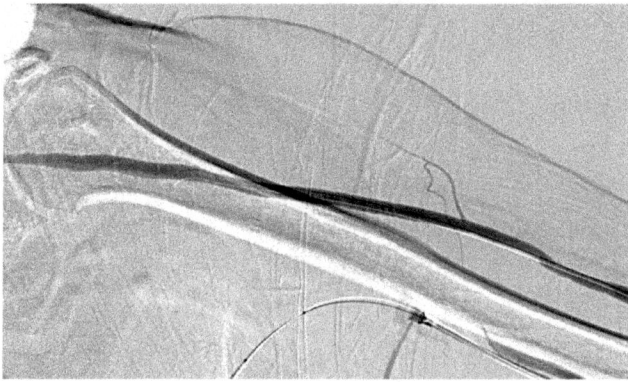

(b)

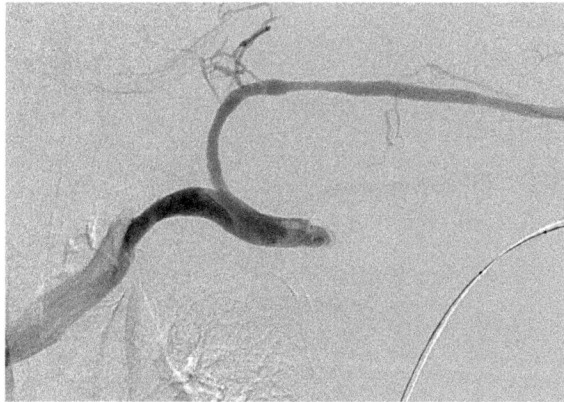

(c)

Fig. 5.   Completion fistulogram after balloon angioplasty.

Even the fistulogram looks bad, crossing of the occlusion may not be very difficult. Attempt with caution, the result could be nice.

Case 2

Madam C, 60 years old, DM, HT, ESRF using left BC AVF for hemodialysis for six years. She has grossly dilated and tortuous course of cephalic vein fistula over the left arm. The venous pressure during dialysis increased to 170–200 mmHg over the last two months. No access flow data was available. Clinically, dilated and tortuous course of cephalic vein fistula was prominently observed with regions of aneurysmal change. Strong pulsation was felt over the fistula but not thrill. Fistulogram and fistuloplasty were arranged.

Due to the very tortuous course of the BC AVF and presence of strong pulsation, the entry site was chosen to be the cephalic vein fistula over the distal arm instead of inflow artery. A 20G IV cannula was used to make the puncture and initial fistulogram is shown as below (Fig. 6):

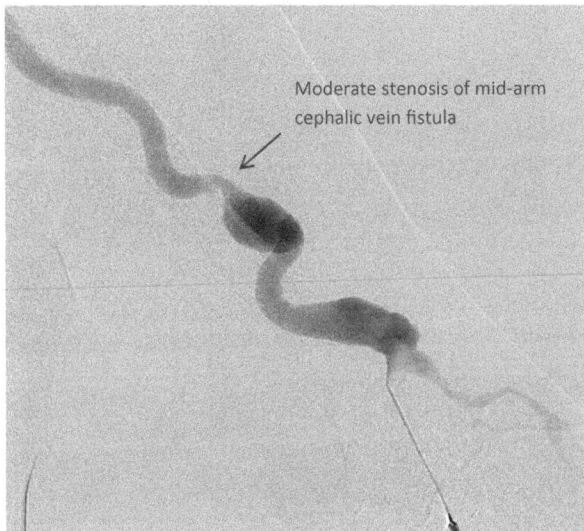

Moderate stenosis of mid-arm cephalic vein fistula

(a)

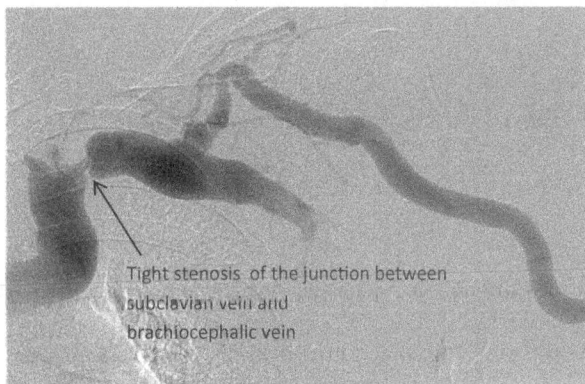

Tight stenosis of the junction between subclavian vein and brachiocephalic vein

(b)

Fig. 6. Fistulogram of the failing left BC AVF.

The cephalic vein fistula and subclavian vein stenosis were crossed and being treated by Reef angioplasty balloon 6 mm/40 and 8 mm/40 (Medtronic Inc. Minneapolis, MN, US) respectively using high pressure (18–22 atm). Residual stenosis was observed but the lumen has significantly improved (Fig. 7).

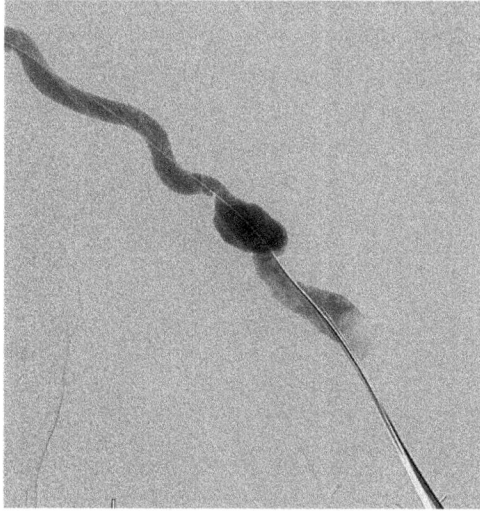

(a)

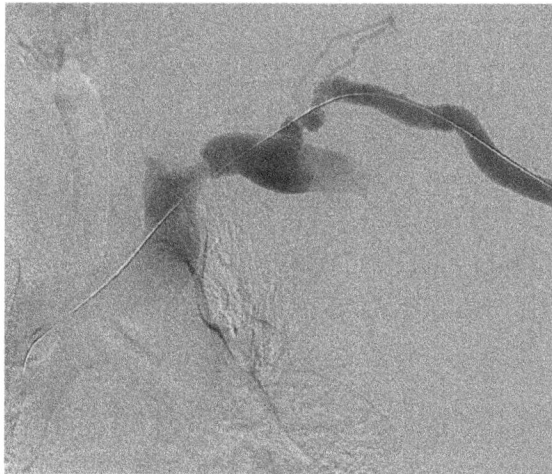

(b)

Fig. 7.   Completion fistulogram after high pressure balloon angioplasty.

This fistula remained functional for another 18 months with one additional balloon angioplasty treatment (eight months later) before it was finally blocked.

If multiple turns are expected, choose a puncture site with a more direct pathway for intervention.

Case 3

Mr L, 69 years old, DM, HT, ESRF using left BC AVF for hemodialysis since four years ago (early 2010). He had juxta-anastomosis cephalic fistula stenosis with balloon angioplasty performed in March 2011. He then defaulted follow-up. He was being admitted to the hospital (April 2012) for failure of dialysis through the left BC AVF. On clinical examination, there was only a weak pulsation over an aneurysmal part of the cephalic vein fistula. A semi-urgent fistulogram was performed.

A 20G IV cannula was inserted into the aneurysmal part of the cephalic vein fistula over the distal arm (antegrade cephalic fistula puncture) to perform a preliminary fistulogram. It showed presence of thrombus and stenosis over the patent cephalic vein segment and total occlusion of cephalic vein fistula over mid-arm (Fig. 8). A 5 Fr sheath was inserted through the same puncture. Heparin 3000unit was given through the sheath.

The cephalic vein occlusion was successfully crossed with 0.035" Glidewire (Terumo Medical Co. Somerset, NJ, US) (Fig. 9). Balloon angioplasty was performed with satisfactory lumen regain except one short segment of persistent recoil (Fig. 10). Stenting was performed for that particular spot using Maris stent 7 mm/40 mm (Medtronic Inc. Minneapolis, MN, US). Post-stenting angiogram showed satisfactory flow and lumen regained (Fig. 11).

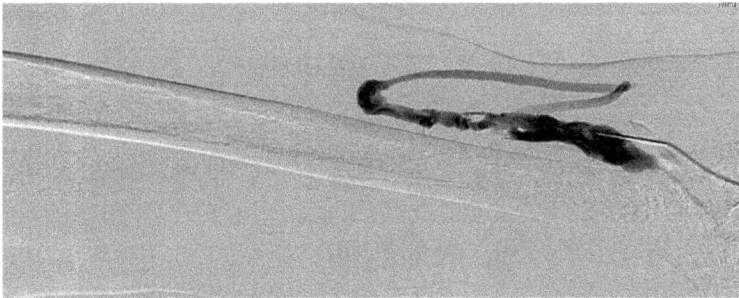

Fig. 8. Fistulogram of the critically failing left BC AVF.

Fig. 9. Angiographic image showing the cephalic vein fistula occlusion was crossed by the guidewire.

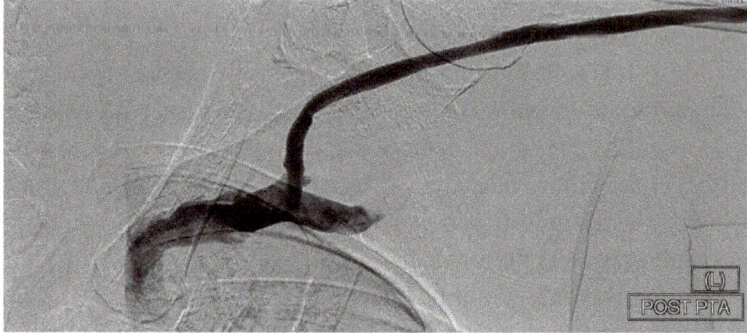

(a)

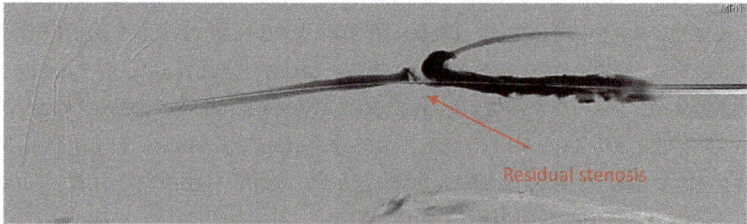

Residual stenosis

(b)

Fig. 10. Angiographic images of the fistula after balloon angioplasty showing patent proximal arm cephalic vein fistula (a) and short segment recoil of venous fistula over mid-arm (b).

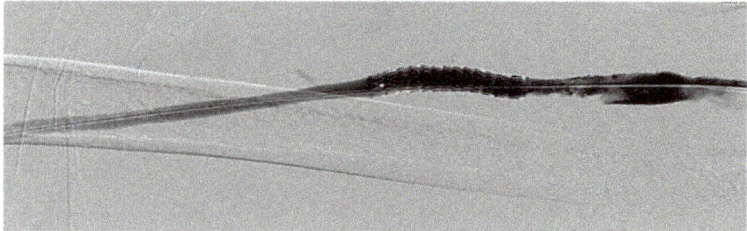

Fig. 11. Post-stenting fistulogram.

As mentioned, since there was a history of juxta-anastomotic stenosis, it is likely to recur and this was confirmed with further fistulogram of the inflow (Fig. 12). A retrograde cephalic vein fistula puncture was then performed on the upper arm to treat the juxta-anastomotic lesion.

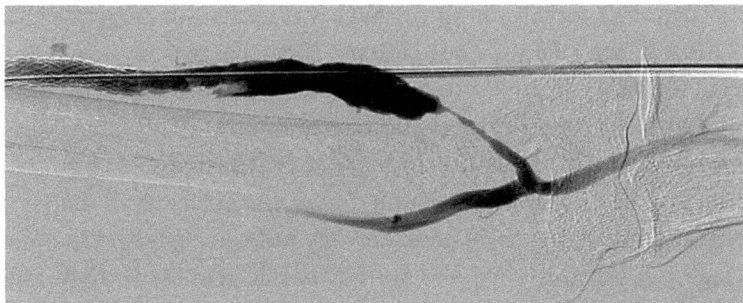

Fig. 12.   Fistulogram showing juxta-anastomotic stenosis.

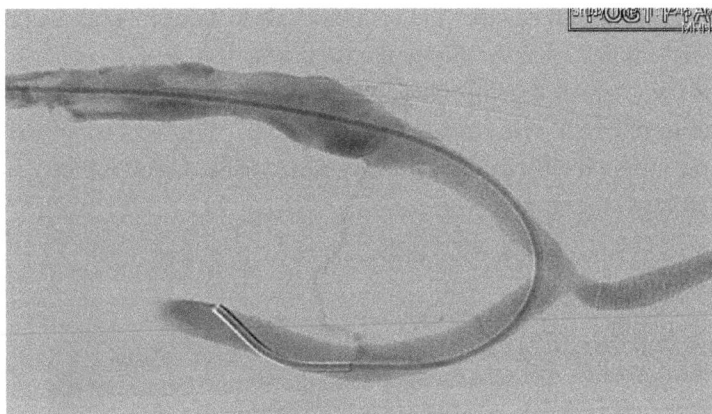

Fig. 13.   Catheter injection completion angiogram after balloon angioplasty of the juxta-anastomotic segment.

Balloon angioplasty was performed to treat this juxta-anastomotic lesion with good result. A Cobra catheter (Cordis Co. Fremont, CA, US) was passed into the brachial artery to perform a good completion angiogram (Fig. 13).

Hemodialysis resumed after this endovascular intervention. Mr L underwent another endovascular intervention four months later. This fistula lasted for 13 months more before it was finally blocked.

An antegrade puncture of cephalic vein fistula provides a direct route to tackle difficult lesions, like this case, long segment occlusion. Assessment of the whole course of the vascular access is important so as not to miss any significant lesion.

Case 4

Madam N, 49 years old, DM, HT, morbidly obese, developed ESRF in year 2008. Vein mapping performed at that time showed cephalic veins in bilateral upper limbs were small in calibre. Left basilic vein over elbow region was 2.8 mm. She was reluctant to have a long wound over the arm. Left forearm loop BB AVG was created in November 2008. She was well using the BB AVG for dialysis until January 2014. There was a gradual reduction of access flow from 1000 ml/min to 700 ml/min. Venous pressure was also noted to be high ~180 mmHg. Pulsation of the AVG was felt on examination. Fistulogram was arranged.

Fistulogram was performed initially through a 5 Fr sheath inserted over the loop of the graft towards arterial limb. Stenosis was noted over the artery-graft anastomosis and in graft (mainly venous limb) (Fig. 14). There was only mild stenosis over the vein-graft anastomosis. Although it looks abnormal, the outflow basilic vein seems to be patent. Balloon angioplasty (Wanda 5 mm/40 mm, Boston Scientific Co. MA, US) of the artery-graft anastomosis was performed with satisfactory result (Fig. 15).

Another 5 Fr sheath was inserted over the loop towards the venous limb of the graft. The guidewire was then parked over the venous limb to treat in-graft stenosis. However, the guidewire was trapped over the elbow region basilic vein.

By changing the angle of the fluoroscopy unit to a cranial-caudal view, a short segment occlusion of the basilic vein was revealed (Figs. 16a, b). The occlusion was crossed with 0.035" Glidewire, supported by straight flush catheter. Balloon angioplasty was performed with good lumen regain (Fig. 17).

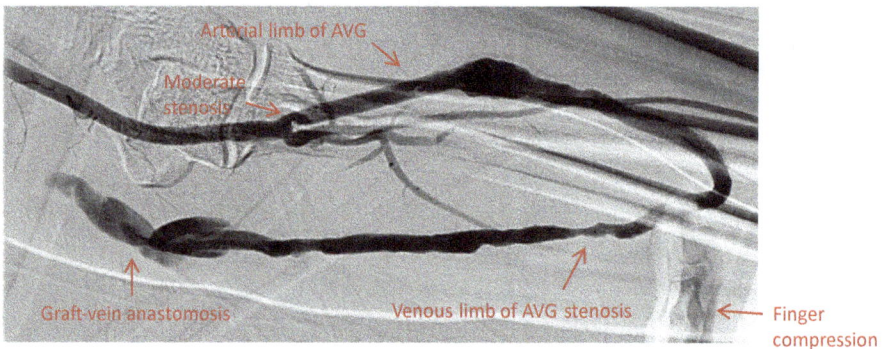

Fig. 14. Fistulogram of the failing left forearm loop BB AVG.

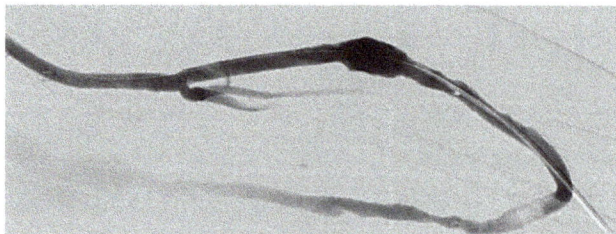

Fig. 15. Arterial limb of the AVG after balloon angioplasty of the artery-graft anastomosis.

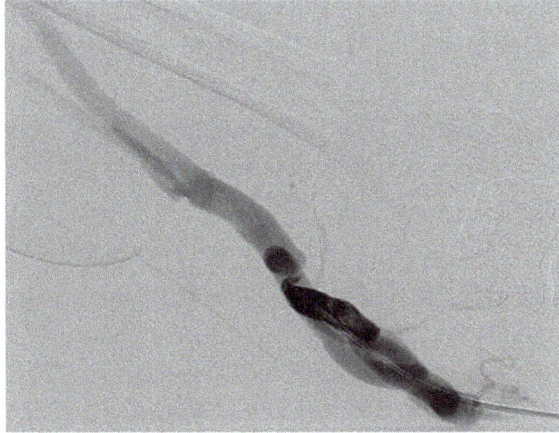

(a)

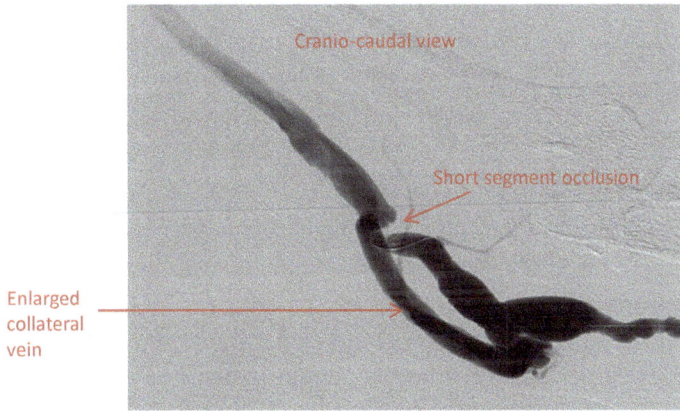

(b)

Fig. 16. Angiogram performed at two different angles reveal a short segment occlusion of the outflow basilic vein.

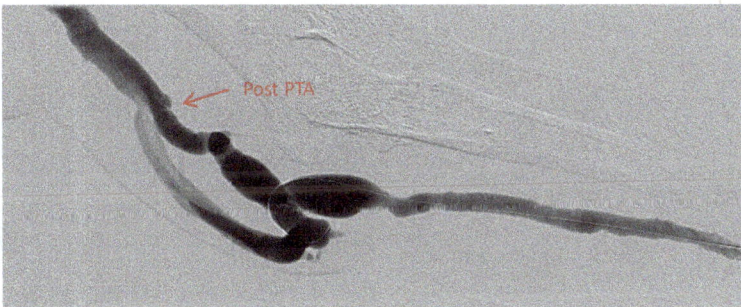

Fig. 17. Completion angiogram after balloon angioplasty of the basilic vein.

Madam N's hemodialysis went well after the intervention. The venous pressure fell to ~120 mmHg after the procedure. The AVG is still functioning well till now.

> *Different angles of fluoroscopy may be needed to view the vessel condition clearly, especially with tortuous vessel or over anastomotic site.*

Case 5

Mr H, 55 years old, DM, HT, chronic smoker, ESRF with hemodialysis commenced four years ago using right BC AVF. He developed diffuse stenosis of the right BC AVF three years ago as shown in the fistulogram (combined image) below (Fig. 18).

The lesions were successfully crossed and balloon angioplasty was performed (Fig. 19). Thrill can be felt well over the BC AVF after fistuloplasty and he continue to use this access for hemodialysis.

Mr H defaulted follow-up for one year and came back to the clinic with difficulty of needling during dialysis. The cephalic vein over his proximal forearm was more prominent than before. Thrill can only be felt around the elbow region and no thrill was palpable over mid- or upper arm. Fistulogram was performed showed re-occlusion of a long segment of the cephalic vein fistula (Fig. 20).

The high grade stenosis over juxta-anastomotic cephalic vein recurred and another tight stenosis was present in the cephalic fistula before a major branch runs into the forearm and connects back into the deep vein. The branch of cephalic vein was well developed and the dialysis nurse had been needling it for "V" cannula. Attempt was made to re-open the blocked right arm cephalic vein fistula but failed. Attention was then shifted to treat the juxta-anastomotic stenosis and the other focal tight stenosis near the branching point.

Good luminal regain and flow were achieved after balloon angioplasty (Fig. 21). Hemodialysis was continued through "A" needling over the distal arm cephalic vein and "V" needling over proximal forearm dilated branch. This fistula is still functional, requiring endovascular intervention every 9–10 months as maintenance therapy. Due to the increased venous drainage through the major branch into the forearm deep vein, his right basilic vein

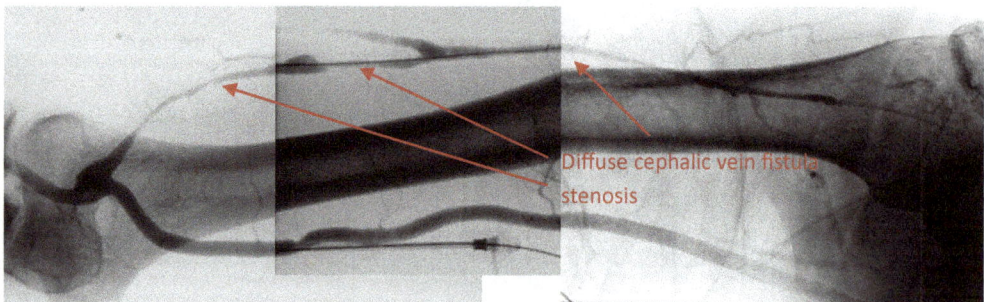

Fig. 18.   Fistulogram of the failing right BC AVF.

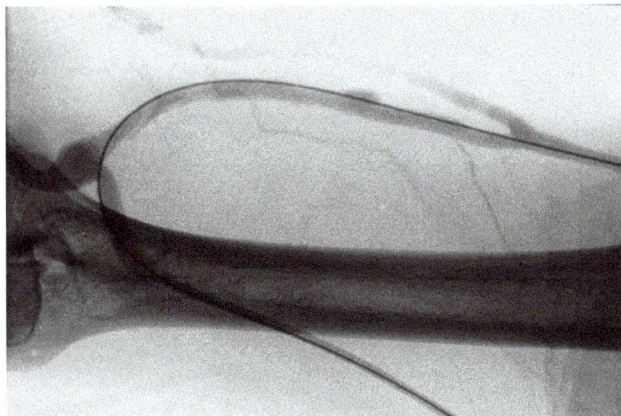

(a)

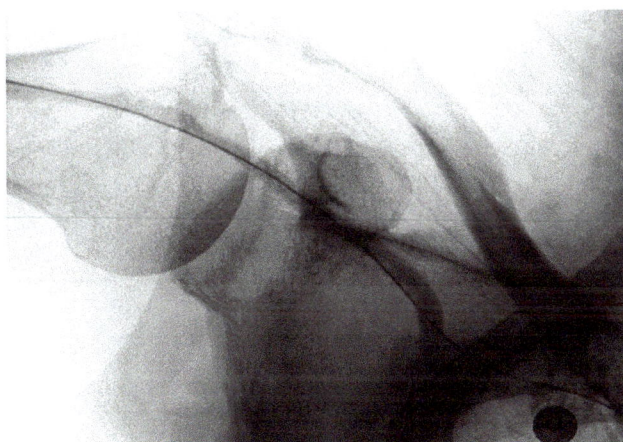

(b)

Fig. 19.   Completion angiogram after balloon angioplasty.

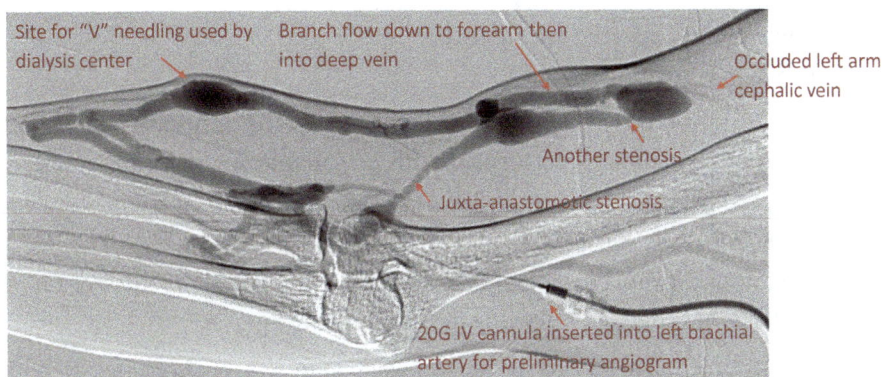

Fig. 20.   Fistulogram of the right BC AVF after recurrence of problem.

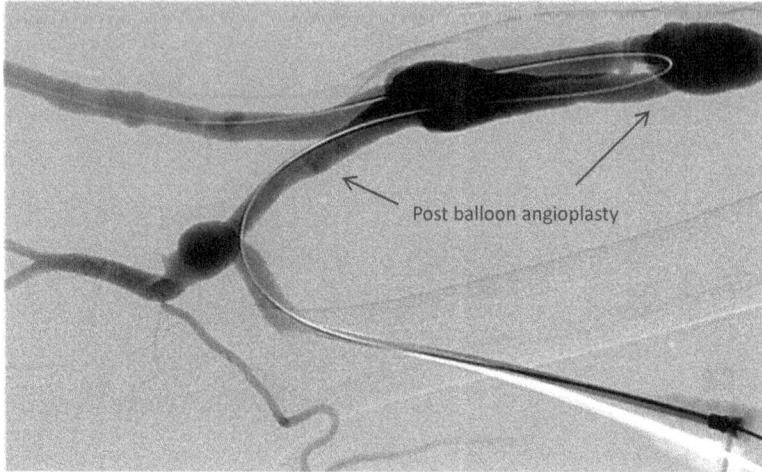

Fig. 21. Completion angiogram after balloon angioplasty.

also enlarges with time. The current plan is to maintain this BC AVF as far as possible. If thrombosis occurs, single stage right arm brachio-basilic AVF creation and BBT would be the next vascular access.

> *The potential of human body adaptation is often beyond our imagination. Clinicians should watch out for the evolving vessel condition of the limb and act to facilitate body's adaptation. That is also one of the fascinating parts of hemodialysis access care.*

Case 6

Mr M, 65 years old, DM, HT. old CVA with left upper limb weakness, developed ESRF since four years ago. His vascular access history is as follwos:

Jan 10 — Tunnelled CVC inserted via right IJV.
Apr 10 — Left BC AVF creation.

Jul 10 — Poor maturation of left BC AVF, fistulogram showed diffuse stenosis of left arm cephalic vein. Balloon angioplasty performed with reasonable angiographic result.

Sep 10 — Re-stenosis and failed maturation of left BC AVF. Left arm basilic vein over the elbow was only 2.2 mm. Mid-arm left basilic vein size was good (~4 mm). There was stiffness and reduced range of movement of his left shoulder joint. During this period of time, right IJV tunnelled CVC changed once due to catheter thrombosis.

Oct 10 — Left arm brachial artery to mid-arm basilic vein AVG was created. Needling for hemodialysis commenced two weeks after the operation. The tunnelled catheter was removed afterwards.

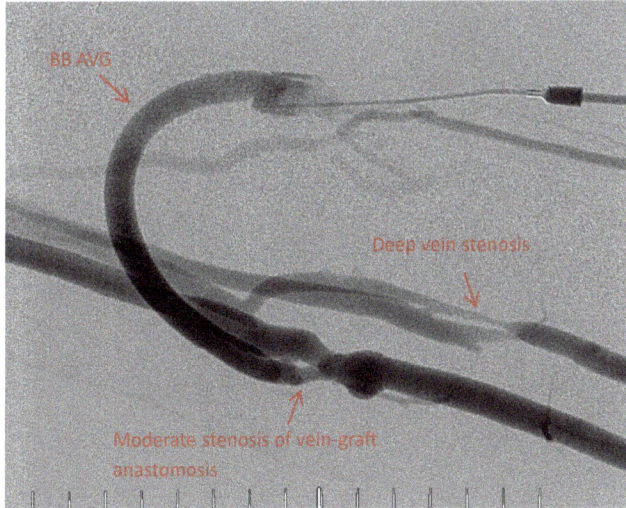

Fig. 22. Left mid-arm fistulogram of the BB AVG.

Oct 11 — Presented back to clinic with left upper limb swelling from hand all the way up to mid-arm. Pulsation of the AVG was felt. Venous pressure on hemodialysis was between 130–160 mmHg. Venous duplex scan was performed for left upper limb to exclude deep vein thrombosis and was negative. Vein-graft anastomosis stenosis was noted in the duplex study. Fistulogram was performed through a puncture of the AVG (Fig. 22) showed graft vein anastomotic stenosis. A segment of brachial vein stenosis was noted.

Venogram further showed another stenotic lesion in proximal arm basilic vein (Fig. 23).

Angioplasty of the vein-graft anastomosis stenosis was performed through the sheath inserted via the graft. However, because of the acute angulation, the guidewire has difficulty in turning proximally to the proximal basilic vein. Percutaneous puncture of the basilic vein around elbow region guided by ultrasound was performed. Balloon angioplasty was then performed to treat the proximal basilic vein stenosis (Fig. 24).

In the presence of upper limb swelling, deep venous system was interrogated. A third access was then made with percutaneous puncture of the brachial vein over cubital fossa under ultrasound guidance and sheath insertion. Stenotic lesions in the mid-arm brachial vein were shown (Fig. 25), crossed and treated (Fig. 26).

Post-intervention, his upper limb swelling subsided. The limb swelling recurred five months later and then another four months later, requiring repeated fistuloplasty and deep vein angioplasty treatment.

> *Multiple punctures to treat various stenotic lesions might be needed. Ultrasound facilitates puncture and access to selective arteries and veins.*

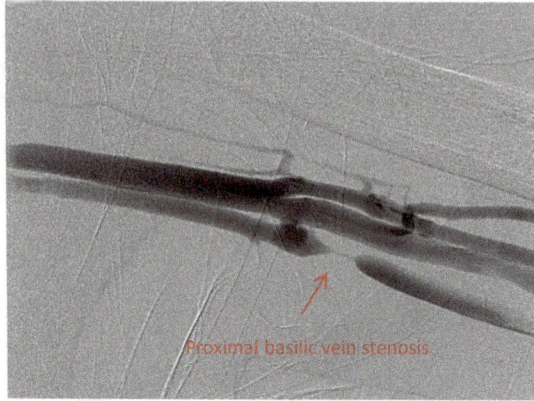

Fig. 23.   Left proximal arm fistulogram.

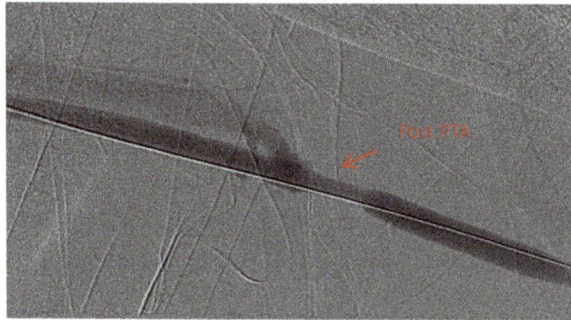

Fig. 24.   Completion angiogram of the proximal basilic vein after balloon angioplasty.

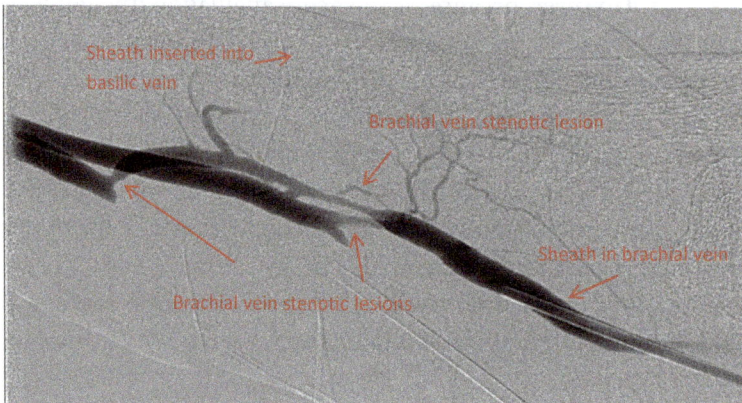

Fig. 25.   Deep vein angiogram showed multiple stenosis of the deep veins.

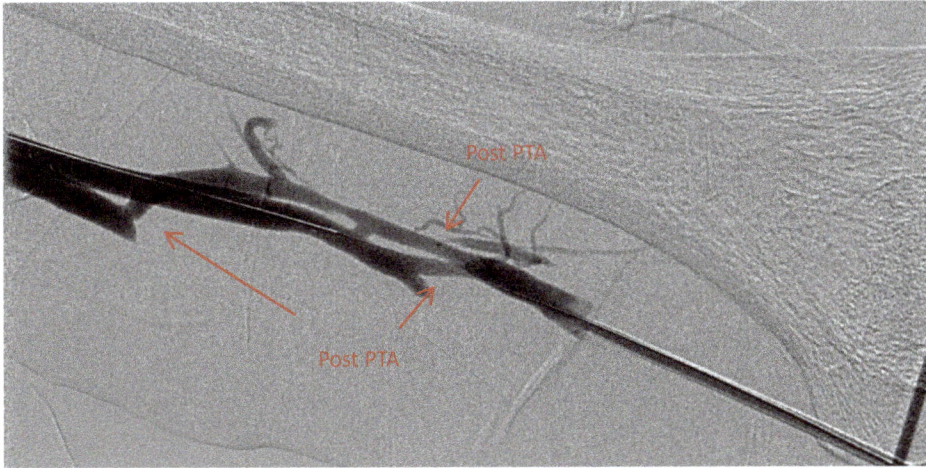

Fig. 26. Completion angiogram of the deep veins after balloon angioplasty.

References

1. Raju AV, May KK, Zaw MH, *et al.* Reliability of ultrasound Duplex for detection of hemodynamically significant stenosis in hemodialysis Access. *Ann Vasc Dis.* 2013, **6**(1): 57–61.
2. Akoh JA. Prosthetic arteriovenous grafts for hemodialysis. *J Vasc Access.* 2009; **10**: 137–147.
3. Bountouris I, Kristmundsson T, Dias N, *et al.* Is repeat PTA of a failing hemodialysis fistula durable? *Int J Vasc Med.* 2014; **2014**: 369687.
4. Tan TL, May KK, Ho P. Outcomes of endovascular intervention for salvage of failing haemodialysis access. *Ann Vasc Dis.* 2011; **4**(2): 87–92.
5. Aftab SA, Tay KH, Irani FG, *et al.* Randomized clinical trial of cutting balloon angioplasty versus high-pressure balloon angioplasty in hemodialysis arteriovenous fistula stenoses resistant to conventional balloon angioplasty. *J Vasc Interv Radiol.* 2014; **25**(2): 190–198.

CHAPTER 9

Endovascular Salvage for Thrombosed Hemodialysis Access

Po Jen Ko and Sung Yu Chu

Indication and Timing

Hemodialysis access is the lifeline for patients on regular hemodialysis treatment. Dialysis-dependent patients are usually put on renal replacement treatment three times a week to keep their normal physiology functions. Access stenosis and access thrombosis are two of the most common access complications that vascular surgeons/interventionists deal with in their daily practice. Once the dialysis graft is found to be acutely thrombosed, action should be taken as soon as possible to maintain or re-establish a functional vascular access for the patient. The procedures could be either salvage of the original access in use or establish a temporary access elsewhere.

The patients need some kind of interventional procedures at least urgently if the vascular access is acutely thrombosed. The procedure should be performed in order not to delay the next scheduled dialysis therapy. In addition, early thrombectomy treatment is said to be associated with higher technical success and potentially may improve midterm access patency.[1] With modern endovascular techniques, the clinical successful rate of salvage procedures for acute dialysis access graft thrombosis, ranges from 70% to 95% in modern literatures,[2,3] are quite satisfactory and at least comparable to its surgical counterparts.[4] The goal of the salvage procedures are to regain adequate blood flow through vascular access to support hemodialysis for the patients and not to jeopardize the normal physiology.

Percutaneous endovascular salvage procedures comprised of two major components: removal of thrombus (thrombectomy) and address of underlying lesion(s). Thrombectomy can be performed as pharmaceutical thrombectomy (lysis of the thrombus by thrombolytic agents), mechanical thrombectomy (removal of the thrombus using balloon or certain devices using mechanical methods) or pharmacomechanical thrombectomy (combination of lytic agents and mechanical methods for removal of thrombus).

Contraindication

Percutaneous salvage procedures for the original access are usually considered as the primary interventional modality to recannalize the acutely thrombosed access. However, for some of the patients, e.g., patients with allergy history to contrast medium, one should choose open thrombectomy (with or without certain type of revision) instead of percutaneous procedures with conventional contrast medium. Percutaneous salvage with CO_2 angiography may be another alternative for patients with the history of contrast allergy. In addition, if the underlying lesion can only be resolved by surgical revision (such as graft pseudoaneurysm/aneurysm, long segmental occlusion of outlet vein, giant fistula aneurysm with heavy thrombus burden, access with profound infection… etc.), surgical treatment should be considered as the priority method compared to percutaneous revascularization.

Percutaneous salvage procedures should not be considered for a clotted newly constructed graft or arteriovenous fistulas (less than 30 days of age) since the concern of angioplasty on unstable anastomosis. In addition, it is too early for those young grafts to develop graft outlet stenosis (the most common cause of dialysis graft thrombosis). Dialysis interventionists should look carefully into the truly underlying cause of the newly constructed graft/fistula failure in order to design a proper way of salvage. The procedure results for those early failure access are usually dismal.[5]

Anesthesia and Adjuvant Medications

Anesthesia methods should also be taken into consideration since it correlates significantly with the procedure-related complications. Chronic renal failure patients are usually so fragile that systemic complications are frequently seen during the perioperative period of access salvage procedures. The possible unfavorable effect of anesthesia agents should be considered cautiously when performing the intervention. Thanks to the minimal invasive nature of percutaneous procedure (with only puncture access needed), the salvage procedure for most of the dialysis grafts can be performed under local anesthesia with infiltration of Xylocaine instead of general anesthesia. Intravenous sedation may be added as well to aid the performance of a percutaneous salvage procedure. Fasting of the patient is not needed unless general anesthesia or sedation is planned due to certain specific reason. For conscious sedation purpose, fentanyl citrate and midazolam hydrochloride can be administered intravenously during dialysis access intervention. Patients are monitored with pulse oximetry, pneumatic blood pressure measurement, and electrocardiography.

Pre-procedure medication with oral anti-platelet agents are usually not necessary. During the procedure, systemic heparinization are not needed due to the relatively short procedure time. Flushing of the introducer sheath occasionally during the procedure with heparinized saline (0.2 ml of heparin in 20 ml of normal saline) to keep the sheath thrombus-free is mandatory during similar procedures. Lytic agents, such as urokinase, are common adjuvant medication given in those percutaneous thrombectomy procedures. Thrombolytic agents are injected into the occluded graft to lyse the intra-graft thrombus (pharmaceutical thrombectomy/catheter-directed thrombolysis) and to aid the proceeding of mechanical thrombectomy. Pure pharmaceutical thrombectomy are seldom used today because of the longer procedure time needed if mechanical thrombectomy is not utilized. Pre-procedure or post-intervention antibiotics are not mandatory. Actually, procedure related local infection or systemic infection are so rare that preventive antibiotics are somewhat redundant.

Pre-Procedure Preparation

Proper diagnosis must be made before jumping into the percutaneous salvage procedure. For both AVG and AVF, vascular ultrasonography is not necessary in terms of establishment of the diagnosis. In fact, physical examination is the most time-saving and effective method of diagnosis. Disappearance of thrill and bruit indicate the absence of active blood flow in the hemodialysis conduit. Absence of pulsatile sensation on palpation of AVG is the indicator of thrombus formation. Occasionally, the inflow side of AVF is still patent but the rest of the fistula thrombosed. Vascular ultrasound would be helpful in confirming the diagnosis in such situations.

Comprehensive history taking regarding the patient's dialysis access should be performed prior to the planning of the procedure. Access type, age, location as well as previous procedures should be identified. Most of the dialysis accesses will not clot without underlying cause. Frequent causes of acute access thrombosis are structural abnormalities in access conduit (aneurysm, stenosis or obstruction somewhere along the circuit), infection, hemodynamic instability, and inappropriate over-compression on the circuit ... etc. Any information to assist the identification of underlying cause of access clotting should not be neglected in order to design the most appropriate and effective salvage method. Underlying cause should be addressed in addition to the removal of the thrombus when dealing with the failed dialysis access treatment in order to achieve a reasonably durable result.

Endovascular Salvage for Thrombosed Graft; Pharmacomechanical Thrombectomy (PMT)

Lacing and maceration by pulse-spray lytic therapy

Currently, infusion thrombolysis alone is seldom used to treat acutely thrombosed grafts because of its time-consuming nature. Adjunctive mechanical components allows faster restoration of flow through thrombosed dialysis grafts. Pulse-spray technique aided PMT had been adopted widely and proposed for more than a decade. It involves the use of high pressure lytic agent spray out through two multiple side-hole catheters inserted toward each other throughout the graft to lace the thrombus during systemic heparinization in dialysis grafts.[6,7] This pulse-spray technique yields reasonable high successful rate and low complication rate. However, the procedure needs specific multiple side-hole pulse-spray catheter and still takes around an hour to wait for the complete dissolution of the clots.

Lyse and wait technique

This particular technique involves infusion of a solution of 250,000 U of urokinase plus 5000 U of heparin through a catheter placed near the arterial anastomosis directed toward the venous end. After 45 minutes of observation, the patient was sent for angiography in the intervention suite and the arterial plug thrombectomy was performed followed by endovascular management any underlying stenosis.[8] Tissue plasminogen (t-PA) can also be used for the lyse and wait technique to substitute urokinase. The immediate technical successful rate of lyse and wait method for dialysis grafts had been reported to be high (above 95%) in literatures with relatively short operation time (usually less than an hour). Primary patency after the procedure was satisfactory (around 4 months).[9,10]

No-wait lysis pharmacomechanical thrombectomy

Ideal thrombectomy procedure should be safe, effective, fast and economical. The classical lyse and wait technique had provided an inexpensive technique with compatible successful rate to thrombectomy using mechanical devices. However, it is still time-consuming. Almehmi *et al.* had published their experience on a "no-wait lysis" approach for prosthetic graft thrombectomy with a shorter procedure time and less radiation exposure.[11] Here we described our modified technique of no-wait lysis PMT for acute AVG thrombosis (Table 1).

Table 1.   Simplified steps of dialysis access PMT.

Basic steps of "no wait" pharmacomechanical thrombectomy (PMT) for blocked AVG
1. Insert antegrade and retrograde sheaths
2. Access outflow veins with antegrade wire
3. Balloon maceration of the thrombus via the antegrade sheath plus negative pressure suction from the sheath
4. Infusion of urokinase solution (250KU in 10 ml saline)
5. Repeat the balloon maceration and suction of thrombus
6. Heparinized saline irrigation
7. Traverse the arterial inlet by wire and Fogarty balloon catheter via the retrograde sheath to extract the arterial plug
8. Angiography of the whole circuit to confirm the flow, discover residual thrombus and underlying vascular lesions
9. Treat any residual thrombus/lesion
10. Remove sheaths, Prolene purse-string suture

Initially, vascular route was established by two 6 Fr sheaths on both end of the graft facing each other (antegrade and retrograde sheaths). The thrombus can be removed by applying negative pressure on either of the sheath. A bolus of urokinase solution (250 KU) can also be injected via the sheath to facilitate the lacing of the graft. If needed, the thrombi maceration could be facilitated by external massage on the graft and the thrombus may be removed by applying negative pressure through the side port of the sheath. The guidewire is then inserted to traverse the venous junction of the graft followed by balloon maceration of the clots within the graft. A noncompliant angioplasty balloon, usually 7 mm or 6 mm in diameter, was used to dilate the whole graft from venous junction through most of the graft length followed by irrigation of the graft by heparinzed saline. Once the graft and its outflow tract are cleared of thrombus, a compliant Fogarty thrombectomy balloon catheter can then be advanced retrogradely through the graft arterial anastomosis to pull back the arterial plug into the graft. At this stage, the active arterial flow should be directed into the graft and the circulation reestablished. Underlying/residual graft stenosis or remaining thrombus fragments are then identified carefully by angiography and addressed by appropriate balloon angioplasty till the lesions have been corrected. With the aid of lytic agent incubation, the fresh clot in dialysis graft is easily macerated and flushed away by irrigation solution. The dialysis circuit can then be re-established easily within a short period. And nothing other than standard wires, sheaths and angipasty balloons are needed. With modern endovascular techniques, PMT is fast, effective, minimal invasive and easy to perform.

Endovascular Salvage for Thrombosis of Arteriovenous Fistula

Treatment for the thrombosed AVF is usually more difficult and technically challenging. This is because of the higher volume of clot formation in the fistula and the irregularity of the fistula anatomy contributed by aneurysm, stenosis and branches. The technique used to do percutaneous salvage for arteriovenous fistula is quite variable, usually customer tailored to each patient's condition. However, the basic principles still applies,[12] two stages of the salvage procedure: (1) clearance of clots; and (2) correction of anatomical cause of thrombosis.

Access to AVF is gained using either one introducer sheath (usually for those with relatively new clots or low clot burden) or two introducer sheaths. If two sheaths are applied, usually 6–8 Fr, they face each other on the fistula segment. The exact sites of sheath introduction should be decided carefully by clinical examination and ultrasound examination. Guidewire should be manipulated into the central vein through the antegrade sheath. The extent of thrombus is then decided by using a 5 Fr, angiography catheter pulling back from the central vein into the venous outlet of the fistula. One should notice that if the wire cannot traverse the fistula outlet into the central vein, the endovascular salvage should be abandoned.

The de-clotting process can be initiated after the definition of clot extent. De-clotting should be conducted from central side retrogradely into the fistula inlet to reduce the chance of pulmonary embolization. The de-clotting maneuver can be conducted by either guiding catheters, catheters with side hole or introducer sheaths. After the de-clotting, any anatomical stenotic segment should be dilated by appropriate angioplasty balloon. Immediate significant recoil of the stenosis should be further addressed by prolonged balloon angioplasty, cutting balloon, stenting, or even direct surgical revision so as to ensure a durable result of the procedure. Some adjuvant maneuvers such as manual declotting without or with ultrasound guidance,[13] suction thrombectomy through a mini-venotomy, or irrigation with lytic solution (250 KU urokinase in 10 ml of normal saline) may sometimes be helpful in difficult procedures.

After recanalization of most of the fistula, the arterial inflow is then addressed by Fogary balloon or angioplasty through the retrograde sheath. After angiography confirms un-obstructed flow from the fistula into the central vein, the sheath is then removed with manual compression or purse-string suture hemostasis.

Mechanical Thrombectomy Devices

Mechanical thrombectomy devices can be divided into direct contact devices, rheolytic devices, rotational devices and ultrasonic. Arrow-Trerotola Percutaneous Thrombectomy Device (Teleflex Inc, NC, US), one of the direct contact devices, is a rotating nitinol basket with a handheld battery which can drive the basket into a rotation speed of 3000 rpm to macerate the thrombus. Rheolytic devices take advantage of the Venturi effect to create a negative pressure gradient to pull in and remove the thrombus. The Angioject catheter (Boston Scienific Co, MA, US) is a typical rheolytic device. Most of the mechanical thrombectomy devices are effective and time-saving. However, higher cost is the Achilles tendon of those mechanical devices.

Post-procedure Consideration

The goal of the salvage procedure for acute dialysis access thrombosis is to restore the active flow in the access and allow the patient to resume hemodialysis therapy as soon as possible. The patient can be put on hemodialysis immediately after the salvage procedure via the original dialysis access and previous puncture sites if no surgical revision involved the puncture site performed.

Post procedure anti-platelet medication and antibiotic prophylaxis are usually not necessary. Since the underlying anatomical lesion had been addressed during the procedure, the pressure gradient of the access usually keeps the circuit with active flow. However, one should be careful in preventing the patient from suffering from hypotension or are hemodynamic unstable to avoid early recurrent thrombosis. In addition, over-compression on the access for hemostasis after each dialysis session should be strictly prohibited as well.

Special Consideration

To ensure favorable outcome of endovascular salvage procedures:

Define the cause of failure

Recanalization of the vascular access means not only clearance of the thrombus but also addressing the underling lesions. It is imperative to speculate the lesion before salvage and define the major cause of the access acute thrombosis during the procedure. Proper diagnosis can lead to an appropriate pre-procedure planning and effective management. One of the most important clues is the history of the access. Underlying anatomical lesions discovered in previous procedures should not be neglected. Careful physical examination of the whole access is the most basic part of diagnosis. Significant stricture (fibrotic cord like structure on palpation), dilated fistula segment and aneurysmal change of artificial graft discovered during inspection should be taken into consideration. Careful inspection of the angiography during the procedure can reconfirm the underlying lesions and lead to further definite treatment. Preoperative vascular ultrasonography sometimes can provide some hint on the underlying access lesions if time and facility permit.

The most frequently discovered lesion for AVGs is the graft venous outlet stenosis. As for the native fistula, stenosis located at the anastomosis swing segment, cephalic arch and lesions up-stream or down-stream to the puncture site aneurysm are all common underlying lesions of AVFs. Central vein stenosis is another underlying cause which should be carefully looked for. For some patients, the major cause of thrombosis may be factors other than anatomical lesions such as hemodynamic instability, over-compression of the circuit, and hyper-coagulation status. All possible factors should be avoided as much as possible after the access salvage procedure.

Address the underlying lesions

One of the advantages of endovascular salvage in contrast to traditional thrombectomy is that it may convey a complete scouting of the whole dialysis circuit, from arterial inlet to central vein. Any hemodynamic significant stenotic lesions which may lead to the failure of the circuit should be addressed during the procedure to ensure the result. Most of the lesions on AVFs found during the salvage procedure can be fixed by endovascular method such as balloon angioplasty during the salvage procedure. If immediate/early recoil, large amount of residual thrombus persists or any extravasation happens, surgical repair should be taken into action to preserve the dialysis access functioning.

AVGs outlet stenosis can also be corrected during the procedure by balloon angioplasty. For those patients with early recoiled graft outlet stenosis (within 3 months of salvage procedure), conventional surgical revision may be needed to ensure more durable result. Covered stent implantation (e.g., Viabahn from W.L. Gore Medical and Associate, AZ, US) to deal with the lesion is a less invasive alternative to surgical revision and may yield a promising longer mid-term access.[14,15]

Monitor and surveillance

Physical examination and clinical evaluation are fundamental skills that are similarly valuable as any surveillance method. To ensure the long-term access functioning access, characteristics such as thrill, pulsatility, hemodialysis parameters (Qb and pressure) should be recorded down and tracked by caregivers carefully. Frequency and type of surveillance methods (access flow, static pressure, Duplex ultrasound study … etc.) adopted for the accesses are dependent on the individual dialysis center and their available resources. Any abnormality of monitor and surveillance results which indicates the recurrence of the anatomical lesions of the access should not be neglected. Pre-emptive treatment, either surgical revision or endovascular intervention, should be considered seriously for patients with difficult vascular access in order to avoid any unexpected recurrent acute access thrombosis.

Avoidance of Possible Complications

Endovascular treatment, in contrast to surgical thrombectomy, is less invasive but not without potential complications. Common complications of endovascular salvage procedures are as below and should be kept in mind at all time:

Access site hematoma

Since there is no flow in the circuit at the beginning of the salvage procedure, gaining access to the dialysis vessel may sometimes be difficult because of the absence of pulsatility. One should be gentle and extremely careful when puncturing the thrombosed graft or fistula. Ultrasound guidance will be of great value when doing the puncture to avoid multiple punctures of the access and thus minimizes the trauma to the vessel during gaining the access and minimizes the subsequent access site hematoma.

Vessel rupture

Similar to other endovascular procedures, interventionist should be gentle and alert when inserting, advancing and inflating any device in the vessel. Keep tracking the tip of the wires and catheters in order to avoid any inadvertant trauma to the vessel. Be relatively conservative when choosing angioplasty balloons for AVF cases, especially those who are undergoing angioplasty for the first time. For those with recurrent lesions, choose the size of balloon according to the medical records of previous intervention. Do no (less) harm is an important rule to follow during any vascular intervention procedure.

Standby covered stents on the shelf is essential for the endovascular salvage procedures. They are important bail-out solutions for unexpected vessel ruptures. Once extravasation of contrast occur, keep the wire in place and inflate the balloon with lower pressure, longer inflation time (usually 5 min, then 10 min) and external gentle manual pressure. Angiography is then followed to check if the rupture being sealed off successfully. If the extravasation persists, go for direct stent grafting with appropriate length and diameter, which usually seals the rupture and maintain the circuit continuity (Fig. 1).

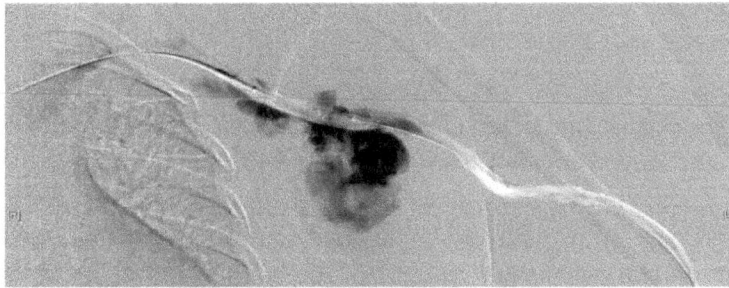

(a)

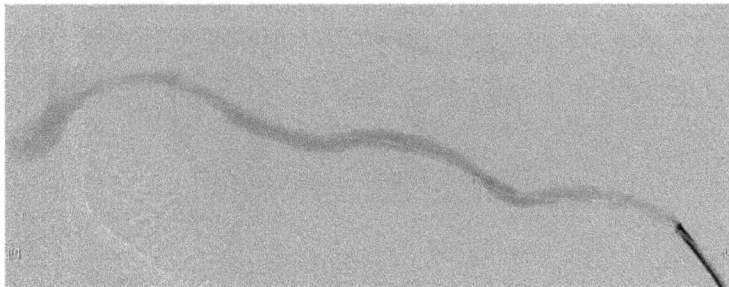

(b)

Fig. 1. (a) An AVG outlet severe extravasation after balloon angioplasty despite of prolonged balloon tamponade. (b) Ruptured vessel repaired by covered stent (Viabahn, W. L. Gore & Associates, Inc. Flagstaff, AZ, US).

Distal arterial embolization

Distal arterial embolization is a serious complication that are not infrequently seen. Sometimes, it can be rather unforgiving. Be sure to be gentle when extracting the arterial plug locating at the arterial inlet of the dialysis access. Slow and gentle passage of wire and balloon catheter through the anastomosis is mandatory to prevent any dislodge of emboli. Completion angiogram may show the presence of thrombus inside the artery. Routine observation of the color and temperature of the patient's fingers post-procedure may help detect any distal arterial embolization early. Once the complication happens, endovascular suction embolectomy, mechanical thrombolysis or surgical embolectomy may be applied to remove the emboli.

Pulmonary embolization

Although rare, the chances of clinical pulmonary embolization cannot be neglected during the procedure.[16,17] Thrombolytic agents such as urokinase can be utilized during the endovascular salvage procedures in order to first, enhance the thrombus breakdown, and second, minimize the chance of pulmonary embolization complication. The total volume of thrombus usually in a graft is small (less than 10 ml in total volume in a 6 mm diameter artificial graft of 20 cm in length), thus the chances of significant clinical pulmonary embolisation is rare. However, a dilated mature AVF, if with aneurysmal change, may harbor significant volume of thrombi that interventionists should try to decrease the volume of the clot (either by repeated suction, irrigation and aspiration, pharmaceutical lysis, or mechanical thrombectomy devices) as much as possible before establishment of the arterial inflow of the circuit. Continuous ECG and pulse oximeter monitor should be applied for every patient undergoing the procedure. An access interventionist should always bear in mind pulmonary embolization. For those patients with limited cardiopulmonary reserve, risk of pulmonary embolization should be vigilantly avoided at all costs.

Illustrative Cases

Case 1

A 74-year-old female patient with left forearm loop BB AVG was presented with acute graft thrombosis. Percutaneous salvage was performed through two access 6 Fr sheaths placed on the graft facing toward each other (Fig. 2).

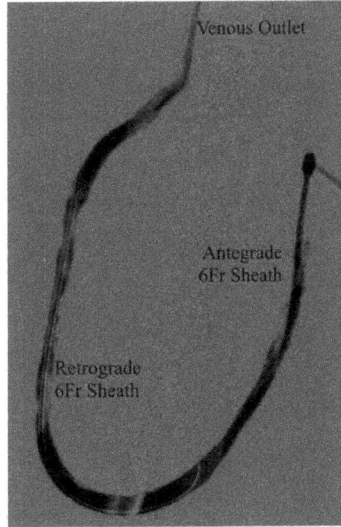

Fig. 2.   Fistulogram showed the position of antegrade and retrograde access sheath position relative to the loop forearm BB AVG.

250 KU urokinase was injected into the graft through the sheath to help with the procedure. A 0.035" Glidewire (Terumo Medical Co. Somerset, NJ, US) was then manipulated to antegradely traverse the graft venous outlet followed by insertion of a non-compliant balloon catheter (6 mm*40 Conquest; BARD Peripheral Vascular Inc. Tempe, AZ, US). The balloon was used to macerate the thrombus all the way from venous outlet to the proximal graft followed by removal of thrombus by applying negative pressure suction through the side port of the sheath. One may also inflat the angioplasty balloon over the venous outlet with low pressure and then retract it towards the arterial limb with syringe aspiration of thrombus through the side port of the sheath simultaneously. Gentle heparinized saline flushing into the graft and then aspirating out was conducted to reassure the clearance of graft and outflow tract. The angioplasty balloon was then reinserted via the retrograde sheath into the feeding artery of the graft to extract the arterial plug located in the graft arterial inlet. At this time, active arteriovenous flow was re-established (Fig. 3).

An angiography was then conducted to scout for any residual lesion. Graft venous outlet stenosis was noted (Fig. 4).

A 7 mm*40 Conquest angioplasty balloon was used to dilate the graft outlet lesion uneventfully at 14 atm for 1 minute (Fig. 5).

Complete angiography (Fig. 6) showed that the loop AVG was successfully recannulized with the graft outlet stenotic lesion addressed.

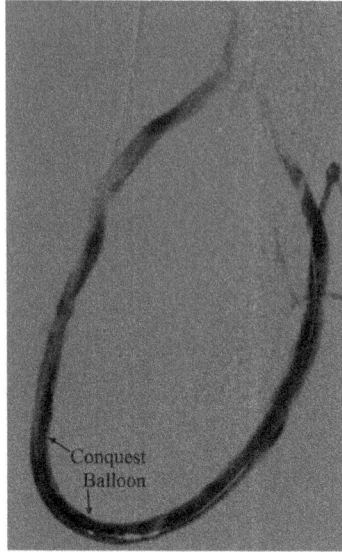

Fig. 3. Angiogram showed re-establishment of flow after arterial thrombus extraction by the Conquest angioplasty balloon.

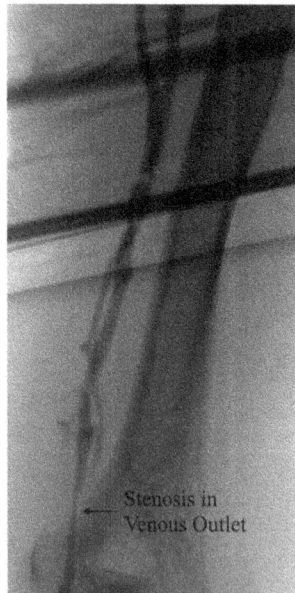

Fig. 4. Completion angiogram showing the stenosis in the venous outlet of the AVG.

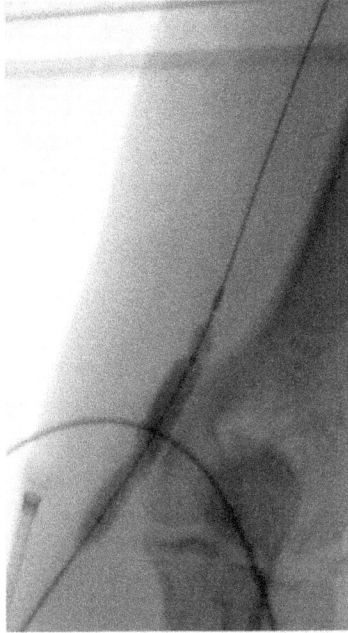

Fig. 5. Balloon angioplasty of the venous outlet stenosis.

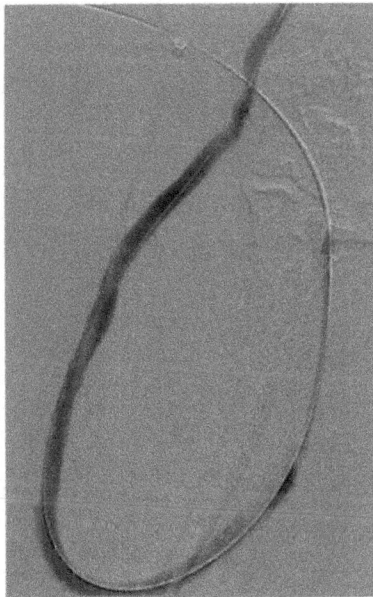

Fig. 6. Completion angiogram showing patent loop BB AVG.

Case 2

A 73-year-old male patient had been put on dialysis via left upper arm BB AVG. A stent graft (7 mm Viabahn) had been placed in the graft outlet 1 year ago for outflow tract stenosis. This time he was admitted for acute access thrombosis.

Two sheaths had been inserted under ultrasound guidance in antegrate and retrograde fashion. Balloon maceration using a 7 mm angioplasty balloon had been performed after intra-graft injection of urokinase solution. Contrast medium had delineate clearly the anatomical lesion located at the venous edge of the stentgraft (Fig. 7).

After clearance of the thrombus by balloon maceration, sheath suction and saline flushing, the arterial plug and thrombus located at the arterial inlet was extracted by over-the-wire thrombectomy catheter (LeMaitre Embolectomy catheter, LeMaitre Vascular, Inc. Burlington, MA, US) (Fig. 8).

Flush angiography can be conducted through the antegrade sheath when the balloon is inflated to show any residual lesion on arterial inlet of the graft (Fig. 9).

The stent graft venous edge stenosis was treated by 7 mm angioplasty balloon uneventfully. The retrograde sheath was removed followed by a completion angiography which had shown recannalization of the graft and resolution of stentgraft edge stenosis (Fig. 10).

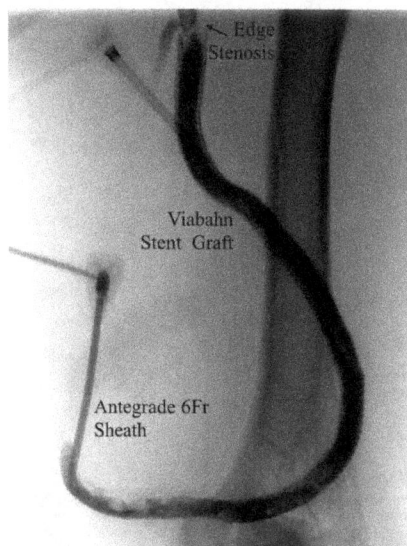

Fig. 7.   Angiogram showing the upper arm BB AVG with antegrade and retrograde sheath inserted and stenosis present over the venous edge of the stentgraft.

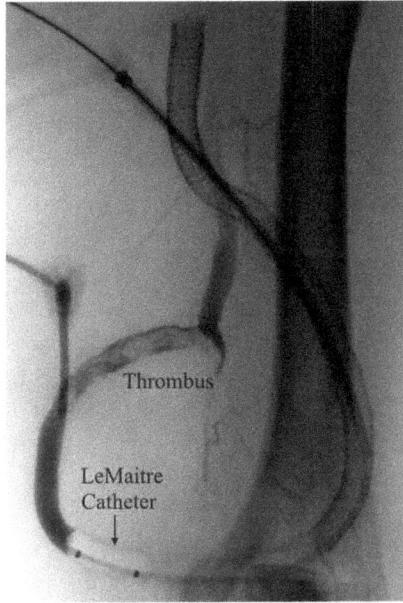

Fig. 8. Fluoroscopy image showing the LeMaitre catheter inside the BB AVG.

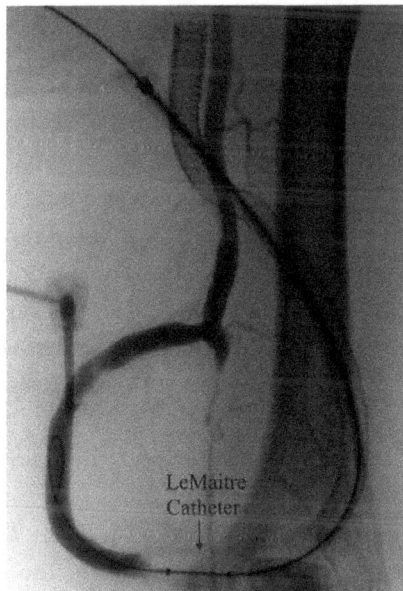

Fig. 9. Angiogram to assess the arterial anastomosis showed no significant stenosis.

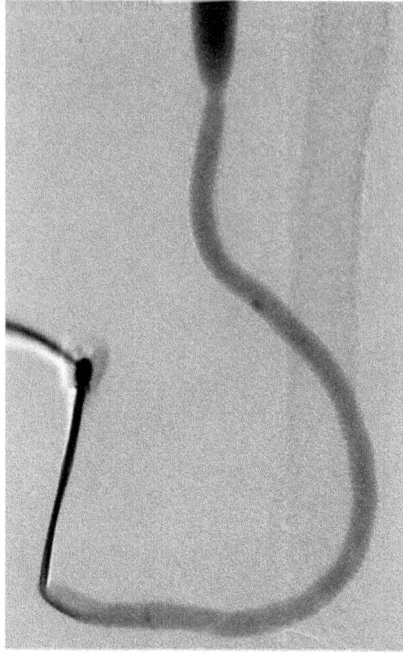

Fig. 10.   Completion angiogram from the antegrade sheath.

Case 3

A 65-year-old female patient had occlusion of BA AVG in her left upper arm. She was allergic to iodinated contrast media. CO_2 was used as contrast media during the salvage procedure. A sheath was inserted into the graft over the arterial side antegradely and CO_2 angiography showed complete occlusion of the graft and reflux of the CO_2 into the feeding brachial artery (Fig. 11). As usual, venous outlet of the AVG was traversed with a wire followed by urokinse inection and balloon maceration.

CO_2 angiography showed underlying multiples stenosis over the graft venous puncture site (Fig. 12) after clearance of the thrombus. The lesions were then treated repeatedly by angioplasty balloon (6 mm*40 mm Conquest; BARD Peripheral Vascular Inc. Tempe, AZ, US).

After recannalization of the venous outflow tract, the arterial inlet was addressed by a retrograde angioplasty balloon successfully.

The arterial anastomosis was patent in this case. With CO_2 angiography, dialysis patient with significant allergy history to traditional contrast media can be percutaneously addressed by double sheath PMT with CO_2 angiography.

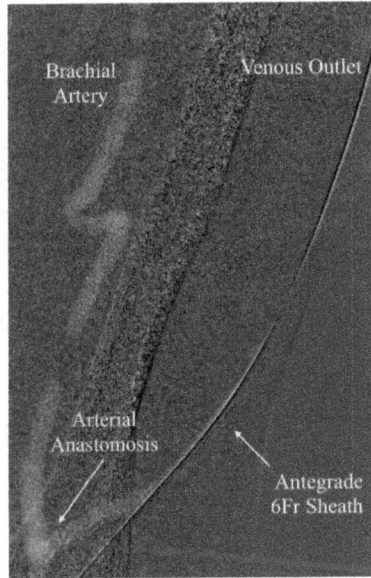

Fig. 11. Initial CO_2 angiogram after antegrade sheath inserted.

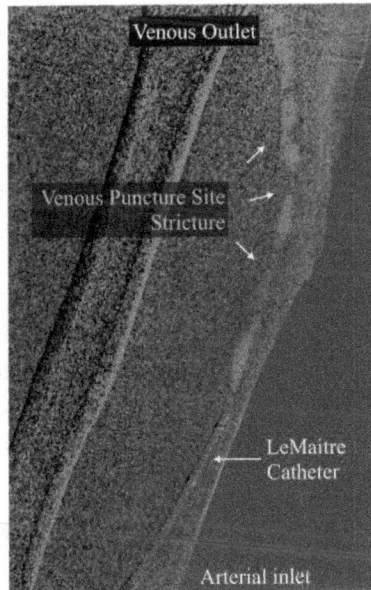

Fig. 12. CO_2 angiogram after thrombus clearance revealed in-graft stenoses.

References

1. Sadaghianloo N, Jean-Baptiste E, Gaid H, *et al.* Early surgical thrombectomy improves salvage of thrombosed vascular accesses. *J Vasc Surg.* 2014; **59**(5): 1377–1384 e1371–1372.

2. Asif A, Merrill D, Briones P, Roth D, Beathard GA. Hemodialysis vascular access: percutaneous interventions by nephrologists. *Semin Dial.* 2004; **17**(6): 528–534.

3. Yang CC, Yang CW, Wen SC, Wu CC. Comparisons of clinical outcomes for thrombectomy devices with different mechanisms in hemodialysis arteriovenous fistulas. *Catheter Cardiovas Interv.* 2012; **80**(6): 1035–1041.

4. Vesely TM, Idso MC, Audrain J, Windus DW, Lowell JA. Thrombolysis versus surgical thrombectomy for the treatment of dialysis graft thrombosis: pilot study comparing costs. *J Vasc Interv Radiol.* 1996; **7**(4): 507–512.

5. Yurkovic A, Cohen RD, Mantell MP, *et al.* Outcomes of thrombectomy procedures performed in hemodialysis grafts with early failure. *J Vasc Interv Radiol.* 2011; **22**(3): 317–324.

6. Bookstein JJ, Fellmeth B, Roberts A, Valji K, Davis G, Machado T. Pulsed-spray pharmacomechanical thrombolysis: preliminary clinical results. *Am J Roentgenol.* 1989; **152**(5): 1097–1100.

7. Valji K, Bookstein JJ, Roberts AC, Davis GB. Pharmacomechanical thrombolysis and angioplasty in the management of clotted hemodialysis grafts: early and late clinical results. *Radiology.* 1991; **178**(1): 243–247.

8. Cynamon J, Lakritz PS, Wahl SI, Bakal CW, Sprayregen S. Hemodialysis graft declotting: description of the "lyse and wait" technique. *J Vasc Interv Radiol.* 1997; **8**(5): 825–829.

9. Vogel PM, Bansal V, Marshall MW. Thrombosed hemodialysis grafts: lyse and wait with tissue plasminogen activator or urokinase compared to mechanical thrombolysis with the Arrow-Trerotola Percutaneous Thrombolytic Device. *J Vasc Interv Radiol.* 2001; **12**(10): 1157–1165.

10. Vashchenko N, Korzets A, Neiman C, *et al.* Retrospective comparison of mechanical percutaneous thrombectomy of hemodialysis arteriovenous grafts with the Arrow-Trerotola device and the lyse and wait technique. *Am J Roentgenol.* 2010; **194**(6): 1626–1629.

11. Almehmi A, Broce M, Wang S. Thrombectomy of prosthetic dialysis grafts using mechanical plus "no-wait lysis" approach requires less procedure time and radiation exposure. *Semin Dial.* 2011; **24**(6): 694–697.

12. Bent CL, Sahni VA, Matson MB. The radiological management of the thrombosed arteriovenous dialysis fistula. *Clinl Radiol.* 2011; **66**(1): 1–12.

13. Huang HL, Chen CC, Chang SH, *et al.* Combination of duplex ultrasound-guided manual declotting and percutaneous transluminal angioplasty in thrombosed native dialysis fistulas. *Ren Fail.* 2005; **27**(6): 713–719.

14. Allon M. Stent graft or balloon angioplasty alone for dialysis-access grafts. *Eng J Medi.* 2010; **362**(20): 1939; author reply 1940.

15. Chan MG, Miller FJ, Valji K, Bansal A, Kuo MD. Evaluating patency rates of an ultralow-porosity expanded polytetrafluoroethylene covered stent in the treatment of venous stenosis in arteriovenous dialysis circuits. *J Vasc Interv Radiol.* 2014; **25**(2): 183–189.

16. Swan TL, Smyth SH, Ruffenach SJ, Berman SS, Pond GD. Pulmonary embolism following hemodialysis access thrombolysis/thrombectomy. *J Vasc Interv Radiol.* 1995; **6**(5): 683–686.

17. Grebenyuk LA, Marcus RJ, Nahum E, Spero J, Srinivasa NS, McGill RL. Pulmonary embolism following successful thrombectomy of an arteriovenous dialysis fistula. *J Vasc Access.* 2009; **10**(1): 59–61.

CHAPTER 10

Surgical Management for Failing and Failed Hemodialysis Access

Jackie P. Ho

Surgical Salvage of Thrombosed Vascular Access

Thrombosis of currently in-use vascular accesses, AVF or AVG, is a common encounter for clinicians managing hemodialysis accesses. Many of these thrombosed vascular accesses are salvageable. Both open surgical or endovascular methods (Chapter 9) can be used to salvage the thrombosed accesses. There is no conclusive evidence to support which approach is more superior.[1,2] The main reasons are:

(1) Majority of thrombosed vascular accesses has underlying hemodynamic cause with stenosis in one or more regions.[3,4] The salvage of thrombosed vascular accesses are composed of two components namely: clearance of thrombus and correction of underlying stenosis.[5] Both components can be managed with open surgical or endovascular treatments. The treatment methods included in various randomized control studies are heterogenous and therefore impossible to produce a conclusive answer.

(2) Lots of advancements in thrombolytic agent, thrombolysis devices and endovascular devices to lysis clots and restore lumen patency occurred over the last 10–15 years. Technology in endovascular treatment may continuously evolve in the near future. A review of surgical versus endovascular salvage of thrombosed vascular access showed that studies performed before 2002 showed better outcomes with surgical treatment but those after 2002 showed similar results with both techniques.[2]

Whether using open thrombectomy or endovascular thrombolysis to clear the clots, once thrombus is being cleared, an angiogram of the vascular access to detect any underlying stenosis is mandatory, either in the same session or soon afterwards if logistically not feasible in the same setting. Correction of the underlying stenosis (using either open surgical or endovascular methods) will significantly improve the secondary patency of the access.[5] In general, centres equipped with good endovascular facilities and comprehensive devices tend to provide endovascular treatment as the first-line therapy for thrombosed vascular accesses. However, open surgery remains an effective treatment to salvage failed

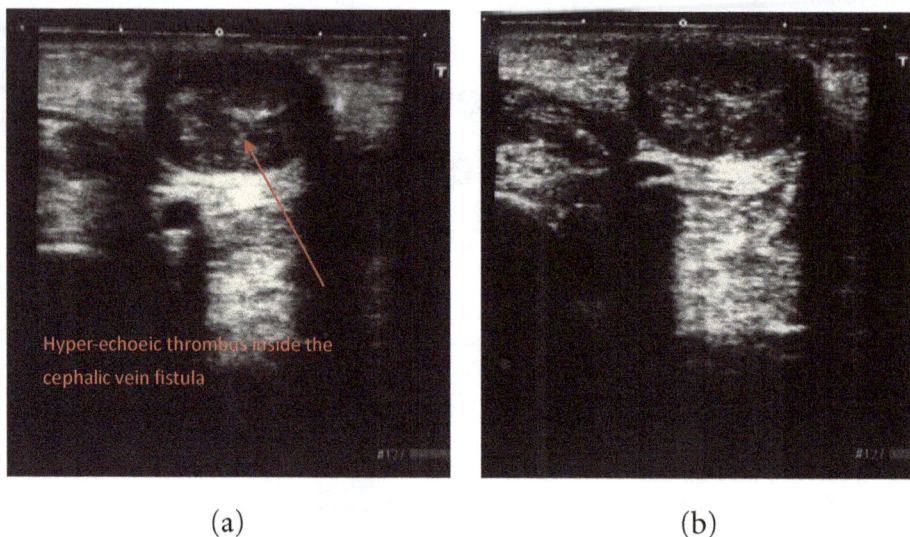

(a)   (b)

Fig. 1.   Ultrasound diagnosis of a thrombosed cephalic vein fistula without (a) and with (b) compression respectively showing hyper-echoeic thrombus and non-compressible feature.

accesses especially when the presentation is delayed, aneurysmal change with large thrombus load or failed recent endovascular treatment. Many centres adopt a combination of open and endovascular approaches for thrombosed vascular accesses. The surgical technique for AVF and AVG salvage is slightly different.

Surgical Thrombectomy for Thrombosed AVG

- Select an incision site that allows thrombectomy catheter to reach both the arterial and venous anastomosis with similar ease, e.g. taking the loop region of a forearm loop AVG.
- The incision is made in a way that minimally occupies potential site of future cannulation (Fig. 2). Therefore, usually the incision is made perpendicular to the path of the graft rather than longitudinal. Alternatively, over the curve part of the graft where it is not suitable for HD cannulation. If the incision was made along the segment of AVG good for cannulation, repeated cannulation over the surgical scar may result in wound breakdown and exposure of graft rendering access loss.
- Avoid making incision near sites of clustering cannulation. Lots of adhesions will be encountered near the cannulation region. Furthermore, the graft may be mechanically weakened and easily torn apart over the frequent needling area.
- The author prefers to give a single dose of prophylactic antibiotic for open AVG thrombectomy procedure.
- Usually a segment of 3–4 cm graft needs to be freed from the subcutaneous tissue to enable vascular clamp application after thrombectomy. If the procedure is performed

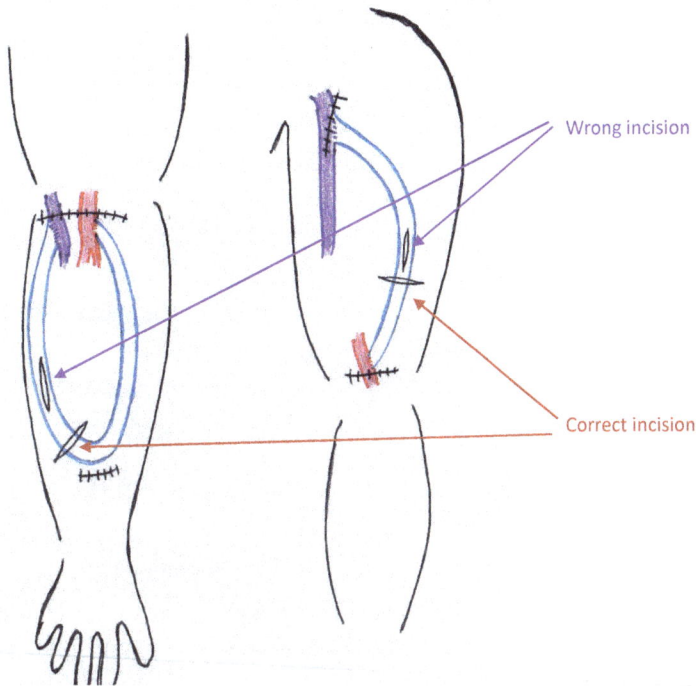

Fig. 2. Diagram showing correct and wrong incisions for graft thrombectomy.

under local anesthesia, ensure the blood pressure of the patient is under control. ESRF patients' blood pressure can be rather labile with anxiety or stress. High blood pressure may lead to excessive bleeding during the thrombectomy process.

- To avoid injury to the inflow arteries by the thrombectomy catheter, one should estimate the distance from the graft incision to arterial-graft anastomosis and avoid passing the thrombectomy catheter too far down the native artery. Judiciously inflate the balloon of the thrombectomy catheter when the balloon is inside the native artery.

- The author prefers to perform thrombectomy from the arterial side first to ensure good inflow is re-established. The incidence of arterial-graft anastomosis stenosis is far less common than graft-vein anastomosis stenosis. One would assume no inflow problem if a strong gush of blood flows out from the arterial limb of the graft after thrombectomy. After thrombectomy of the venous limb, one may insert a large calibre sheath (e.g. 7 or 8 Fr) directly into the opened graft and control bleeding with a snare or double throw vessel loop. Angiogram can then be performed through the sheath. Further treatment for stenosis detected can then be performed (either endovascular angioplasty ± stenting or surgical patch angioplasty). A large calibre sheath enables a variety of endovascular devices to be used for treatment of any stenosis. Using this method, one could also use wire and catheter skill to open a total occlusion of the venous anastomosis in the absence of backflow.

Surgical Thrombectomy for Thrombosed AVF

- Previously, thrombosed native vein AVF was considered not beneficial to salvage because of the low success rate. Subsequently, many clinicians reported[6-9] encouraging success rate in AVF salvage and satisfactory secondary patency.
- The principles of incision placement to avoid frequent needling site and minimally occupy potential cannulation site also apply to AVF salvage.
- The site, the extent of thrombosis and the thrombus load varies in different failed AVFs. On the easy end of the spectrum, it may be only a partially obstructive localized thrombus present next to a stenotic site of fistula or after a problematic cannulation. A small incision made over the thrombosed segment under local anesthesia followed by correction of the underlysis stenosis will be able to salvage the fistula (Fig. 3). Depending on the site of the high grade stenosis that is causing the failure and collateral vessel situation, the thrombus may present right from the AV anastomosis and extending to a various length into the venous fistula. In large diameter AVFs with existing aneurysmal changes, large volume of fresh and old thrombus may exist together and render thrombus removal more challenging. The worst end of the spectrum would be failed AVFs with multiple stenotic sites, tortuous and grossly aneurysmal change of fistula and extensive hard thrombus felt over the whole pathway of fistula.
- Ultrasound study of the thrombosed AVF by the operator is helpful in estimating the difficulty of thrombus removal, clinical decision to proceed for salvage procedure, counselling the patient on expectation and making surgical decision.
- Cull *et al.*[9] reported a surgical technique to manage failed AVF by making an incision near the AV anastomosis. The most common sites of native vein AVF are AV anastomostic sites and juxta-anastomotic region. The adherent clot over the AV anastomosis can be fished out using thrombectomy catheter, milked out or pulled out using forceps or artery forceps. The thrombus inside the venous fistula can also be milked out with external compression.
- Once the clots are cleared, assessment of the site of stenosis using fistulogram and correction of stenosis are essential to good outcomes. The fistulogram can be performed either via venous fistula opening or a separate percutaneous access from the inflow artery or outflow vein, depending on the location of the AVF and relative position of the stenotic areas and the fistula opening.
- Options of correction of underlying stenosis for AVF include endovascular angioplasty ± stenting, surgical revision of anastomosis and interposition graft. There is no RCT comparing angioplasty and surgical revision of anastomotic and juxta-anastomotic stenosis after AVF thrombectomy. However comparative studies of failing AVF treatment showed similar overall patency after angioplasty or surgical revision although the restenosis rate is higher with angioplasty.[10,11]
- Tourniquet over proximal inflow artery may help to reduce blood loss during open thrombectomy.

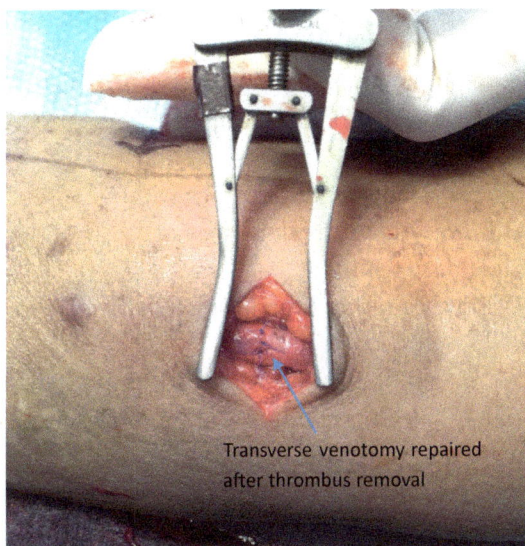

Transverse venotomy repaired
after thrombus removal

Fig. 3. Localized thrombus in a BC AVF was removed through a small transverse incision made over the fistula under local anesthesia.

General Issues of Surgical Salvage for Failed Vascular Access

- Thrombus becomes more adherent with time. For both AVFs and AVGs, particularly AVFs, the outcomes are better with earlier thrombus clearance.[12] Therefore, once access failure is noted by patient or the dialysis nurses, patient should seek treatment immediately and salvage procedure should commence as soon as possible.
- Potential complications of open vascular access thrombectomy include failure to salvage the access, bleeding, perforation, arterial embolization, clinically symptomatic pulmonary embolization, wound and access infection. These potential complications should be discussed during pre-operation counselling. These potential complications also apply to AVG thrombectomy.
- In some situations, after successful thrombus removal and treatment of hemodynamic obstruction and flow restoration, there may still be some non-flow limiting residual thrombus present inside the vascular access. 24 to 48 hours of therapeutic anti-coagulation therapy (heparin infusion) may help to dissolve the residual clots. Caution needs to be paid in cannulation of those vascular accesses during the anti-coagulation therapy to prevent hematoma formation.
- After thrombectomy and treatment of underlying stenoses of the hemodialysis access, it is important to assess the success through review of hemodialysis parameters including arterial and venous pressure, and access flow. Early fistulogram and additional necessary (endovascular or surgical) salvage procedures are indicated if those parameters are grossly abnormal.

Surgical Procedures for Failing Vascular Access

The aim is to maintain a functional hemodialysis access for a patient with failing vascular access and to avoid the need of central venous catheter. These procedures include:

(1) Surgical correction of persistent stenotic regions without vascular access downtime;
(2) Creation of a new AVF over the upstream of an existing access to avoid vascular access downtime (Type I secondary fistula);
(3) Creation of a new AVF over another limb (Type II secondary fistula).

In all these situations, a thorough evaluation of the mode of access failure, various options of salvage, arterial and venous condition of the access limb and other limbs is the key to optimal management and outcome.

Surgical Correction of Persistent Stenotic Regions without Vascular Access Downtime

When there is persistent stenosis of the fistula which is resistant to endovascular therapies, one may consider surgical procedures to overcome the stenosis. For AVF, usually the stenotic areas are AV anastomotic site, justa-anastomotic area, or frequent needling site. The surgical methods used include proximal neo-anastomosis, interposition graft to bypass the stenotic region, excision of stenotic segment and primary anastomosis for very focal tight fistula stenosis, patch repair using vein, processed tissue or synthetic material. Georgiadis *et al.*[7] reported similar primary patency of RC AVF salvage using proximal neo-anastomosis and short interposition PTFE graft. Mallik *et al.*[13] reported the primary patency of RC AVF after proximal neo-anastomosis created (for both failure to mature and failing in-used fistula) was 78.5% and 54.9% at 1 and 3 years respectively. Murphy *et al.*[14] in their early report of the outcome of elbow AVFs salvaged with surgical method (excision and re-anastomosis, patch repair with vein and interposition PTFE graft) achieved an overall 70% success rate.

In a favorable condition, e.g. AV anastomotic stenosis with a healthy juxta-anastomotic fistula and an inflow artery situated next to one another, proximal neo-anastomosis would be a preferable approach as it does not involve usage of synthetic graft which is costly and bears a higher risk of infection. Interposition PTFE graft would be needed if the suitable segment of venous fistula and arterial inflow segment for anastomosis are located far apart. Localized excision and primary anastomosis is suitable for focal resistant tight stenosis of juxta-anastomotic segment or any part of venous fistula that is not the frequent needling site. For longer stenotic segment, patch repair would be required. Dissection and patch repair of the stenotic segment resulting from frequent needling is difficult due to extensive adhesion around the region and unhealthy nature of the venous fistula. Bypass for this kind of stenotic segment would be better than patch repair.

In some BC AVFs, cephalic arch high grade stenosis is the cause of access dysfunction. Surgical procedures for this condition could be cephalic transposition and venovenostomy, venovenostomy to basilic vein, stenotic segment resection, and cephalic-jugular vein bypass graft. Wang *et al.*[15] noted that angioplasty prior to surgical repair of the cephalic arch stenosis is associated with poorer patency after surgery. It would be difficult for clinician to suggest direct surgical repair without any attempt of endovascular treatment for a de novo cephalic arch stenosis, which is much less invasive to patient. However, if post-angioplasty show significant re-coil immediately, or early recurrence of significant stenosis, early conversion to surgical treatment for patients with low surgical risk should be considered.

For AVG, surgical treatment of failing access is recommended when endovascular intervention fails.[16] In AVG, usually the vein-graft anastomosis is the site of stenosis. The stenosis may extend into the outflow vein proximally for a variable length.[17] Occasionally, skip lesion may present along the outflow vein. If the venous anastomotic stenosis is localized and the extension of stenosis into the outflow vein is a short segment, patch repair of the anastomosis using either vein, processed tissue or synthetic patch could be considered. However, if there is a long segment of outflow vein stenosis or extensive skip lesions, interposition graft may be considered. Alternatively, if the outflow vein is of good quality and calibre, secondary AVF creation using the arterialized outflow vein will be a more favorable option (discussed below).

Potential complications associated with the surgical salvage of failing AVF and AVG include early re-occlusion, steal syndrome, bleeding, wound (and, or graft) infection.

Creation of a New AVF Using the Outflow Vein of an Existing Access to Avoid any Vascular Access Downtime (Type I Secondary Fistula)

In failing RC AVF resistant to endovascular therapy with extensive fistula aneurysmal change, diffuse stenosis or poor skin condition of the forearm, and, good size and quality of the outflow vein, creation of a new AVF upstream may provide a functional access without any downtime which avoids the need of a central venous catheter. If there is an adequate length of the outflow vein that is good for immediate cannulation (>10 cm long, >6 mm in diameter and <6 mm under the skin), an end-to-side anastomosis of the outflow vein with the more proximal inflow artery is preferred together with ligation of the failing RC AVF to avoid steal syndrome. However, if the outflow vein condition is still marginal, a side-to-side anastomosis between the outflow vein and the more proximal arterial inflow could be used to preserve the failing RC AVF for current cannulation while waiting for the secondary AVF to mature. As more than one site of the arterial system of the limb were tapped to support vascular access, risk of steal syndrome is increased. A detailed assessment of the arterial system is mandatory before the procedure. Judicious sizing of the AV anastomosis during the surgery is required.

Case 1

Mr N, 59 years old, had been using his right RC AVF for seven years. The access flow was noted to reduce from >800 ml/min to <300 ml/min. Clinically, the thrill was only felt over the distal forearm and not palpable over the proximal forearm. The fistula near the anastomotic area felt hard. There were two areas of aneurysmal change (frequent cannulation sites) over mid- and proximal forearm.

Fistulogram (Figs. 4 and 5) was performed and showed tortuous distal radial artery with focal high grade stenosis just proximal to the AV anastomosis. Another segment of high grade stenosis was present over the mid-forearm segment of the fistula. Heavy calcification was noted around the AV anastomosis region.

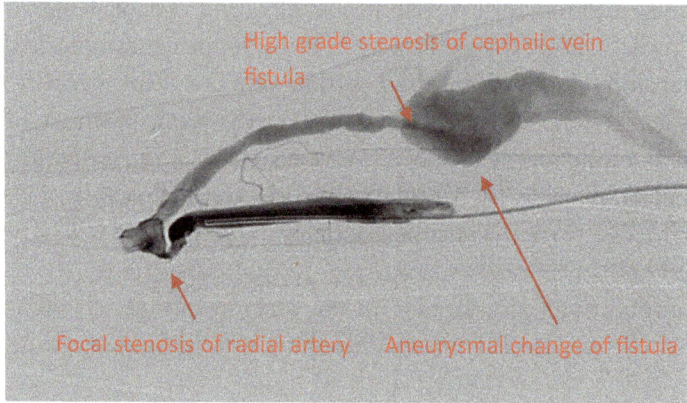

Fig. 4.   Fistulogram showing tortuous and focal stenosis of the radial artery and a segment of stenosis over cephalic vein fistula.

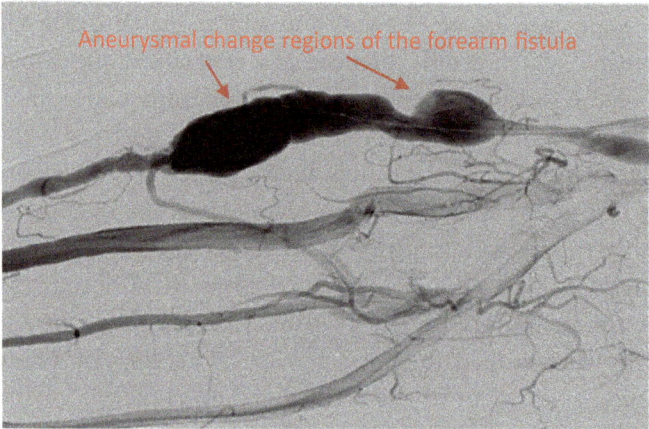

Fig. 5.   Two aneurysmal change areas of cephalic vein fistula over mid- and proximal forearm.

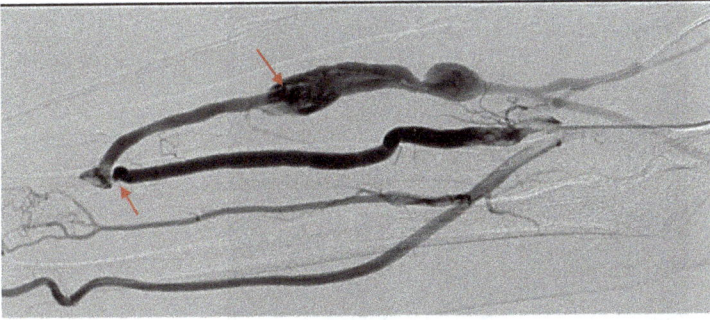

Fig. 6.   Completion angiogram after balloon angioplasty of the radial artery and fistula.

Great difficulty was encountered while passing the guidewire through the stenosis. Both antegrade right brachial access and retrograde proximal forearm fistula access were used. Eventually the radial artery lesion was crossed with 0.014" Hydro ST wire (Cook Medical Inc. Bloomington, IN, US), supported by straight 90 cm CXI catheter (Cook Medical Inc. Bloomington, IN, US) and fistuloplasty was successfully performed with Splinter NC balloon 2.75/20 mm (Medtronic Inc. Minneapolis, MN, US). The fistula lesion was crossed via retrograde approach and angioplasty performed using Sterling balloon 6 mm/40 (Boston Scientific Co. MA, US). Completion angiogram showed lumen regain over the fistula lesion but radial artery stenosis re-coiled (Fig. 6).

The access flow after fistuloplasty was improved to 450 ml/min. However, his access flow was reduced again to <300 ml/min two months later. A quick ultrasound study performed in the clinic showed persistent distal radial artery stenosis and recurrence of fistula stenosis. The cephalic vein fistula of the right forearm drained to the antebrachial vein, which mainly supplies to the basilic vein and also cephalic vein of the arm. The size of cephalic vein in the distal and mid arm was between 5.2 mm and 5.8 mm. The right basilic vein size was good at about 7 mm but deeper underneath the subcutaneous tissue. The ulnar artery was patent and the brachial artery measured 5 mm in diameter. In view of the heavy calcification around the stenotic sites and rapid recurrence of stenosis after fistuloplasty, surgical revision of the fistula was recommended and Mr N agreed with the procedure.

Ultrasound study was performed in the operating theatre to locate a region where the antebrachial vein is in close proximity to the distal brachial artery (Figs. 7 and 8)

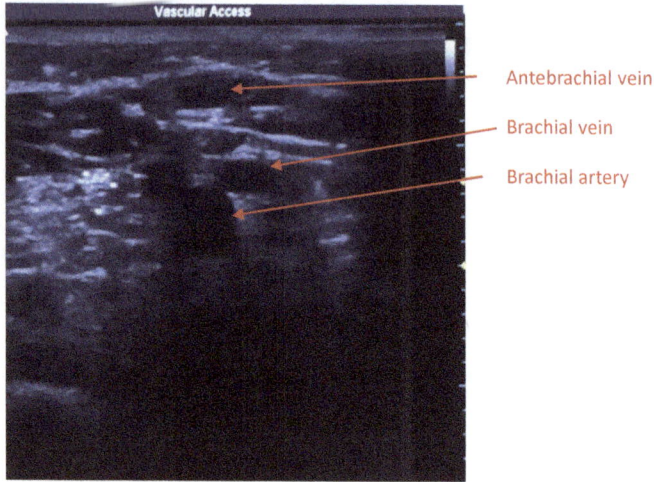

Fig. 7.   Ultrasound study showing the right antebrachial vein superficial to the distal brachial artery.

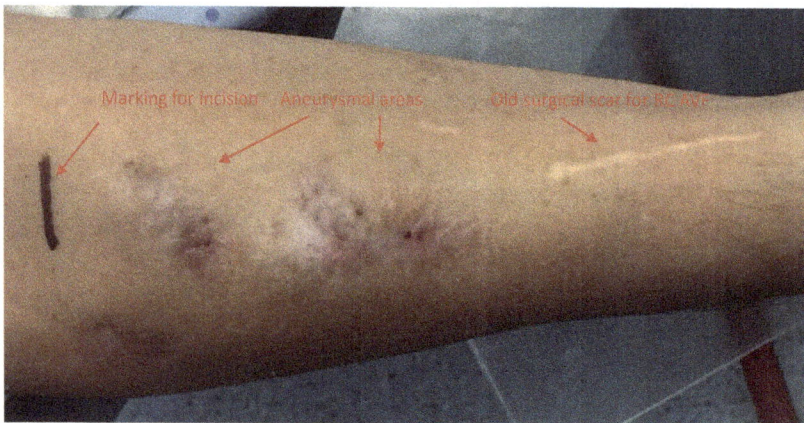

Fig. 8.   Marking for the incision where the antebrachial vein was in close proximity to the right distal brachial artery.

Side-to-side anastomosis was made between the antebrachial vein and the distal brachial artery. Post-operation, good thrill was felt over the proximal forearm aneurysmal part of the fistula, cubital fossa, as well as both cephalic and basilic vein of the arm. There was no feature of steal syndrome. "A" cannulation was sited over the distal arm cephalic vein fistula (measured 5.6 mm immediately after operation). The "V" cannulation was sited over the proximal forearm aneurysmal part (Fig. 9). No CVC was required for Mr N.

Please also see Chapter 12 Case 2 for another illustration.

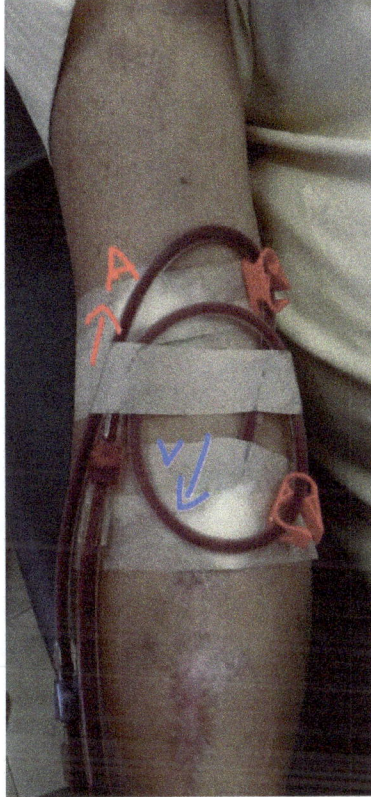

Fig. 9.   Photo is provided by the dialysis center after cannulation of Mr N's fistula after the revision surgery.

Creation of a New AVF over Another Limb (Type II Secondary Fistula)

When there are uncorrectable problems over the existing vascular access limb that further new access creation is considered to be not possible or at high risk, e.g. symptomatic central vein occlusion or significant arterial disease not correctable by endovascular interventions, a new vascular access can be created over the opposite upper limb or the lower limb.

In some conditions, the native superficial vein on one limb is exhausted but there is good quality superficial vein over the opposite limb. In most situations, the opposite upper limb is the dominant hand. Proper counselling should be provided to the patient for better adaptation. Besides standard clinical and ultrasound assessment for regular vascular access creation, clinicians also need to find out the history of CVC insertion and its duration in the opposite side, and if indicated, proper imaging evaluation of the central vein patency of the limb planning for new vascular access creation prior to actual operation.

Case 2

Mr A, 62 years old, with DM, HT, ESRF, IHD, sleep apnoea, left hand dominance and had right BC AVF created three years ago. Before this permanent vascular access, he had a long period of tunnelled CVC insertion to both side IJV. He also had right brachiocephalic vein angioplasty and stenting performed two years ago for persistently high venous pressure. He defaulted follow-up for some time and recently presented with right upper limb swelling (Fig. 10).

Fistulogram showed occlusion of the in-stent segment of right brachiocephalic vein and the subclavian vein. The occluded in-stent right brachiocephalic vein was crossed via a femoral approach. Crossing of the occluded subclavian vein was failed with both antegrade (through the fistula) and retrograde (through right femoral vein) endovascular approach (Fig. 11).During the procedure, the connection between right jugular and brachiocephalic vein was opened (Fig. 12) and this provided some relief of the venous drainage of the right upper limb. There was mild reduction in the swelling of his right arm after the procedure. A new access was planned as there was no long lasting solution for the right side.

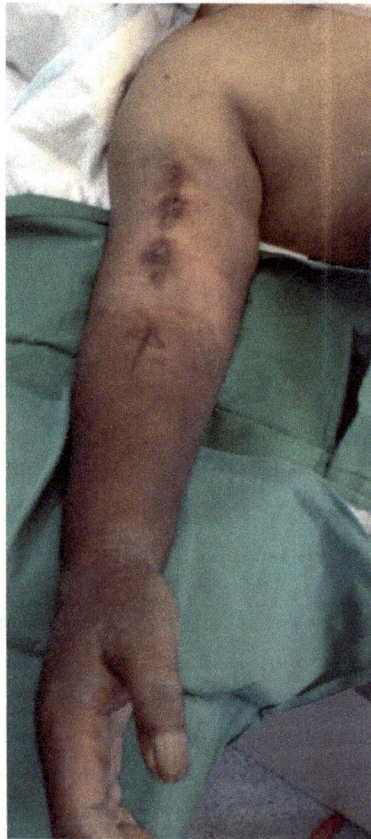

Fig. 10.   Clinical photo of Mr N's right upper limb with symptomatic central vein obstruction.

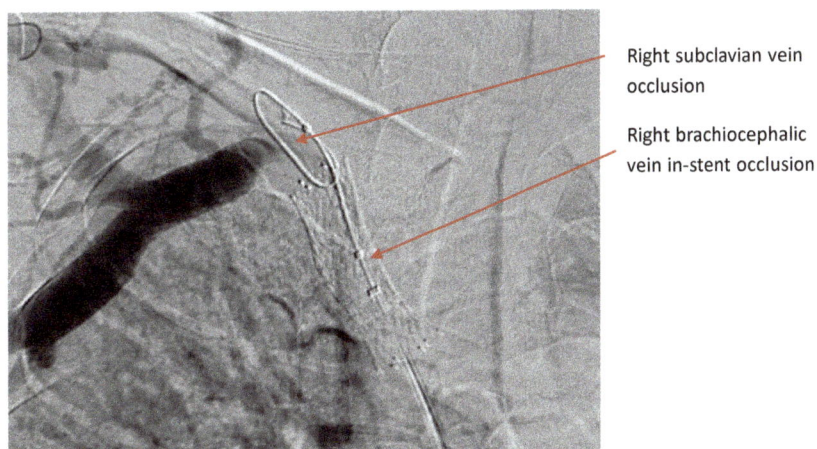

Fig. 11.   Fistulogram showing right brachiocephalic and subclavian vein occlusion. Guidewire failed to cross the subclavian segment.

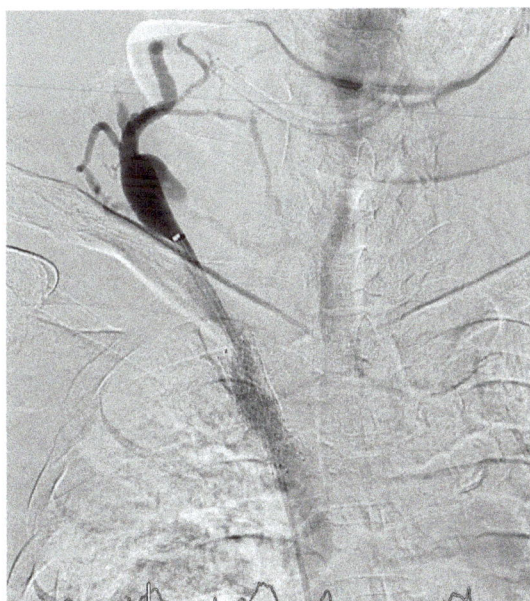

Fig. 12.   The connection between right jugular vein and brachiocephalic vein was crossed and angioplasty performed.

Mr A's left upper limb median cephalic vein was of good size and left BC AVF was planned. Based on his long-term CVC history, one needs to rule out left central vein obstruction before creating fistula on the left upper limb. Left upper limb venogram performed showed a segment of high grade stenosis of the left brachiocephalic vein (Fig. 13). Angioplasty followed by stenting (Smart stent 12 mm, Boston Scientific Co. MA, US) of the left brachiocephalic vein extending into the SVC was performed (Fig. 14). Left BC

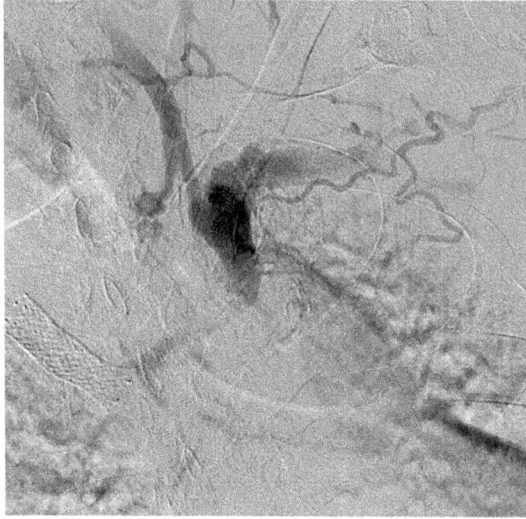

Fig. 13.   Venogram of Mr A showing left brachiocephalic vein high grade stenosis.

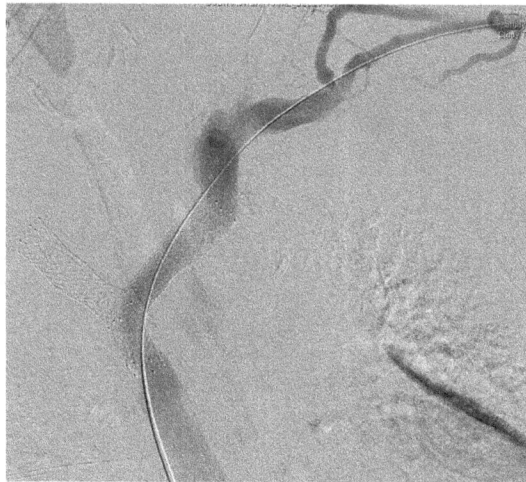

Fig. 14.   Central venogram performed after angioplasty and stenting of the left brachiocephalic vein and upper SVC.

AVF was then created. It matured eight weeks after creation. The right upper limb swelling gradually worsened and the cannulation of the fistula became difficult. Right BC AVF was ligated off and left BC AVF was used for hemodialysis.

Another option for failed vascular access that is not salvageable is to create a new AVG using rapid access graft (e.g. GORE Acuseal, Atrium Flixene™, Nicast AVflo™ etc.), when there is no good native vein AVF option. Usually, cannulation of these grafts for hemodialysis can be commenced between 24 and 72 hours after creation. However, if there is edema of the limb after access creation, cannulation may have to be deferred. Currently, only small scale studies are available to predict the long-term performance of these grafts.[19,20] More evidence is needed for consideration of wider usage.[21]

References

1. Green LD, Lee DS, Kucey DS. A metaanalysis comparing surgical thrombectomy, mechanical thrombectomy, and pharmacomechanical thrombolysis for thrombosed dialysis grafts. *J Vasc Surg.* 2002; **36**: 939–945.

2. Tordoir JH, Bode AS, Peppelenbosch N, *et al.* Surgical or endovascular repair of thrombosed dialysis vascular access: Is there any evidence? *J Vasc Surg.* 2009; **50**: 953–956.

3. Dougherty MJ, Calligaro KD, Schindler N, *et al.* Endovascular versus surgical treatment for thrombosed hemodialysis grafts: A prospective, randomized study. *J Vasc Surg.* 1999; **30**: 1016–1023.

4. Poulain F, Raynaud A, Bourquelot P, *et al.* Local thrombolysis and thromboaspiration in the treatment of acutely thrombosed arteriovenous hemodialysis fistulas. *Cardiovasc Intervent Radiol.* 1991; **14**: 98–101.

5. Liu YH, Hung YN, Hsieh HC, *et al.* Surgical thrombectomy for thrombosed dialysis grafts: comparison of adjunctive treatments. *World J Surg.* 2008; **32**: 241–245.

6. Ponikvar R. Surgical salvage of thrombosed arteriovenous fistulas and grafts. *Ther Apher Dial.* 2005; **9**: 245–249.

7. Georgiadis GS, Lazarides MK, Lambidis CD, *et al.* Use of short PTFE segments (6 cm) compares favorably with pure autologous repair in failing or thrombosed native arteriovenous fistulas. *J Vasc Surg.* 2005; 76–81.

8. Lipari G, Tessitore N, Poli A, *et al.* Outcomes of surgical revision of stenosed and thrombosed forearm arteriovenous fistulae for haemodialysis. *Nephrol Dial Transplant.* 2007; **22**: 2605–2612.

9. Cull DL, Washer JD, Carsten CG, *et al.* Description and outcomes of a simple surgical technique to treat thrombosed autogenous accesses. *J Vasc Surg.* 2012; **56**: 861–865.

10. Tessitore N, Mansueto G, Lipari G, *et al.* Endovascular versus surgical preemptive repair of forearm arteriovenous fistula juxta-anastomotic stenosis: analysis of data collected prospectively from 1999 to 2004. *Clin J Am Soc Nephrol.* 2006; **1**(3): 448–454.

11. Napoli M, Prudenzano R, Russo F, *et al.* Juxta-anastomotic stenosis of native arteriovenous fistulas: surgical treatment versus percutaneous transluminal angioplasty. *J Vasc Access.* 2010; **11**(4): 346–351.

12. Sadaghianloo N, Jean-Baptiste E, Gaid H, *et al.* Early surgical thrombectomy improves salvage of thrombosed vascular accesses. *J Vasc Surg.* 2014; **59**(5): 1377–1384.

13. Mallik M, Sivaprakasam R, Pettigrew GJ, *et al.* Operative salvage of radiocephalic arteriovenous fistulas by formation of a proximal neoanastomosis. *J Vasc Surg.* 2011; **54**: 168–173.

14. Murphy GJ, Saunders R, Metcalfe M, *et al.* Elbow fistulas using autogeneous vein: patency rates and results of revision. *Postgrad Med J.* 2002; **78**: 483–486.

15. Wang S, Almehmi A, Asif A. Surgical management of cephalic arch occlusive lesions: are there predictors for outcomes? *Semin Dial.* 2013; **26**(4): E33–E41.

16. Schild AF. Maintaining vascular access: the management of hemodialysis arteriovenous grafts. *J Vasc Access.* 2010; **11**: 92–99.

17. Bachledaa P, Utikala P, Kocher M, *et al.* Arteriovenous graft for hemodialysis, graft venous anastomosis closure — current state of knowledge. Minireview. *Biomed Pap Med Fac Univ Palacky Olomouc Czech Repub.* 2014; **158**: 1–4.
18. Slayden GC, Spergel L, Jennings WC. Secondary arteriovenous fistulas: converting prosthetic AV grafts to autogenous dialysis access. *Semin Dial.* 2008; **21**(5): 474–482.
19. Schild AF, Schuman ES, Noicely K, *et al.* Early cannulation prosthetic graft (Flixene™) for arteriovenous access. *J Vasc Access.* 2011; **12**(3): 248–252.
20. Aitken EL, Jackson AJ, Kingsmore DB. Early cannulation prosthetic graft (Acuseal) for arteriovenous access: a useful option to provide a personal vascular access solution. *J Vasc Access.* 2014; **15**(6): 481–485.
21. Inston NG, Jones R. Devices in vascular access: is technology overtaking evidence? *J Vasc Access.* 2014; **15**(2): 73–75.

CHAPTER 11

Strategies for Central Vein Obstruction

Kyung J. Cho

Central vein obstruction remains a significant source of morbidity and mortality in hemodialysis patients. It can usually be suspected on the basis of the clinical presentation and a history of central line placement. Central vein obstruction causes swelling, discomfort and pain of the extremity associated with vascular access and dialysis. Obstruction of the superior vena cava (SVC) causes superior vena cava syndrome characterized by facial swelling, dyspnea and neck vein distention. Endovascular approach is the preferred method for treating symptomatic central vein obstruction. Selection of the proper recanalization equipment, and development of a facile catheterization and recanalization technique are essential in obtaining a successful recanalization of central vein obstruction. This chapter discusses the anatomy, imaging workup and strategies for central vein obstruction in hemodialysis patients.

Anatomy

The central venous anatomy of the upper extremity is less complicated than the arterial anatomy with fewer variations, but a thorough knowledge of its relationship with the adjacent critical organs and arteries (innominate and pulmonary arteries, and aorta). The central veins of the upper extremity are generally demonstrated by contrast-enhanced computed tomographic venography (CTV) and magnetic resonance venography (MRV) (Fig. 1) as well as by injecting contrast medium into the peripheral vein of the upper extremity. CO_2 is a useful contrast agent for central venography of the upper extremity, particularly in the presence of central vein occlusion.

The subclavian vein is the continuation of the axillary vein extending from the lateral border of the first rib to the sternal end of the clavicle, where it is joined by the internal jugular vein to form the brachiocephalic vein. It generally lies anterior and inferior to the subclavian artery. The tributaries of the subclavian vein are the external jugular vein, the anterior jugular vein, and occasionally the cephalic vein. At its junction with the internal jugular vein, the right subclavian vein drains the right lymphatic duct, and the left subclavian vein receives the thoracic duct.

The right brachiocephalic vein is a short vein, about 2.5 cm in length, beginning at the junction of the internal jugular vein and the subclavian vein behind the sternal end of the clavicle. The right brachiocephalic vein courses anterior and to the right of the innominate artery. It joins the left brachiocephalic vein to form the SVC. The left brachiocephalic vein is about 6 cm in length and courses behind the manubrium anterosuperior to the aortic arch and anterior to the innominate artery, the left common carotid artery, and left subclavian artery. Both the brachiocephalic veins drain the vertebral, internal thoracic, and inferior thyroid veins. The left brachiocephalic vein drains the left superior intercostal veins, the thymic vein and pericardial veins.

The superior vena cava originates with the junction of the right and left brachiocephalic veins. It usually begins at the level of the first intercostal space to the right of the innominate artery and the ascending aorta. It is about 2 cm in diameter and 7 cm in length. About half of its length lies within the pericardial sac, covered by serous pericardium. The azygos vein drains into the posterior aspect of the SVC just caudal to the junction of the right and the left brachiocephalic veins.

A left sided SVC is the most common congenital venous anomaly in the SVC. In a majority of cases, it is accompanied by a normal right-sided SVC (called SVC duplication) but it rarely occurs as an isolated left-sided SVC.

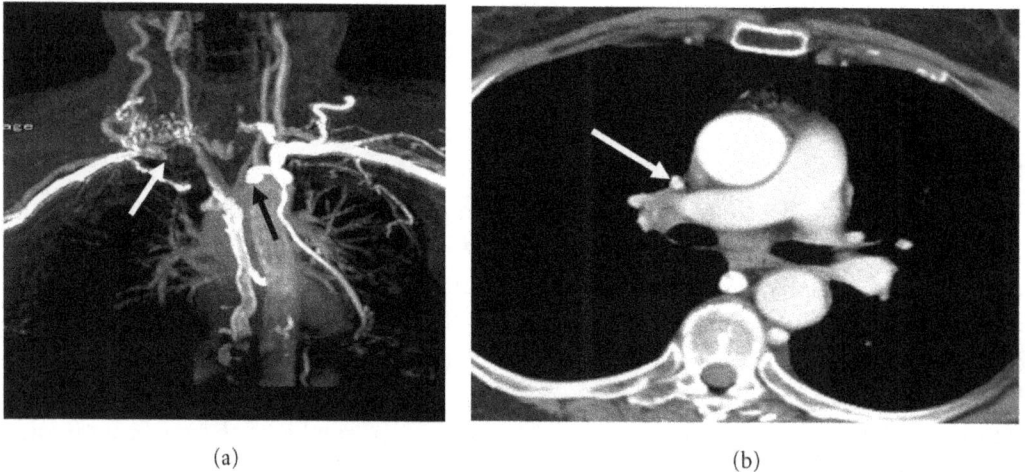

(a)   (b)

Fig. 1. SVC syndrome in a 67-year-old woman. (a) MRV was performed with the injection of dilute gadolinium contrast into both peripheral IV lines showing occlusion of the right subclavian (white arrow) and left central brachiocephalic (black arrow) veins extending to the SVC. Collateral circulation is seen through the azygos and left superior intercostal vein. (b) Axial CT section at the level of SVC shows SVC occlusion (arrow). Both brachiocephalic veins were cannulated through the SVC from the femoral approach, and after PTA, left brachiocephalic vein was stented with a 12 mm × 4 cm Wallstent while the SVC was stented with a 16 mm × 4 cm Wallstent.

Imaging Workup

Cross sectional imaging of the neck and chest should be obtained to demonstrate the important vascular anatomy in the planning of an endovascular procedure for central vein obstruction. The potential target inflow and outflow vessels (proximal and distal to the occlusion) must be identified to facilitate the procedure. The imaging must demonstrate the subclavian and innominate arteries, the thoracic aorta and right pulmonary artery adjacent to the junction of the SVC and right atrium when recanalizing the SVC occlusion. Ultrasonography is done to evaluate the dialysis feeding artery, arterial anastomosis, outflow veins, and the occlusion prior to intervention. Both internal jugular veins are also evaluated by ultrasonography as the jugular vein is often used for catheter introduction. Sites for catheter introduction in the outflow vein, the extent of occlusion, and its relationship with the critical anatomy should be determined prior to the start of the endovascular intervention. Subclavian vein recanalization is usually performed from the basilic or cephalic vein accessed at the mid arm. The femoral vein or the great saphenous vein is accessed for the evaluation of the central extent of the occlusion and placement of a loop snare for targeting during sharp recanalization.

Sharp recanalization of the SVC occlusion usually requires access from the internal jugular and femoral veins (dual venous access to both ends of the occlusion).[1] The brachial vein is also accessed to demonstrate the extent of the venous occlusion and collaterals. In patients with SVC syndrome, a single subclavian or internal jugular vein should be established in continuity with the SVC to relieve the SVC syndrome. CO_2 is a useful contrast agent to outline central vein occlusion in kidney disease patients with residual renal function or iodinated contrast allergy. Its low viscosity allows CO_2 injection through a small bore needle placed in a peripheral vein. The viscosity also allows filling of the central veins through collateral veins including the contralateral jugular and brachiocephalic veins, and the SVC.

Access and Venography

Prior to recanalization of a central vein obstruction, a fistulogram is performed as the initial evaluation of the fistula, outflow and central veins. Percutaneous access to the outflow vein, femoral and jugular veins utilizes the Seldinger technique. The micropuncture set is commonly used to access the veins. Ultrasound guided venipuncture is preferred. The access site is selected to allow relatively short and straightforward pathway of the wire, catheter and special devices to tackle the central vein occlusion. When the puncture site is selected, the area is prepped and draped in sterile fashion. In a patient with an AVF of the upper extremity, the outflow vein (the basilic or the cephalic vein) should be punctured towards the axillary vein approximately 5–10 cm away from the arterial anastomosis of the AVF.

The venous puncture technique is as follows: the 21gauge needle is advanced under ultrasound guidance until blood flows through the needle. A 0.018" guidewire is inserted through the needle into the vein. The passage of the wire should be smooth into the axillary vein. After removal of the needle, the coaxial catheter is introduced, and the inner 3 Fr dilator is removed. A 0.035" Safe-T-J guidewire is advanced through the outer dilator to the axillary vein. Over the wire, a 6 Fr or 7 Fr sheath is introduced, and venography is performed to visualize the outflow and central veins. Either iodinated contrast medium or CO_2 (20–30 cc/sec) may be used. Various CO_2 delivery systems can be used to avoid air contamination. When visualization of the fistula/or graft and arterial anastomosis is indicated, contrast medium is injected with compression of the outflow vein or application of a pressure cuff above the puncture site inflated at a pressure above 250 mmHg. For the evaluation of the arterial inflow from the brachial and axillary arteries, a second access is usually needed and a catheter can be placed near the arterial anastomosis or inside the feeding artery. The Seldinger micropuncture technique is used for catheterization of the femoral vein, and a 7 Fr long introducer sheath is advanced to the brachiocephallic vein. Occasionally the brachial vein on the opposite arm is catheterized to facilitate recanalization of the central vein occlusion.

Conventional Recanalization Technique

The technique for recanalization of central vein stenosis or occlusion is as follows: After catheterization of the outflow segment of the AVG or AVF and/or the femoral vein, 7 Fr sheaths are inserted. Pressure is measured in the outflow vein or right atrium, and a venogram is obtained. The catheter is then advanced to the point of the occlusion and a repeat venogram is performed to visualize the anatomy of the lesion and associated collaterals. The occlusion is then crossed using a Kumpe catheter (Cook Medical Inc., Bloomington, IN) as support and 0.035" hydrophilic guidewire. Once the occlusion has been traversed, angioplasty is performed using the upper limb vein approach. When stent placement is indicated, an exchange Glidewire with stiff shaft (Terumo Medical Corp. Somerset, NJ, US) is advanced through the obstruction into the SVC where the wire is snared using a loop snare (Amplatz GooseNeck Snare, ev3 Endovascular, Inc., Plymouth, MN) from the femoral vein. Over the through-and-thorough wire, a 5 Fr vertebral or HIH Glide catheter is advanced through the obstruction into the outflow vein. Over the new exchange wire such as an Amplatz Super stiff wire, balloon angioplasty is performed on the subclavian vein occlusion using a 10 and 12 mm diameter angioplasty balloon such as Conquest balloon (Bard Peripheral Vascular, Tempe, AZ, US) or Mustang balloon catheter (Boston Scientific Corp., Marlborough, MA, US). A 12–14 mm diameter balloon catheter is used to dilate the brachiocephalic vein occlusion and a 14–16 mm diameter balloon catheter used to dilate the SVC lesion. If angioplasty fails due to vessel recoil, a metallic stent is deployed.[2–4] The stent diameters used for the subclavian and brachiocephalic veins, and the SVC are usually 12 mm, 14 mm, and 16 mm, respectively (Figs. 2 and 5).

Recanalization Steps

- Assess the AVF or AVG or outflow (basilic, cephalic or brachial vein) veins using ultrasound.
- Using a micropuncture set under ultrasound guidance, insert a 6 Fr, 5 or 7 cm long sheath.

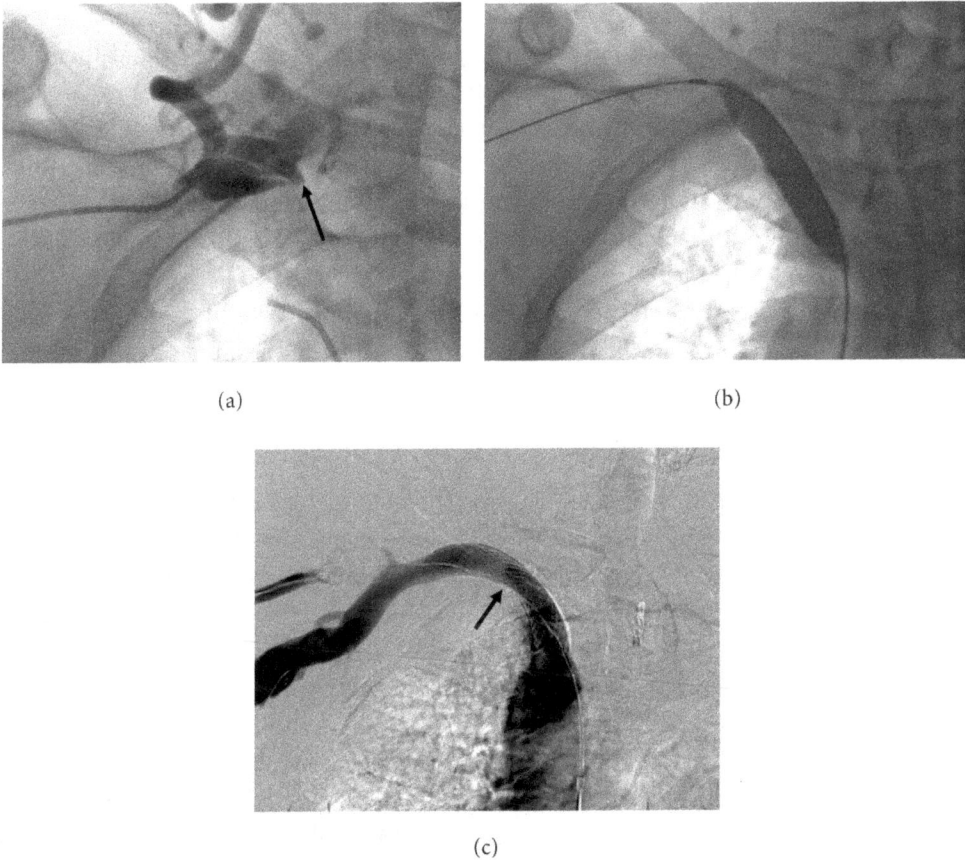

(c)

Fig. 2. Right subclavian and brachiocephalic vein occlusion in a patient with right arm AVF and swelling. (a) After catheterization of right basilic vein under ultrasound guidance, a venogram shows occlusion of the right subclavian and brachiocephalic veins (arrow) with numerous collateral veins. (b) After percutaneous catheterization of the right femoral vein, an 8 Fr, 55 cm sheath was advanced to the right atrium, and then a 5 Fr HIH catheter was advanced to the brachiocephalic vein. From the basilic vein access with a 5 Fr sheath a 5 Fr Kumpe catheter was advanced to the point of the occlusion (arrow) and a venogram was performed. Using the stiff end of the guidewire, the occlusion was crossed, and the catheter was then advanced to the right atrium. From the femoral vein approach the occlusion was dilated to 12 mm, (c) 14 mm × 40 mm Wallstent (Boston Scientific Corp.) was deployed from the axillary vein to the brachiocephalic vein. After post dilation to 14 mm, a completion venogram showed a widely patent stent (arrow).

- Perform a fistulogram to assess the outflow and central veins. Use CO_2 as contrast agent in patients with residual renal function or contrast allergy.
- Assess the arterial anastomosis if necessary with contrast injection through the sheath after inflating a blood pressure cuff above the tip of the sheath to 300 mmHg.
- Advance a 5 Fr catheter to the point of occlusion, and perform a venogram.
- Insert a 7 Fr sheath in the femoral vein.
- Advance a 5 Fr catheter to the point of the occlusion, and perform a venogram to assess the central side of the occlusion.
- Advance a 0.035" hydrophilic guidewire intermittently through the occlusion.

- Use appropriate support catheter to support the wire crossing the occlusion.
- Once the guidewire has been passed through the occlusion, advance the catheter over the guidewire into the SVC, and confirm the catheter position with contrast injection.
- Advance the sheath introducer into the brachiocephalic vein, and then inject contrast medium through the side port of the sheath into the occluded segment while leaving the guidewire in place in the SVC to document intraluminal passage of the guidewire.
- Select a proper size of balloon catheter for angioplasty. Use a smaller balloon for the initial dilatation.
- Place a stent if vessel recoil results in incomplete dilation.
- Completion venogram with the injection of contrast medium in the outflow vein.

Sharp Recanalization

When recanalization of the occlusion using the standard catheter and hydrophilic guidewire fails, sharp recanalization can be performed using one of the specially designed puncture devices.[5-11] A simple, commonly used technique is crossing the occlusion by intermittently cautiously moving the stiff end (back end) of a 0.035" or 0.018" hydrophilic guidewire forward through the occlusion. If this fails, a specially shaped puncture needle is used to cross the occlusion. The puncture needles that have been used for sharp recanalization include the Ross modified Colapinto needle (Cook Medical, Inc., Bloomington, IN, US) (Fig. 3a), the Rösch-Uchida transjugular liver access set (Cook Medical, Inc., Bloomington, IN, US) (Fig. 3b), the BRK Transseptal needle (St. Jude Medical Inc., St. Paul, MN, US) (Fig. 4),[1,5] PowerWire RF Guidewire (Baylis Medical Company Inc., Montreal, QC Canada), and the Outback-LTD re-entry catheter (Cordis Corp, Fremont, CA, US).[12] The modified Ross Colapinto needle is a 16 gauge assembled in a 10 Fr catheter which is introduced through a 10 Fr sheath. The 30° needle tip curve facilitates the puncture along the course of the occluded segment. The Rösch-Uchida transjugular access set that is used for portal vein access for a TIPS procedure, may be used for sharp recanalization. The trocar stylet (0.038") within a 5 Fr catheter combined with the stiffening cannula within the 10 Fr catheter facilitates puncture of the chronic fibrotic occlusion of the vein. The BRK Transseptal needles that are generally used for a puncture in the interatrial septum, can be used for sharp recanalization of the occlusion. It consists of a luminal puncture needle (accepting 0.018" guidewire), a solid stylet and a dilator. The distal section of the needle is curved to facilitate the direction of the needle away from the critical structure towards the loop snare target positioned near the central end of the occlusion (Fig. 6). The system is advanced through a 7 Fr sheath. The PowerWire RF Guidewire may be used to cross the occluded venous segment that is difficult to negotiate with a standard or sharp recanalization. Recanalization is achieved by delivery of radiofrequency energy through the atraumatic radiopaque tip of the RF guidewire. The Outback-LTD re-entry catheter has also been successfully used for revascularization of an occluded brachiocephalic vein using the same technique employed for re-entry of the true lumen from the subintimal channel.[12]

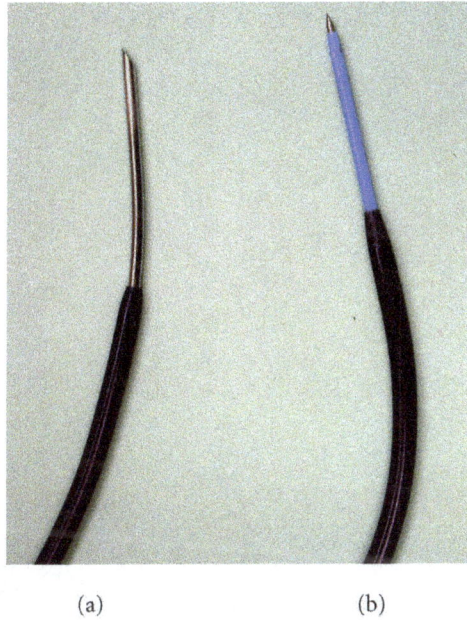

Fig. 3. The Ross modified Colapinto needle (a) and Rösch-Uchida Transjugular liver access set (b) for sharp recanalization. The 10 Fr catheter is used to advance the 16 gauge Ross Modified Colapinto needle. The needle and catheter assembly is advanced through a 10 Fr sheath. The Rösch-Uchida set consists of a 21 gauge stylet in the 5 Fr catheter inserted through a 14 gauge stiffening cannula which is passed through a 10 Fr catheter. The needle and cannula assembly is introduced through a 10 Fr sheath.

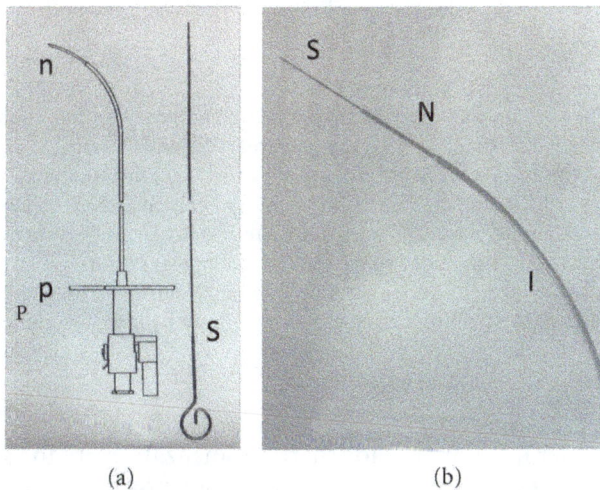

Fig. 4. BRK Transseptal Needle used for sharp recanalization of central venous occlusion. (a) The arrow base-plate (p) on the needle shield indicates the direction of the needle curve. The stylet (S) is used to facilitate the insertion of the needle into the needle shield. The 18 gauge needle with the beveled tip (n) is used for puncture. (b) Close-up view of the needle showing the stylet, the needle (N) and introducer dilator (I).

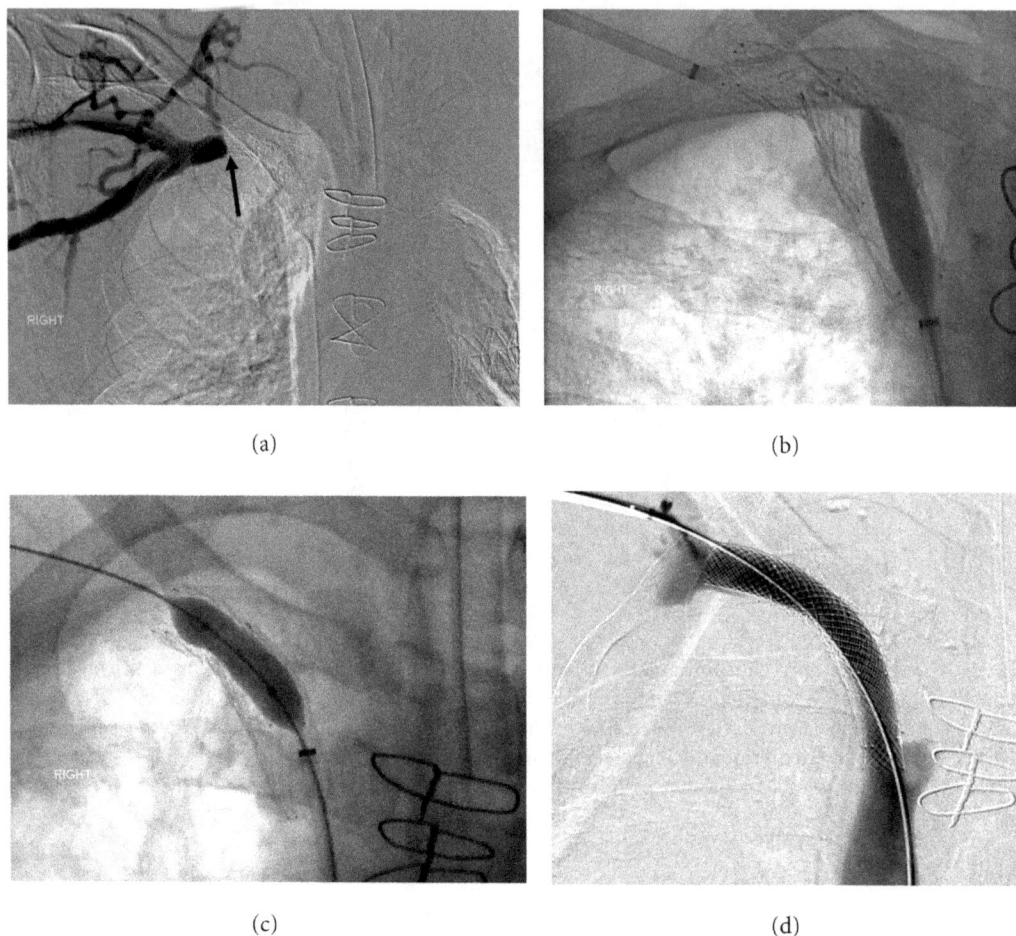

(a)   (b)

(c)   (d)

Fig. 5.   A 76-year-old man with a right upper extremity AVF and arm swelling following stent placement for subclavian vein occlusion. (a) After accessing right brachial vein with placement of a 7-Fr sheath, a venogram shows occlusion of the subclavian stent (arrow). A 9 Fr × 70 cm introducer sheath was introduced in the right common femoral vein and advanced into the superior vena cava. (b) A balloon is placed within the central portion of the stent, and then a 0.035" stiff Glidewire and Rosen guide were used to cross the occlusion. (c) Balloon angioplasty of the occluded subclavian vein stent was performed using a 12 mm high pressure balloon catheter. (d) After placement of a 12 mm × 60 mm Wallstent within the right subclavian vein stent and dilation of the stent to 12 mm utilizing Atlas angioplasty balloon, a completion venogram shows patent stent.

The sites for catheter introduction for sharp recanalization are the same as those used for the standard recanalization and balloon angioplasty. The basilic and femoral or great saphenous veins are the usual sites for access to perform recanalization. After placing a 6 Fr sheath in the basilic vein and a 7 Fr sheath in the femoral or great saphenous vein, the catheters are positioned on either side of the occlusion, and venography is performed in the anteroposterior and oblique projections to assess the length and course of the occluded segment. A loop snare is placed at the central end of the occlusion for targeting the central vein, sharp recanalization is performed with the stiff end (back end of the wire) of the guidewire and if it fails, one of the specially designed puncture needles is passed from either the basilic vein or the internal jugular vein to cross the occlusion.

The technique of sharp recanalization involves passing the stiff end of the guidewire through the catheter positioned on the proximal point of the occlusion. The guidewire is advanced gently using an intermittent forward motion. Once the guidewire is passed through the occlusion, it should pass smoothly into the brachiocephalic vein and then into the SVC. If any resistance is encountered, the guidewire should be withdrawn and the process should then be repeated at a slightly different angle of puncture. Once free passage of the wire into the SVC is achieved, the catheter is advanced into the central vein, and contrast injection is made to confirm the position of the guidewire in the brachiocephalic vein and SVC. An exchange hydrophilic guidewire is passed from the basilic access to the SVC where the wire is snared and brought out through the femoral sheath. This facilitates passage of a balloon catheter through the occlusion from the femoral access.

If the stiff end of the hydrophilic guidewire cannot cross the occlusion from either the basilic or femoral approach, the BRK Transseptal Needle is used to puncture the occlusion. When recanalizing right subclavian vein occlusion from the axillary vein, the needle should be directed anteriorly towards the loop snare target positioned at the junction of the right internal jugular and brachiocephalic veins to avoid arterial injury.

When recanalizing the left subclavian vein occlusion, an attempt is made to access the left mid upper arm basilic vein to insert a 7 Fr x 45 cm sheath. Then a 7 Fr x 45 cm sheath is inserted in the femoral vein, and advanced to the left brachiocephalic vein. A 15 cm diameter loop snare is opened at the confluence of the left internal jugular vein and the subclavian occlusion. The relationship between the occluded segment, and the left subclavian and internal mammary arteries is carefully evaluated to avoid inadvertent puncture of the artery. The direction of the needle is evaluated by taking AP and multiple oblique projections. Using the loop snare as a target, the BRK1 needle with the stylet in place is advanced into the loop snare target. After withdrawing the stylet, a V-18 guidewire is advanced through the left brachiocephalic vein and the SVC into the right atrium. The needle pass is tested with a 5 Fr IVUS or with the injection of contrast medium to assess any evidence of vessel perforation. The snared wire is pulled out from the femoral sheath, establishing a through-and-through wire. Initially, the recanalized vein is dilated with a small diameter balloon catheter (usually 4–6 mm in diameter) to facilitate passage of the catheter and sheath. If no contrast extravasation is seen, repeat dilation is performed with an 8 to 10 mm diameter balloon catheter. After reconfirmation of the intraluminal pass without extravasation, final balloon dilatation is performed with a 12 mm balloon catheter.

Recanalization of the brachiocephalic vein and SVC occlusion is performed similarly with the use of a smaller balloon, initially, and followed by dilatation of the brachiocephalic vein to 14 mm and the SVC to 16 mm. The commonly used high pressure balloon catheters are Conquest (Bard peripheral vascular, Tempe, AZ, US) and Mustang (Boston Scientific

Corp., Marlborough, MA, US). Initially, cutting balloon angioplasty (8 mm x 20 mm cutting balloon, Boston Scientific Corp., MA, US) may be used to dilate the lesion and followed by dilation with a larger balloon catheter. Atlas (Bard peripheral vascular, Tempe, AZ, US) is used to dilate the brachiocephalic vein to 14 mm and the SVC to 16 mm.

After balloon dilation, stent placement is usually required to maintain patency of the recanalized segment. Both balloon expanding stent (Palmaz stent, Cordis Corp, Fremont, CA, US or iCAST, Atrium Medical Corp., Hudson, NH, US) and self-expanding stents (Wallstent, Boston Scientific Corp., Marlborough, MA, US) have been used to stent the brachiocephalic vein and SVC occlusion. When stenting brachiocephalic vein, the stent should extend just distal to the junction of IJV and SCV to maintain patency of both veins. Covered stents are used to treat symptomatic perforation.[13] When both the brachiocephalic vein and the SVC are to be stented, the segment of SVC draining the azygos vein should remain uncovered.

After completion of angioplasty and stenting, a venous pressure gradient is re-measured across the lesion, and then a completion venogram is performed. If residual stenosis with a significant pressure gradient is present, repeat balloon dilation is performed using a larger diameter balloon. If the lesion recoils after balloon dilation, stent deployment is indicated.

Recanalization of the occluded stent is performed using the same technique as that used for recanalization of the central vein occlusion. If the peripheral end of the stent has extended into a tributary vein, recanalization should be performed through its interstices. Once the stent has been crossed through the interstices from the axillary vein, the stent and interstices are dilated with a high pressure balloon catheter. Stent placement is often necessary to maintain the patency of the stent.

Sharp Recanalization Steps

- Use the same access technique as for standard recanalization.
- Advance a 7 Fr sheath to the point of occlusion from the femoral vein or great saphenous vein.
- Place a 15 mm or 20 mm loop snare at the central side of occlusion.
- Advance a 7 Fr sheath to the proximal point of the occlusion from the basilic vein for subclavian vein occlusion.
- Advance a 7 Fr or larger sheath from the jugular vein to the point of the brachiocephalic or SVC occlusion.
- Fluoroscopy in AP and oblique projections for the best alignment between the puncture needle and loop snare.
- Advance the needle for a short distance to puncture the proximal occlusion.
- Advance a guide wire through the needle.
- Advance a 4 Fr or 5 Fr catheter through occlusion into the central veins.
- Confirm the intraluminal position of the catheter with contrast injection.
- Inject contrast medium into the occluded segment to confirm no extravasation.
- Advance the sheath through the occlusion.

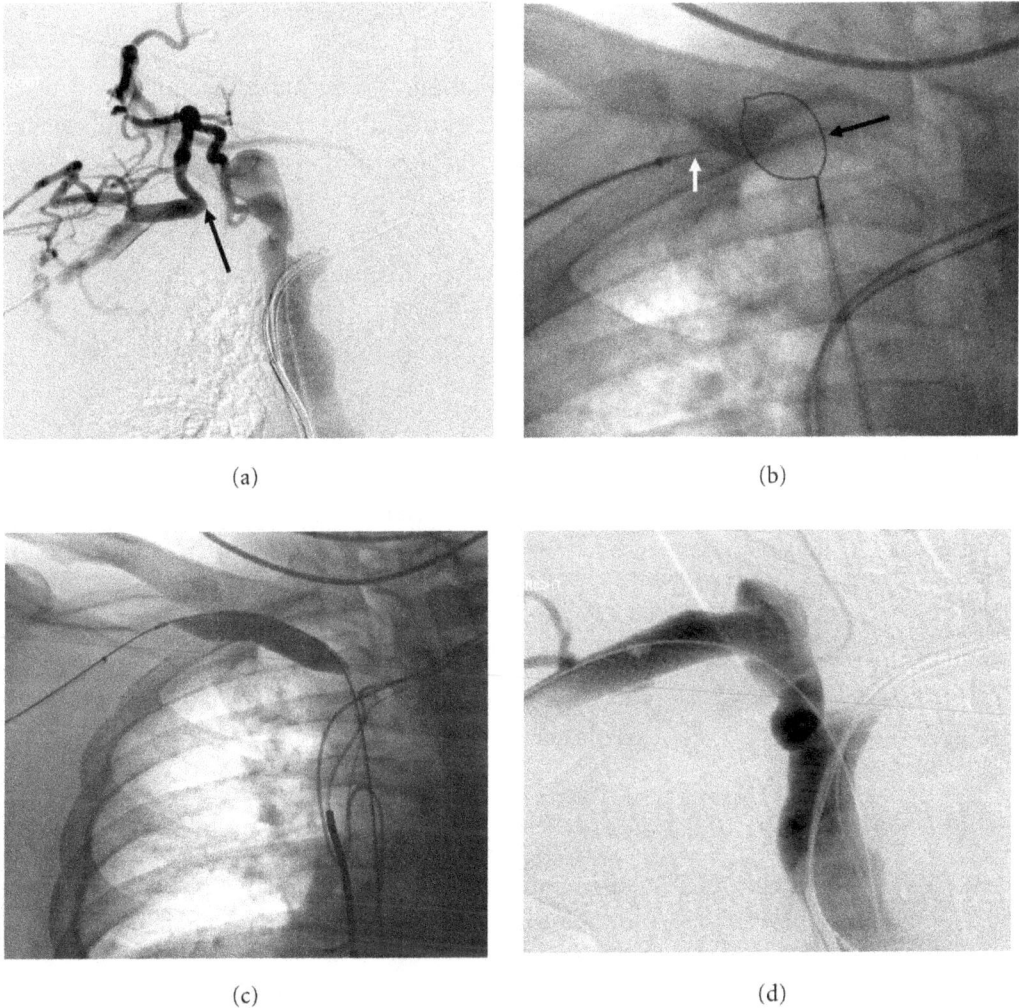

(a)

(b)

(c)

(d)

Fig. 6.   Sharp recanalization of right subclavian vein occlusion in a patient with right arm AVF and arm swelling. (a) Right upper extremity venogram from the basilic vein shows right subclavian vein occlusion (arrow) with reconstitution of the brachiocephalic vein by collateral circulation. (b) After unsuccessful attempts to cross the occlusion from the basilic vein using a vertebral catheter, and the soft and rigid ends of a 0.018" and 0.035" hydrophilic guidewire, sharp recanalization was performed using the BRK needle. The needle was advanced through a 6 Fr sheath from the basilic vein, and then punctured towards the loop snare placed in the patent central part of the brachiocephalic vein under fluoroscopy. The needle passage was directed (act as target for the sharp canalization) anteriorly to avoid arterial injury. The wire was snared and pulled out through the femoral vein sheath. (c) Over the 0.035" Amplatz superstiff guidewire, balloon dilation was performed sequentially using a 4 mm, 12 mm and 14 mm balloon catheters. (d) After placement of a 14 mm × 4 cm Wallstent, a completion venogram shows patent subclavian vein. Subsequently his arm swelling was resolved.

Courtesy of Dr. D. Williams, University of Michigan Frankel Cardiovascular Center.

- Intravascular ultrasound of the recanalized occluded segment if available.
- Initially, dilate the occlusion using a smaller balloon.
- Confirm no extravation with contrast injection into the occluded segment.
- Dilate the occlusion with larger balloons (10 mm and 12 mm balloon for subclavian occlusion).
- Place a stent if the vessel recoils after PTA (14 mm diameter stent for subclavian vein occlusion).
- Obtain completion venogram with the injection of contrast medium into the outflow vein of AVF.
- Treat concommittent AVF or AVG hemodynamically significant stenosis if present.

Potential Complications

The most serious complication of central vein recanalization is perforation resulting in hemothorax, mediastinal hematoma, pericardial tamponade, retrosternal discomfort, arrhythmia, balloon rupture, and stent migration. Delayed complications include in-stent restenosis, in-stent occlusion, and stent migration. Fluoroscopy and electrophysiology mapping guidance can be used to prevent pericardial perforation during sharp recanalization of SVC occlusion. Reinterventions are usually performed with balloon dilation, cutting balloon angioplasty, or stent placement.

Patient Monitoring during Procedure

Close monitor of patients' vital signs (mentioned in Chapter 6) is mandatory during central vein intervention. Watch out for any chest pain experienced by the patient signifying vessel perforation.

Postprocedure Management

Postprocedure care is essentially the same as for balloon angioplasty of venous lesions. This author prefers anticoagulation for several months (longer or often life-time for the patients with hypercoagulability) and administration of a baby aspirin ± cilostazol (Pletal). Venographic follow-up is performed 4 months after the recanalization. Restenosis and re-occlusion is rather common.

Summary

Endovascular recanalization with balloon dilatation and stent placement of central vein obstruction is a safe and effective treatment for central vein obstruction in hemodialysis patients. It will prolong the durability of dialysis access. Preoperative cross-sectional imaging and a facile endovascular technique are essential in obtaining a successful recanalization. Dual venous access to both ends of the occlusion as well as multiple different angle of fluoroscopy are commonly required. Meticulous intra- and post-procedure monitoring is mandatory. Frequent re-interventions are usually necessary in order to maintain the patency of the recanalized central vein obstruction and the functioning dialysis access.

References

1. Farrell T, Lang EV, Barnhart W. Sharp recanalization of central venous occlusions. *JVIR.* 1999; **10**: 149–154.
2. Verstandig AG, Bloom AI, Sasson T, Haviv YS, Rubinger D. Shortening and migration of Wallstents after stenting of central venous stenosis in hemodialysis patients. *Cardiovasc Intervent Radiol.* 2003; **26**: 58–64.
3. Lorenz JM. Use of stents for the maintenance of hemodialysis access. *Semin Intervent Radiol.* 2004; **21**(2): 135–140.
4. Surowiec, SM, Fegley AJ, Tanski WJ, *et al.* Endovascular management of central venous stenoses in the hemodialysis patient: results of percutaneous therapy. *Vasc Endovascular Surg.* 2004; **38**(4): 349–354.
5. Honnef D. Wingen M, Gunther RW, Haage P. Sharp central venous recanalization by means of a TIPS needle. *Cardiovasc Interv Radiol.* 2005; **28**: 673–676.
6. Haage P, Günther RW. Radiological intervention to maintain vascular access. *Eur J Vasc Endovasc Surg.* 2006; **32**: 84–89.
7. Levit RD, Cohen RM, Kwak A, Shlansky-Goldberg, Clark TW, *et al.* Asymptomatic central venous stenosis in hemodialysis patients. *Radiology.* 2006; **238**: 1051–1056.
8. Kim YC, Won JY, Choi SY, *et al.* Percutaneous treatment of central venous stenosis in hemodialysis patients: long-term outcomes. *Cardiovasc Intervent Radiol.* 2009; **32**(2): 271–278.
9. Ozyer U, Harman A, Yildirim E, Aytekin C, Karakayali F, Boyvat F. Long-term results of angioplasty and stent placement for treatment of central venous obstruction in 126 hemodialysis patients: a 10-year single-center experience. *AJR Am J Roentgenol.* 2009; **193**: 1672–1679.
10. Kundu S, Central venous obstruction management. *Semin Intervent Radiol.* 2009; **26**: 115–121
11. Athreya S, Scott P, Annamalai G, Edwards R, Moss J, Robertson I. Sharp recanalization of central venous occlusions: a useful technique for haemodialysis line insertion. *Br J Radiol.* 2009; **82**: 105–108.
12. Anil G, Taneja M. Revascularization of an occluded brachiocephalic vein using Outback-LTD re-entry catheter. *J Vasc Surg.* 2010; **52**: 1038–1040.
13. Anaya-Ayala JE, Smolock CJ, Colvard BD, Naoum JJ, Bismuth J, *et al.* Efficacy of covered stent placement for central venous occlusive disease in hemodialysis patients. *J Vasc Surg.* 2011; **54**: 754–759.

CHAPTER 12

Prevention, Diagnosis and Management of Steal Syndrome

Jackie P. Ho

Dialysis associated steal syndrome (DASS) is a situation when tissue perfusion distal to the vascular access become insufficient, related to blood being diverted to support the access.[1] It is one of the most dreadful complications of vascular access creation. The severity and implication of ischemia can be classified[2] into:

Grade 1. **Mild** (cool extremity with few symptoms but demonstrable abnormal vasculature by flow augmentation with fistula occlusion), no treatment needed;

Grade 2. **Moderate** (intermittent ischemia only during dialysis/claudication), intervention sometimes needed;

Grade 3. **Severe** (ischemic pain at rest/tissue loss) (Fig. 1), intervention mandatory.

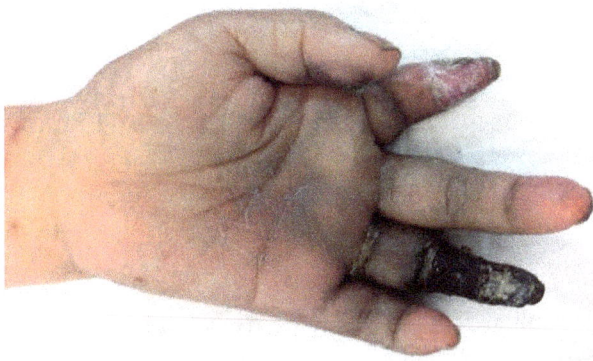

Fig. 1. DASS with gangrenous change of the left ring finger.

DASS may present early or late. DASS often present early for AVG while DASS in AVF tends to present later when the fistula is dilated and become mature.[3]

179

When a vascular access was created on a limb, the ultimate perfusion reaching the distal extremity of that specific limb is influenced by:

(1) **Original blood supply to the limb** — depends on the size and patency of the whole arterial system of the limb;
(2) **Amount of blood diverted to the access** — depends on the size of the AV anastomosis, diameter of fistula or conduit, and the flow resistance relative to the arterial outflow;
(3) **Body's adaptation ability** — depends on the capacity of arterial or collateral dilatation in response to the reduced flow resistance.

Prevention and treatment of DASS involve attention and decision-making in all stages of vascular access creation as summarized in the diagram below.

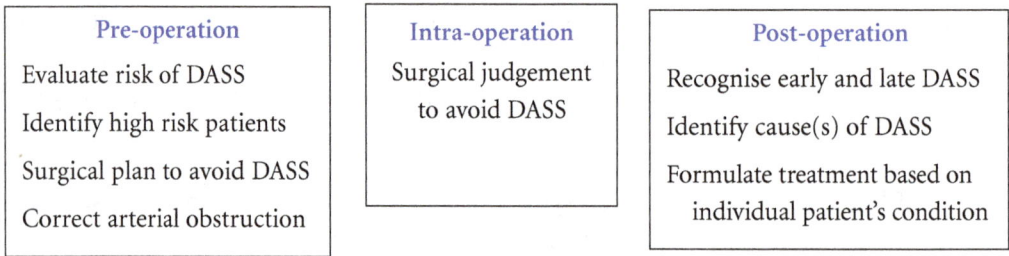

Pre-operation	Intra-operation	Post-operation
Evaluate risk of DASS	Surgical judgement to avoid DASS	Recognise early and late DASS
Identify high risk patients		Identify cause(s) of DASS
Surgical plan to avoid DASS		Formulate treatment based on individual patient's condition
Correct arterial obstruction		

Prevention of DASS

Overall incidence of DASS is estimated between 4% and 10%. DASS is found to be more prone in patients with diabetes, female gender, brachial artery as inflow, side-to-side anastomosis, peripheral vascular disease and multiple previous vascular accesses.[4,5] In Asia, diabetes-induced nephropathy is the main cause of kidney failure.[6] Access surgeons working in high diabetic incidence regions have to pay extra attention to prevent, detect and manage DASS. It is interesting to note that brachial artery-based vascular accesses have the highest incidence of DASS (can be as high as 20%). The incidence of DASS using radial[7] or axillary artery as inflow is less comparatively. Thus on creating access on brachial artery, clinicians have to exercise stringent judgement on AV anastomosis size.

(1) *Pre-operative assessment and planning*

Checking the radial and ulnar pulses (dorsalis pedis and posterior tibial if lower limb access is planned) and Allen's test are the basic arterial assessment. If brachial, radial and ulnar pulses are all present but weak, one has to compare the brachial pressure of both upper limbs to detect subclavian artery stenosis.

Routine pre-operative duplex vascular assessment is becoming the standard of care in many centers. The study protocol should include the assessment of axillary, brachial, ulnar

8.0mm

5.6mm

6.0mm

5.6mm

5.6mm

PSV ratio
7.3
(50–99% Stenosis)

PSV ratio
4.4
(50–99% Stenosis)

PSV ratio
2.8
(50–99% stenosis)
at distal radial
artery)

4.2mm

3.9mm

3.9mm

4.6mm

4.6mm

4.1mm

Venous tributary noted in distal segment
of forearm cephalic vein
4.5mm

Patient with upper limb arterial
disease high risk of DASS

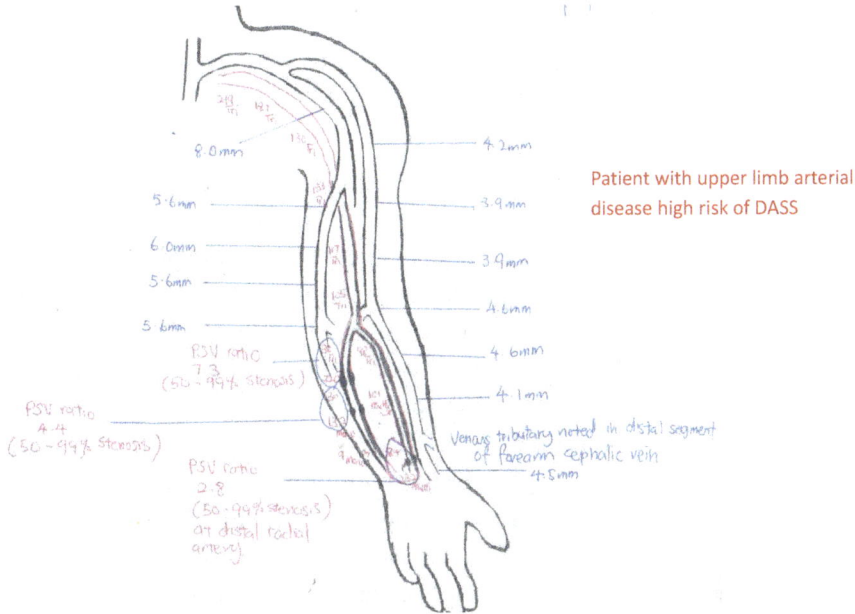

Fig. 2. Pre-operative duplex assessment of left upper limb arterial and venous system.

and radial artery size, degree of calcification, waveform and any hemodynamic significant obstruction of the limb planned for vascular access (Fig. 2).

If the physical findings are equivocal or suspicious of arterial disease, a pressure study of Digital Brachial Index (DBI) could provide an objective parameter for evaluation. Patients with lower DBI have higher risk of developing DASS but there is no DBI threshold that can well predict steal.[8,9]

In high-risk patients with significant multi-level arterial disease and without another better option limb, one may consider catheter based arteriogram to assess the arterial lumen and perform balloon angioplasty (± stenting) to correct the obstruction prior to access creation. If patient is not yet hemodialysis dependent, CO_2 angiogram could be an alternative (see Chapter 6). This interventional procedure can also be done in the same session of vascular access creation if the healthcare facility and expertise allow.

In patients with mild arterial disease (e.g. absent ulnar pulse only but Allen's test normal), one may proceed with the planned access surgery with proper counselling of DASS risk to patient, as well as meticulous control of the size of arterial anastomosis during operation. Nonetheless, it is challenging to achieve an anastomosis size just good enough to support vascular access flow and yet not taking away too much arterial flow to the hand and fingers.

In situations where arterial obstruction is not correctable or the correction is unsustainable, and there is no better other limb option, proximalization of vascular access could be a workable solution. Jennings[10] *et al.* reported 30 high-risk patients with non-palpable brachial, radial and ulnar pulses underwent transposed AVF creation using axillary artery as inflow. None of the patients developed DASS after the access creation.

(2) *Intra-operative judgement*

A quick ultrasound scan immediately before surgery is helpful in many circumstances. If the brachial artery over planned surgical site is only about 2 mm on ultrasound assessment, the procedure should either be cancelled or converted to use more proximal inflow.

Vascular access using brachial inflow has the highest incidence of DASS in the upper limb. The size of arteriotomy on brachial artery for anastomosis should be kept ≤4 mm if the brachial artery size is small (~3 mm), absent either radial or ulnar pulse, or with large size outflow vein (>3 mm). Subjective judgement of arteriotomy size may not be accurate. Measure with a sterile paper ruler will help achieve accurate sizing of the anastomosis. If an AVG was planned, tapered graft 4–6 mm or 4–7 mm can be used to minimize size mismatch between arteriotomy and graft. Be aware that vasospasm may affect the intra-operative brachial artery size judgement. Ultrasound assessment before any surgical dissection gives a more accurate account of the brachial artery size. Trial of vasodilator (e.g. Papaverine) local infusion or intravenous injection (e.g. Glyceryl trinitrate) could be given to reverse any vasospasm.

Diagnosis of DASS

Diagnosis of DASS is primarily clinical based on signs and symptoms. Duplex and pressure study provide additional hemodynamic information to help make the diagnosis.

Immediately after the vascular access surgery, the finger color, nail bed capillary refill, radial and ulnar pulses should be checked. If uncertain, one can use the pulse oximeter (readily available in operating theatres) to check for finger oxygen saturation and perfusion status.

Some remedy procedures can be done immediately in the operating theatre. For example, re-open the wound and re-do anastomosis if the AV anastomosis is considered too big. If a surgeon suspects occlusion of distal artery after clamping, re-explore the wound and reverse the obstruction could salvage the situation.

The typical clinical features of DASS include pallor, absent distal pulses, coldness, pain or paraesthesia over hand and finger, weakness in hand grip, and tissue loss (ulcer/gangrene) of fingers. Patient may experience pain of the fingers during hemodialysis in Grade 2 DASS. Occasionally, it can be difficult to differentiate DASS from nerve injury with only paraesthesia symptom over the hand and fingers in early post-operation period.

Finger pressure or DBI provides information on perfusion pressure which may help to establish diagnosis in equivocal cases. Duplex study of the arterial system and the flow around AV anastomosis provide important information to confirm diagnosis, reveal the cause of DASS and aid planning of treatment.

Delay onset of DASS is not uncommon. The signs and symptoms of DASS should be assessed for all vascular access patients during their clinic follow up.

Arteriogram of the affected upper limb is a useful investigation. It may reveal situations such as any obstructive lesion along the whole upper limb arterial system, too big size AV anastomosis, preferential flow into the fistula (signifying high flow situation), multiple venous outflows and generalized small size of the arteries. Endovascular intervention to treat any identified arterial obstruction or local anesthetic procedure to ligate one of the outflow veins can be performed in the same session.

Treatment of DASS

Goal of treatment: *Reverse the ischemia symptom of the limb and preserving the vascular access.*

Ligation of the vascular access is the simplest way to reverse DASS. However, ligation will certainly sacrifice one valuable access. Furthermore, in ESRF patient with DASS history, subsequent vascular access creation would still carry high DASS risk.

For Grade 1 DASS, observational management can be adopted. Usually, the mild ischemic symptom may improve with time due to collateral recruitment. Nonetheless, close monitoring is warranted. Educate the patient to recognize steal syndrome so that the patient himself or herself could report early if symptom gets worse.

For Grade 2 and 3, intervention or surgery is required to reverse or minimize the adverse effect of ischemia.

There are numerous methods available to treat DASS (listed in table below) suitable for different causes and conditions.

Endovascular intervention for arterial obstruction
Ligation or embolization of distal radial artery
Banding of vascular access
Ligation of one of the outflow veins
Revision to narrow down anastomosis
Revision using distal inflow (RUDI)
Proximalization of arterial inflow (PAI)
Distal revascularization and interval ligation (DRIL)
Distal revascularization without interval ligation

There are few important questions clinicians need to answer when planning the treatment strategy

(1) What is/are the causes

This could be proximal artery obstruction causing inflow problem, radial or ulnar artery obstruction, palmer arch and digital artery disease (distal vessel disease), excessive retrograde flow into radial artery, generalised small sized arteries relative to a large sized fistula, too big AV anastomosis, high flow fistula access, several outflow vein of fistula access, or a combination of the above.

(2) Severity of DASS

Grade 2 symptom indicate the perfusion deficiency of hand and fingers is not very severe. Treatments that just tilt the hemodynamic balance over are likely sufficient in correcting the symptom. Whereas in cases with significant tissue loss, more drastic change of blood flow (often requiring more extensive surgery) is required.

(3) Patient's general medical condition

Whether patient is suitable to undergo extensive surgery?

Each treatment method regarding its application and limitation will be discussed as follows:

Endovascular intervention for arterial obstruction

Subclavian artery stenosis or occlusion is not uncommon, left side more often than right. With widely adoption of pre-operative routine duplex study, many of the proximal inflow problem would have been identified before access creation. Sometimes, this can still be missed. The lesion can be tackled by brachial retrograde ± femoral retrograde approach. The outcome of endovascular balloon angioplasty ± stenting for subclavian artery is rather durable.

Calcified and stenosed radial and ulnar arteries are more common in elderly and diabetic patients. The lesion can be tackled with brachial antegrade approach punctured under ultrasound guidance just proximal to the antecubital skin crease level (where the brachial artery is rather superficial to skin and with a reasonable distance from brachial bifurcation). To treat the diseased radial and ulnar arteries, 0.014" wire and angioplasty balloons are preferred. Balloon angioplasty could effectively improve the radial or ulnar artery flow but restenosis is common.

After endovascular intervention, even with complete resolution of DASS, regular monitoring of DASS recurrence and surveillance studies of the arterial system are indicated. Repeated angioplasty might be needed to maintain vessel patency.

Ligation or embolization of distal radial artery (for RC AVF)

Retrograde flow of blood over radial artery distal to the RC AVF anastomosis is common. However, in patients with small vessel disease of the digital arteries, DASS is diagnosed with excessive shunting of blood from the ulnar artery through palmer arch back into radial artery and the fistula, together with poor opacification of the digital arteries recognized in angiogram. To test the effects of distal radial artery occlusion, one may manually compressed the distal radial artery distal to RC anastomosis and repeat the angiogram to observe for digital artery flow. The retrograde flow can be stopped by surgical ligation of distal radial artery or deploying an embolization coil into it (Fig. 3).[11]

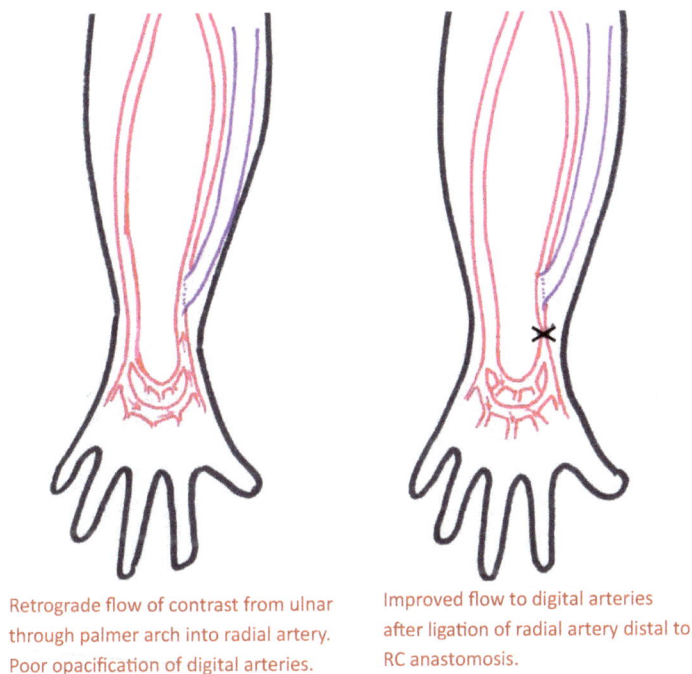

Retrograde flow of contrast from ulnar through palmer arch into radial artery. Poor opacification of digital arteries.

Improved flow to digital arteries after ligation of radial artery distal to RC anastomosis.

Fig. 3. Schematic diagram of blood flow condition before and after ligation of radial artery distal to the anastomosis.

Banding of vascular access and ligation of one of the outflow veins

Both methods are tackling high flow situation of the fistula either due to large luminal diameter or multiple outflow (e.g. in side-to-side AV anastomosis, or median antebrachial vein supplying both cephalic and basilic veins). High flow into vascular access can be diagnosed by duplex study. More accurately, in angiogram, the contrast will flow predominantly into the fistula instead of the arteries distal to the AV anastomosis. Digital compression of the fistula will improve contrast flow into the distal arteries.

To increase the resistance and reduce the flow into a large lumen venous fistula, various methods can be used including suture plication on one side of the venous fistula near the anastomosis followed by ePTFE banding (Fig. 4a),[12] or minimally invasive limited ligation endoluminal-assisted revision (MILLER) banding (Fig. 4b).[13] Intra-operative flow monitoring[14] to assess fistula flow reduction and radial, ulnar artery flow increase is advisable to evaluate the effectiveness and prevent over-do which can cause fistula failure. Banding is more likely to resolve DASS of Grade 2 severity or milder Grade 3. Its flow redistribution effect may not be good enough for those with significant tissue loss.

For multiple outflow veins, ligation of one of the outflow veins will help increase the flow resistance and reduce access flow.

Sometimes, concomitant distal arterial obstruction will be detected during angiogram assessment. Endovascular intervention for the arterial obstruction can be performed in the same session with the outflow procedure to help redistribute the flow into the hand.

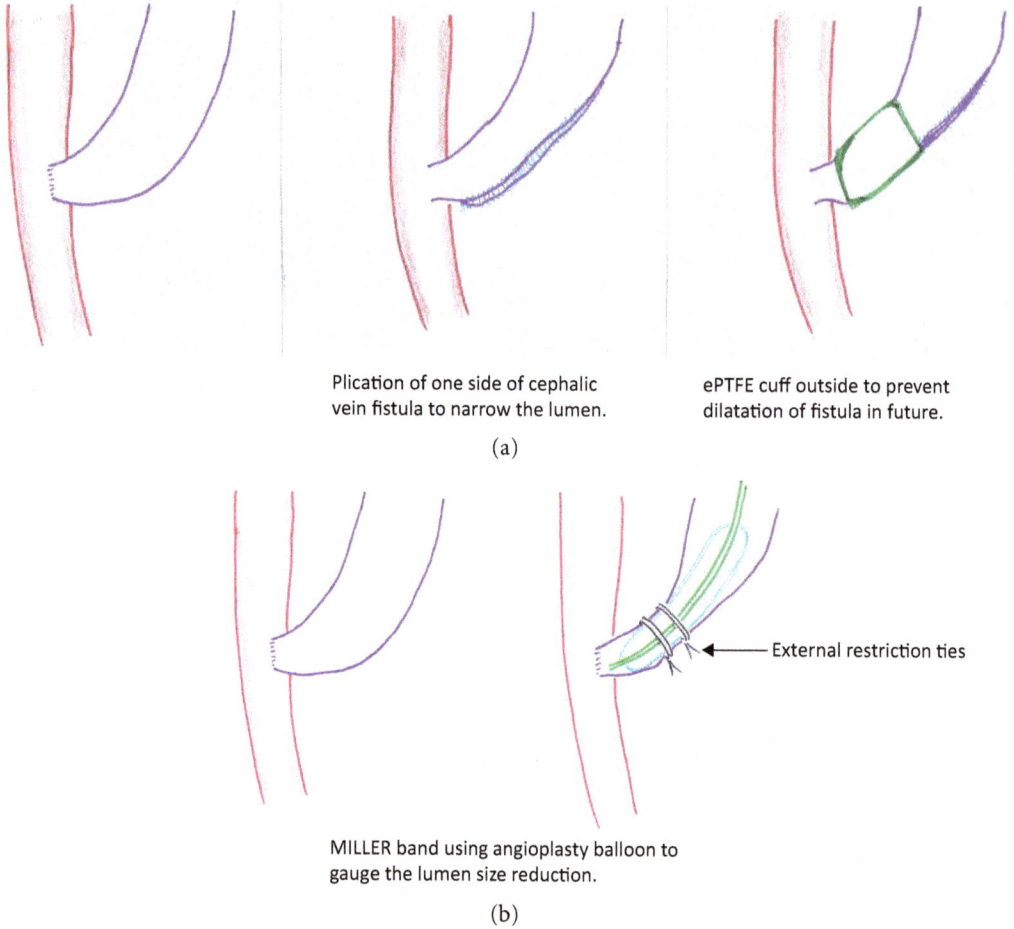

Plication of one side of cephalic
vein fistula to narrow the lumen.

ePTFE cuff outside to prevent
dilatation of fistula in future.

(a)

External restriction ties

MILLER band using angioplasty balloon to
gauge the lumen size reduction.

(b)

Fig. 4.   Schematic diagram of various method to increase the outflow resistance of the fistula.

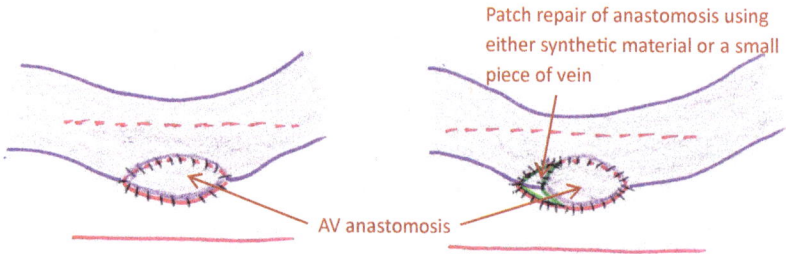

Patch repair of anastomosis using
either synthetic material or a small
piece of vein

AV anastomosis

Fig. 5.   Schematic diagram of reducing the size of AV anastomosis.

Revision to narrow down anastomosis

AV anastomosis may have been created larger than usual in heavily calcified artery or judgement issue. The size of the anastomosis can also be evaluated in angiogram. Revision of anastomosis to a smaller size will reduce the proportion of flow into the fistula (Fig. 5). However, this method may not be effective for natively small brachial artery (<3 mm).

Revision Using distal inflow (RUDI), proximalization of arterial inflow (PAI), distal revascularization and interval ligation (DRIL), distal revascularization without interval ligation

These four methods involve different magnitude of anastomosis revision or bypass operation. Principle of each operation is shown in the diagrams below using BC AVF as an example.

RUDI[15] moves the anastomosis distally from brachial artery to more distal vessel (proximal radial or ulnar artery 2–3 cm distal to the brachial bifurcation) so as to divert more flow back to the arterial system (Fig. 6). It requires patent radial and ulnar artery and palmer arch. The flow of the palmer arch vessel should be checked with Doppler or Duplex with the new inflow vessel clamped. A segment of vein, preferably with diameter smaller

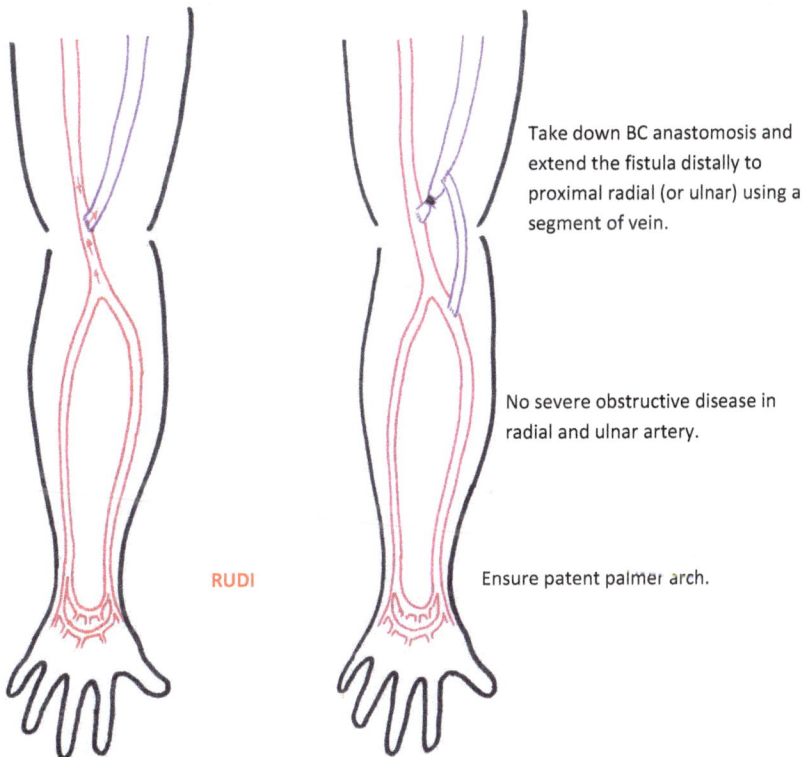

Take down BC anastomosis and extend the fistula distally to proximal radial (or ulnar) using a segment of vein.

No severe obstructive disease in radial and ulnar artery.

Ensure patent palmer arch.

RUDI

Fig. 6.   Schematic diagram of revision using distal inflow RUDI.

than the original vein fistula, is needed to extend the fistula to the proximal radial or ulnar artery. This vein can be obtained from branch of cephalic vein, forearm reasonable sized basilic vein, transposed saphenous vein or segment of brachial vein. RUDI is not suitable in patients with extensive obstructive disease of the radial and ulnar artery, and those with severe small vessel disease inside the hand.

PAI[16] involves ligation of the existing brachial anastomosis and connects the original fistula with a synthetic graft to the axillary artery (Fig. 7). Proximal artery inflow for access creation has better adaptation ability and will increase the overall flow into the whole upper limb. It is useful for patients with diffusely diseased small vessel of the upper limb or generalized small sized arteries. However, it is virtually converting an AVF into an AVG. The advantage of better patency of native vein fistula will be reduced. PAI may not be a good option if the fistula is a precious and good quality one, especially if the patient has rather long life expectancy.

Distal Revascularization[17] with or without Interval Ligation involves a proximal to distal brachial artery bypass using a segment of vein to divert proximal arterial flow to the distal vessels (Fig. 8). Usually, reversed saphenous vein is harvested as the bypass conduit. Synthetic graft is less preferred. It is well shown to improve flow and symptoms[17,18] to the extremity although the need for interval ligation is still controversial.[19] Since it is a bypass of the artery, the original fistula created is not being affected. Satisfactory patency of the fistula has been reported. There is concern of causing arterial damage during bypass procedure and worsening of ischemia if blockage of this bypass occurs. Satisfactory graft patency has been reported from specialized centers.[19,20] Nonetheless, DRIL is a major

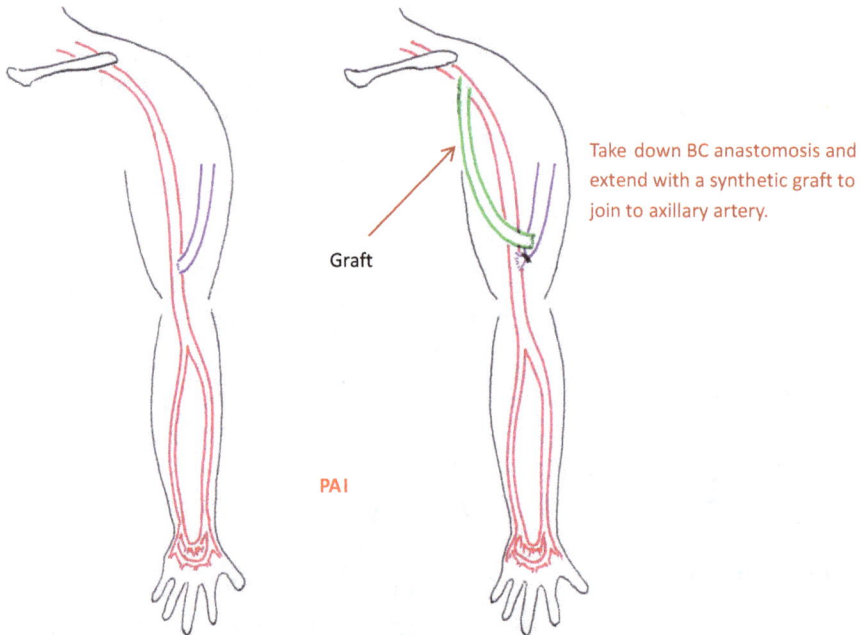

Take down BC anastomosis and extend with a synthetic graft to join to axillary artery.

Graft

PAI

Fig. 7.   Schematic diagram of proximalization of arterial inflow PAI.

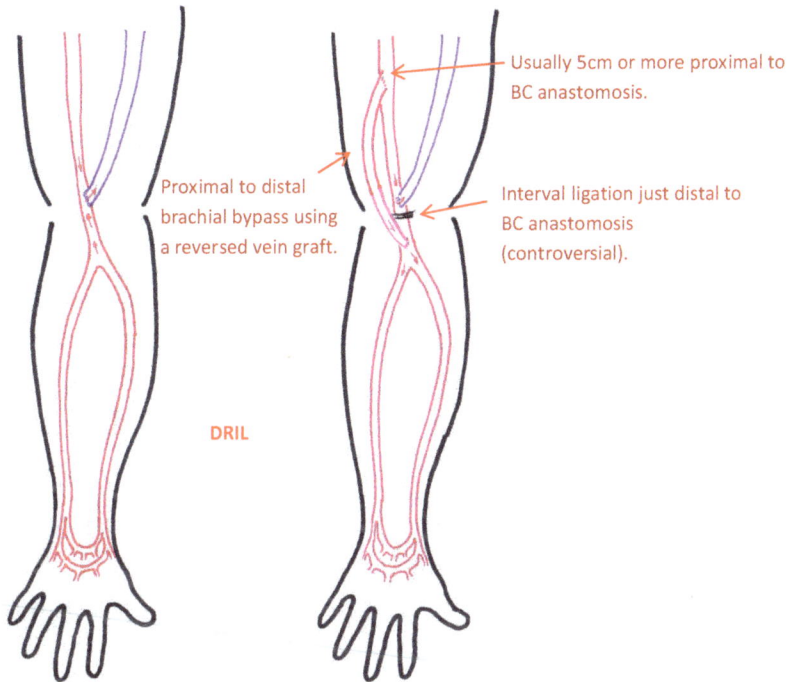

Usually 5cm or more proximal to BC anastomosis.

Proximal to distal brachial bypass using a reversed vein graft.

Interval ligation just distal to BC anastomosis (controversial).

DRIL

Fig. 8.   Schematic diagram of revascularization and interval ligation DRIL.

surgery that potentially leads to substantial morbidities.[20] It is a good option for younger, medically healthier patients with reasonable brachial artery condition. DRIL may not be a favorable option for patients that are medically frail, with small and calcified brachial artery or extensive severe obstruction of hand and forearm arteries.

Extreme conditions

Some patients are medically at high risk, with severe systemic atherosclerosis involving multi-level arteries of the upper limb. Ligation of the vascular access and converting to CVC dialysis would be a safer option.

In other situations, there are medically frail patients on long-term hemodialysis that come back with multiple extensive finger gangrene. As a result, the functional hand cannot be preserved. Some clinicians suggest ligation of the vascular access and perform necessary amputation or debridement of the hand wound. I take another point of view. If there is no functional hand that can be salvaged, it is better to preserve the hemodialysis access for the patient. With caution, I would keep the access and manage the hand wound with necessary dressing or surgical amputation. Ligation of vascular access has to be considered if the severity of ischemia prevent proper wound healing.

A flow-chart of the pre-, intra- and post-operative strategies that access surgeons can apply to avoid and treat DASS is illustrated in Fig. 9.

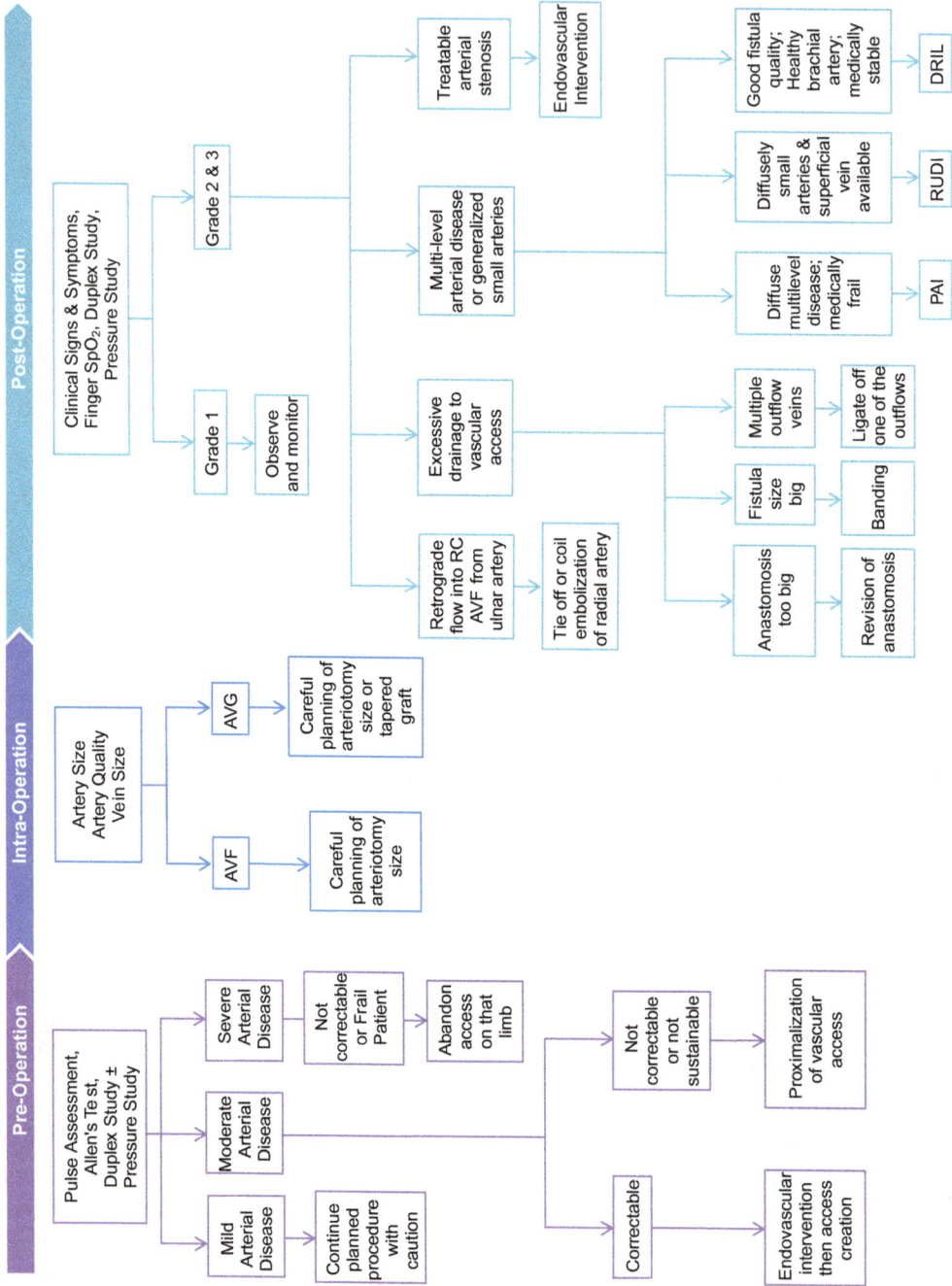

Fig. 9.   Flow-chart illustrating the pre-, intra-, and post-operation measures to prevent and treat DASS.

Case 1

Mr G, 70 years old, DM and DM nephropathy, HT, IHD, peripheral arterial disease. He was already using right IJV CVC for hemodialysis for one month when he attended the first clinic visit for access creation. Left radial and brachial artery pulse was good but ulnar pulse was absent. Allen's test was negative. The left forearm cephalic vein is of good size.

Left RC AVF was planned. Intra-operatively, the left radial artery was heavily calcified but quality of cephalic vein was good. RC AVF was created. Post-operation, left fingers color pinkish and there was no pain or numbness of the hand. Mr G came back for follow-up two weeks later and complained of left middle and ring fingers pain and numbness. Angiogram of the left upper limb was performed via brachial artery antegrade approach (Fig. 10) using a 4 Fr sheath inserted under ultrasound guidance.

Mr G's finger symptom resolved after the ulnar artery angioplasty. He defaulted follow-up and returned to clinic three months later, presenting with left middle finger 2 mm gangrene patch together with left big toe gangrene. Left RC AVF was matured and in-use but dialysis centre reported in-sucking condition over the arterial needle. Left upper and lower limb angiogram was carried out.

The left lower limb angiogram showed anterior and posterior tibial artery obstruction. Balloon angioplasty was performed and in-line flow to the foot was established. Post-intervention, left dorsalis pedis and posterior tibial pulses were palpable.

Left upper limb angiogram via brachial approach is being shown in Fig. 11a. Balloon angioplasty to both radial near AV anastomosis and ulnar artery was performed (Fig. 11b).

Post-second intervention, his left finger lesion healed. Dialysis via the left RC AVF was successful with access flow 800 ml/min. The CVC catheter was removed afterwards. Several sessions of debridement was performed to remove the gangrenous tissue over his left big toe distal part and the wound healed after two months of care. He was being followed up regularly every two months for any recurrence of hand symptom.

> *Endovascular intervention provides a minimally invasive solution to improve arterial blood flow. However, the patency may not be durable especially for smaller sized distal vessels.*

(a)

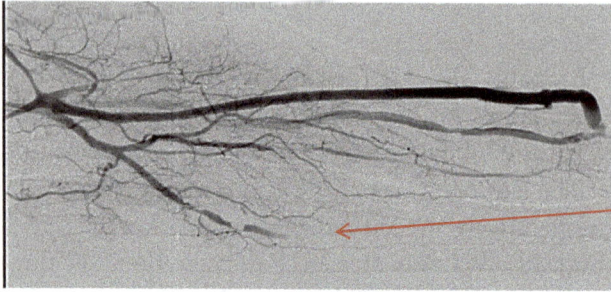

Long segment of high grade stenosis and CTO along left ulnar artery.

(b)

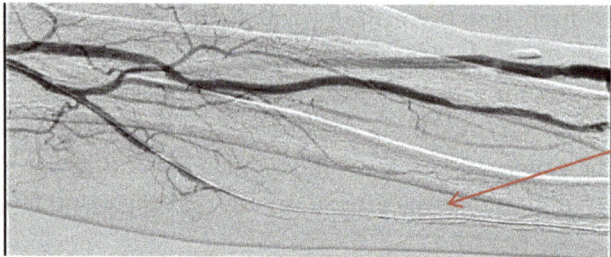

Left ulnar artery CTO crossed by 0.014" Command Wire (Abbott Laboratories. Abbott Park, IL, US).

(c)

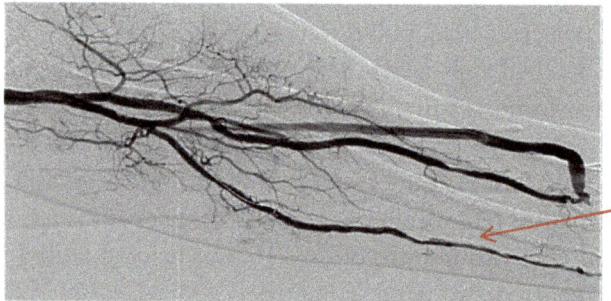

Post Bantam Alpha (2mm/80mm) balloon angioplasty (BARD Peripheral Vascular Inc. Tempe, AZ, US), flow in ulnar artery improved.

(d)

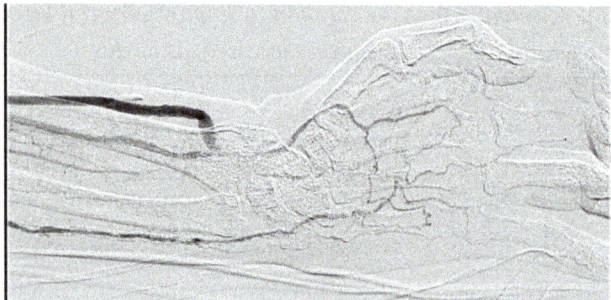

Completion angiogram showed presence of small vessel disease. But the digital vessel perfusion was improved.

Fig. 10. Angiogram of Mr G's left upper limb (a), the ulnar artery angioplasty (b & c) and the arterial outflow after ulnar artery angioplasty.

(a)

Stenosis of left radial artery proximal and distal to the anastomosis

Restenosis of left ulnar artery

(b)

Post-balloon angioplasty of both ulnar and radial artery

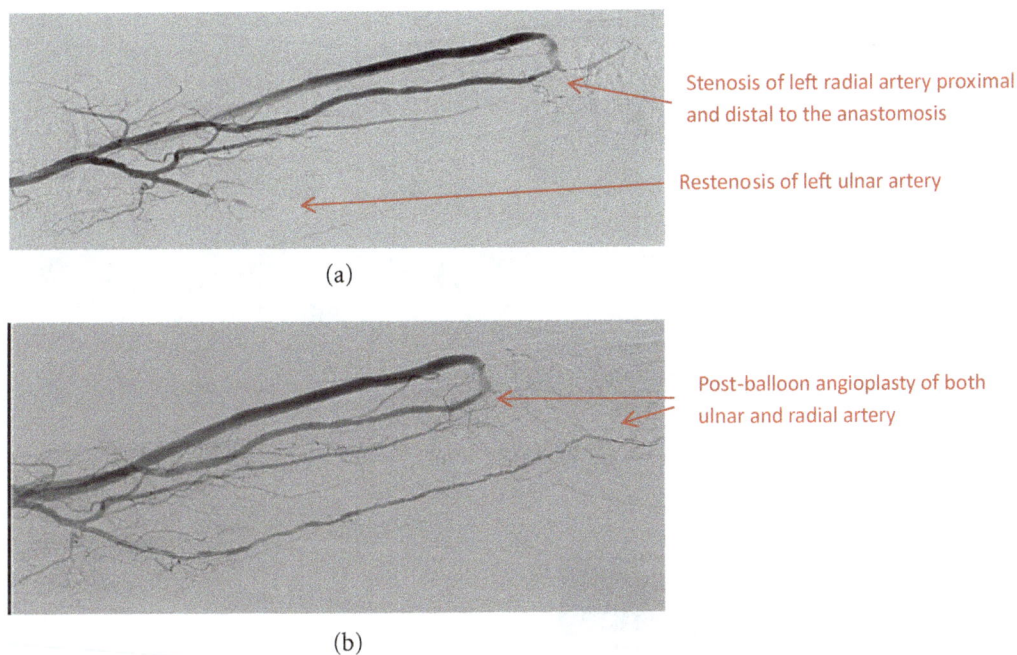

Fig. 11. Angiogram of left upper limb before (a) and after (b) angioplasty.

Case 2

Mr R, 46 years old, DM, HT and DM nephropathy. He started hemodialysis four years ago using right RC AVF. The fistula was well except the access flow was dropping over the last six months. Fistulogram was performed as shown in Fig. 12.

Angioplasty of the AV anastomosis and cephalic vein fistula was performed but the improvement was mild. He was scheduled for revision of right upper limb AVF four weeks later but readmitted few days before the scheduled date with a thrombosed distal RC AVF. Side-to-side anastomosis was made between median antebrachial vein and distal brachial artery so as to preserve part of the arterialized forearm fistula. Post-operation, Mr R continued his hemodialysis using the AVF with "A" needle over forearm fistual and "V" needle over the arm cephalic vein (Fig. 13). His right hand was well and he was discharged two days after the surgery.

Three months later, Mr R returned to clinic and complain of right index finger pain on dialysis and a small gangrene patch over the tip of his index finger (Fig. 16a). Angiogram was performed using right femoral retrograde puncture. Arch aortogram and selective angiogram showed no proximal arterial lesion.

Angiogram over the elbow and forearm region showed preferential contrast flow into the fistula (Fig. 14). Both radial and ulnar arteries were not well opacificed. The angiogram was then repeated with manual compression over the AV anastomosis (Fig. 15a). The radial artery was noted to be thrombosed. The ulnar artery was patent without significant stenosis but the flow was slow.

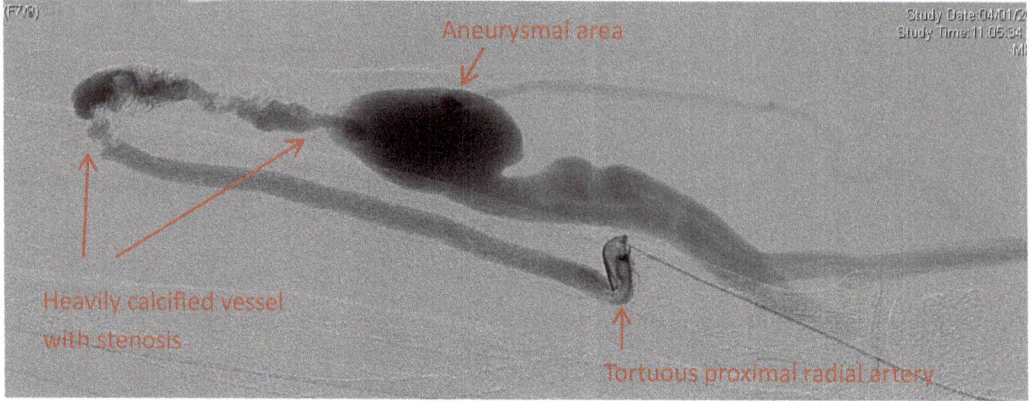

Aneurysmal area

Heavily calcified vessel with stenosis

Tortuous proximal radial artery

Fig. 12.   Fistulogram of right RC AVF.

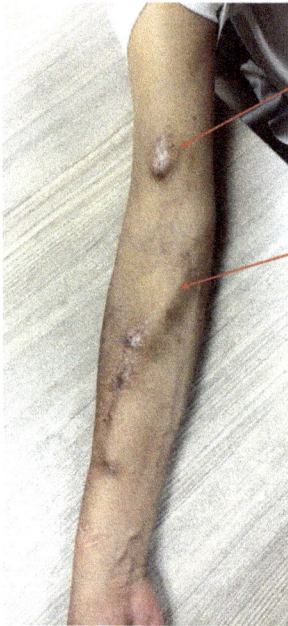

Dilated arm cephalic vein fistula for "V" cannulation

Side to side anastomosis between antebrachial vein and distal brachial artery. Proximal forearm dilated cephalic and antebrachial vein preserved for "A" cannulation after operation.

Fig. 13.   Clinical photo of Mr R's right upper limb fistula 6 weeks after access salvage operation.

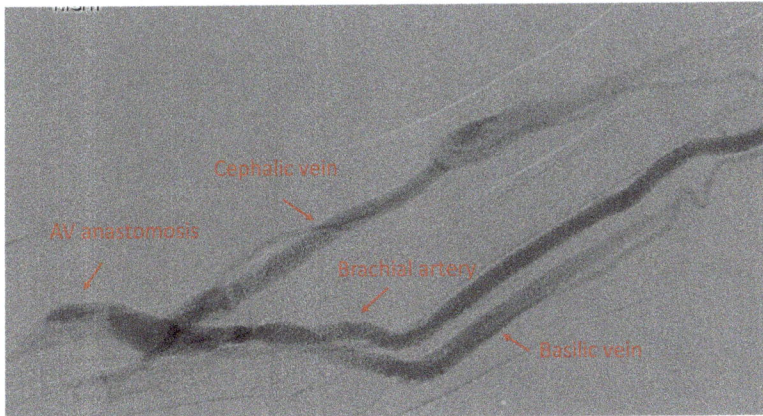

Fig. 14. Fistulogram of the left upper limb showing preferential flow of contrast into both cephalic and basilic vein of the arm. There was minimal flow through radial and ulnar arteries.

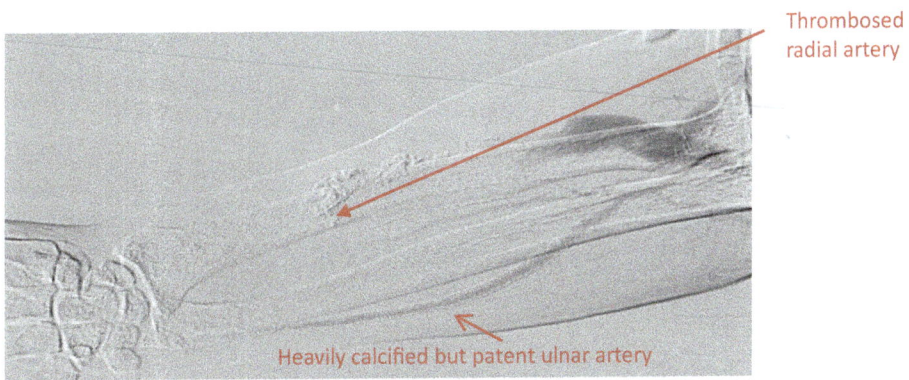

(a)

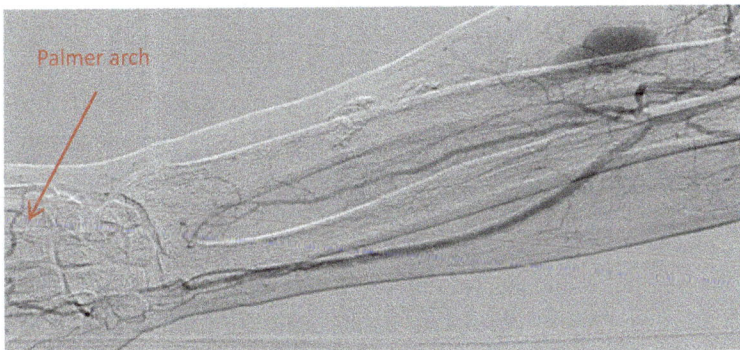

(b)

Fig. 15. Angiogram with antebrachial vein to distal brachial artery anastomosis compressed (a) and after ligation of the basilic vein (b).

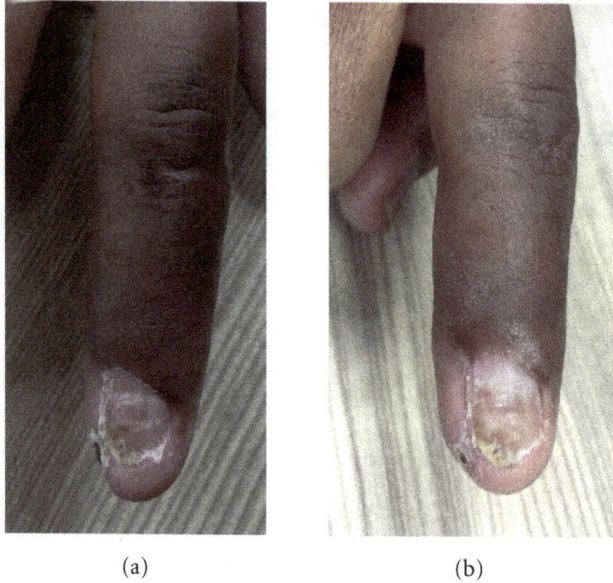

(a)   (b)

Fig. 16.   Right index finger gangrene patch before (a) and four weeks after (b) ligation of basilic vein.

The connection between the median antebrachial vein and basilic vein was divided surgically using a small incision. Angiogram was repeated. This time, right hand palmer arch was displayed better on angiogram (Fig. 15b). Mr R's finger pain subsided after the procedure. Pulse oximeter applied on the right index finger yield 100% saturation. His finger gangrene patch shrunk in size and healed in two months (Fig. 16b).

> *Cut off one of the outflow veins to help redistribute blood flow to the hand.*

Case 3

Mr N, 62 years old, DM and DM nephropathy, HT, IHD. He used his left RC AVF for hemodialysis for three years with recent balloon angioplasty of the AV anastomosis a month ago. He was admitted to hospital because the hemodialysis session cannot be completed due to clogging of the dialysis line. Clinically, the thrill over his RC AVF was weak. Hardening of a segment of cephalic fistula was felt near the anastomosis. The left forearm cephalic vein drained through the median antebrachial vein to the basilic vein. The size of basilic vein was about 5 mm over the left arm on ultrasound study. Cephalic vein in the left arm was not visible.

To provide an early functional fistula and to avoid the need of CVC, left distal brachial artery to basilic vein AVF was created in a side-to-side manner together with transposition of the arterialized basilic vein. The brachial artery was diffusely, heavily calcified. A slightly bigger than usual arteriotomy was made. After the operation, Mr N had good flow in both the forearm cephalic as well as arm transposed basilic vein. Moderate swelling of the left forearm was noted. Mr N started to experience left hand numbness on day 2 post-operation. The ulnar pulse was palpable but weak. Radial pulse was not palpable. Left upper limb angiogram was performed using right femoral retrograde approach. The angiogram showed preferential flow of contrast into grossly dilated basilic vein and fore-arm fistula. The radial and ulnar arteries were poorly opacified (Fig. 17). With the basilic vein to brachial artery anastomosis compressed, angiogram of the radial and ulnar arter-ies showed no significant stenosis (Fig. 18).

Mr N was taken back to the operating theatre. Two procedures were done to reduce the fistula flow: (1) ligation of the stenosed RC AVF; (2) re-open the cubital fossa wound to band down the size of the basilic vein proximal and distal to the anastomosis (to ~4 mm diameter) (Fig. 19).

A CVC was inserted via right IJV for hemodialysis while waiting for the forearm swelling to subside and transposed basilic vein to mature. His numbness resolved after the banding surgery. The CVC catheter was removed three weeks later. He used proximal forearm fistula for "A" needling and transposed basilic vein for "V" needling. The fistula lasted till Mr N's demise due to acute myocardial infarction three years later.

Secondary fistula on the same upper limb with a failing original AVF will increase the risk of steal syndrome. Clinicians have to balance the benefit of early functional fistula and DASS risk. A combination of procedure can help restroe the balance of blood flow to both the hand and the fistula.

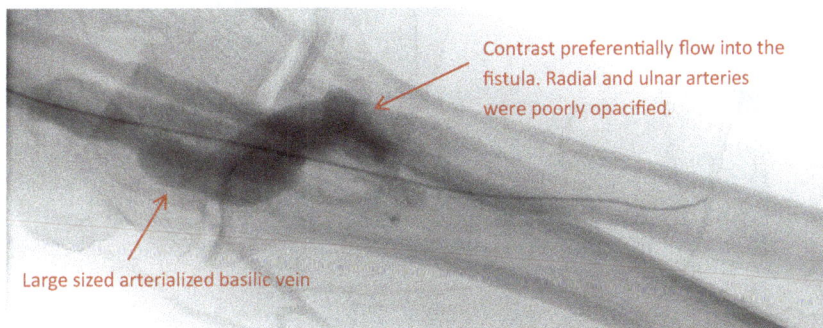

Contrast preferentially flow into the fistula. Radial and ulnar arteries were poorly opacified.

Large sized arterialized basilic vein

Fig. 17. Contrast injection from catheter parked in axillary artery showed preferential flow of contrast into the basilic vein.

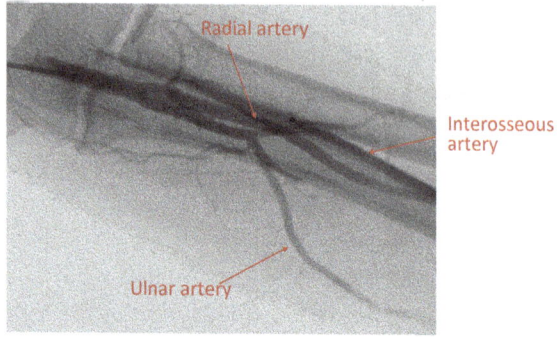

(a)

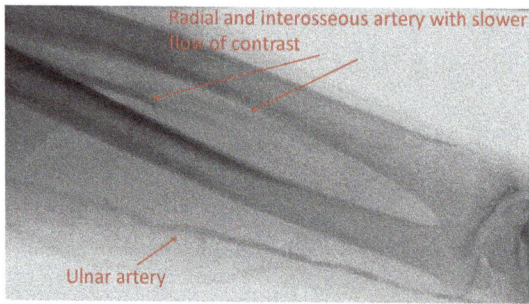

(b)

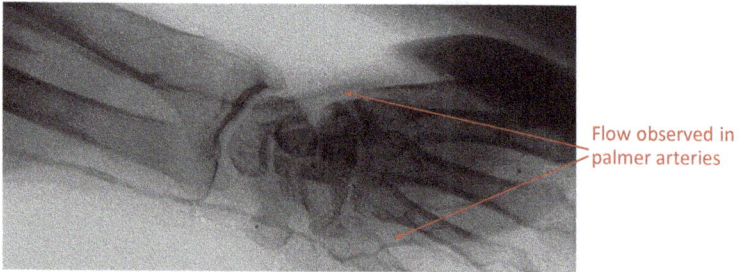

(c)

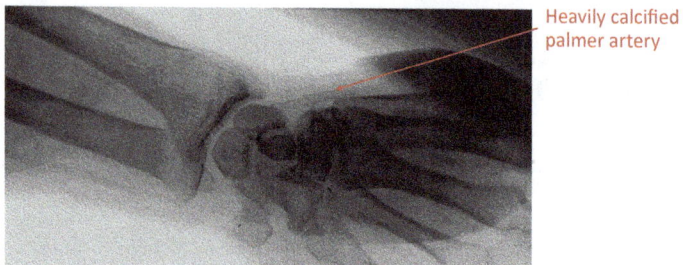

(d)

Fig. 18.   Angiogram of left forearm (a, b) and hand (c) with a long (110 cm) straight flush catheter injecting contrast in the distal brachial artery together with compression over AV anastomosis. Fluoroscopic image of the left hand showed heavily calcified palmar arch (d).

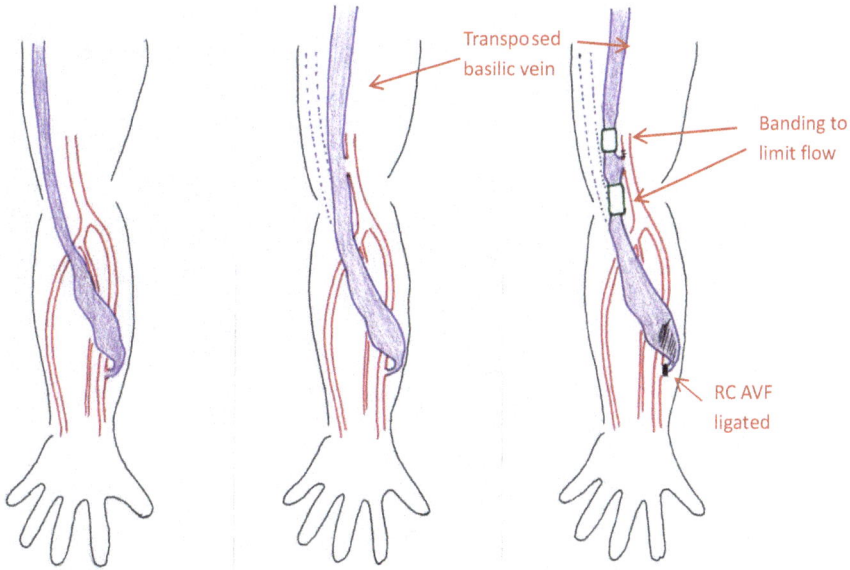

Fig. 19. Schematic diagram of the second operation for Mr N to reverse DASS.

Case 4

Madam A, 62 years old, DM and DM nephropathy, HT, IHD presented to the vascular access clinic in October 2009. She had already started hemodialysis using a right IJV tunnelled CVC for three months. She is right handed. Her body height was 150 cm with a body weight of 41 kg. Bilateral upper limb vein mapping was available as shown in Fig. 20.

Clinically, there was no visible cephalic vein over both upper limbs. Both the radial and ulnar pulses were palpable over the left wrist. A detailed discussion was made with Madam A about the options of two stages BB AVF/BBT, and BB AVG. Eventually, BB AVG was decided to minimize CVC duration.

The operation was performed in December 2009. Upon surgical dissection, a brachial artery of ~2 mm was noted. Further assessment confirmed Madam A has early bifurcation of brachial artery. Over the cubital fossa, there were both radial and ulnar arteries. The ulnar artery was slightly bigger (2 mm) than the radial artery and was taken as the inflow. The basilic vein was 2.8 mm. A tapered ePTFE 4–6 mm Gore-tex graft (W.L. Gore, Flagstaff, AZ, US) was used for forearm loop AVG creation. Post-operation, the AVG flow was good but patient complained of left hand numbness. Her symptom got worse with development of weak hand grip. A bypass between the ulnar artery above and below the elbow without interval ligation was performed using reversed GSV (2.5 mm diameter). Post-operation, the numbness and weakness of her left hand gradually resolved.

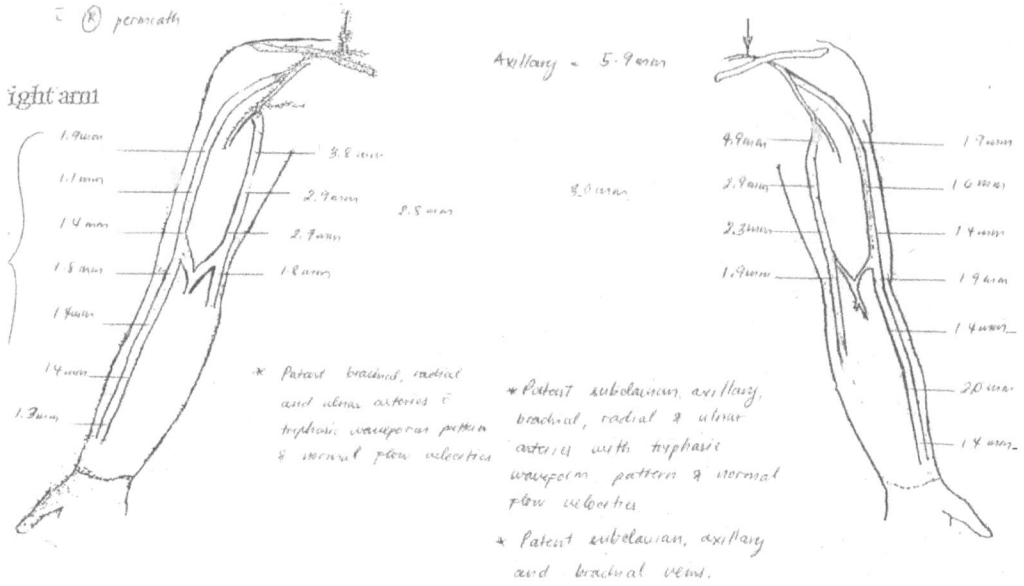

Fig. 20.   Pre-operative duplex assessment of bilateral upper limb arterial and venous system.

She was readmitted in April and November 2010 for blocked left forearm loop AVG. Angiojet (Boston Scientific Co. MA, US) thrombolysis was performed to restore flow. Stenosis was found over the arterial anastomosis, venous anastomosis, in-graft and mid-arm basilic vein. Angioplasty of all the lesions and stenting for mid-arm basilic vein stenosis were performed in April. Eventually, the forearm AVG was not salvageable in November 2010 even after thrombolysis and angioplasty.

In December 2010, another AVG was created for her. This time, ultrasound was performed in the theatre to locate the brachial bifurcation. An incision over upper arm was made over the site marked to dissect out proximal radial and ulnar artery. The proximal ulnar artery diameter was about 3.5 mm, whereas radial about 2.5 mm. Intra-operatively, sterile Doppler probe was prepared to assess the arterial flow in the hand and wrist region during the operation. Then the proximal basilic vein was also dissected out (~4 mm). The snuff box radial artery and wrist ulnar artery flow were checked with the proximal ulnar artery temporarily clamped. Only a mild signal reduction of the distal ulnar artery was noted and the snuff box radial artery signal was maintained. Proximal ulnar artery to basilic vein loop (arm) AVG using tapered ePTFE 4–7 mm Gore-tex graft (W.L. Gore, Flagstaff, AZ, US) was created (Fig. 21). Madam A did not experience any steal syndrome after this operation.

Graft thrombosis occurred in March 2012 and resolved by thrombectomy and balloon angioplasty of the in-graft and vein-graft anastomosis stenosis. However, thrombosis recurred six weeks later. A new ring supported ePTFE graft Distoflo 6 mm (Impra; Bard Inc, Tempe, AZ, US) was used as jump graft to connect between the arterial limb of the graft to the axillary vein in April 12 (Fig. 22).

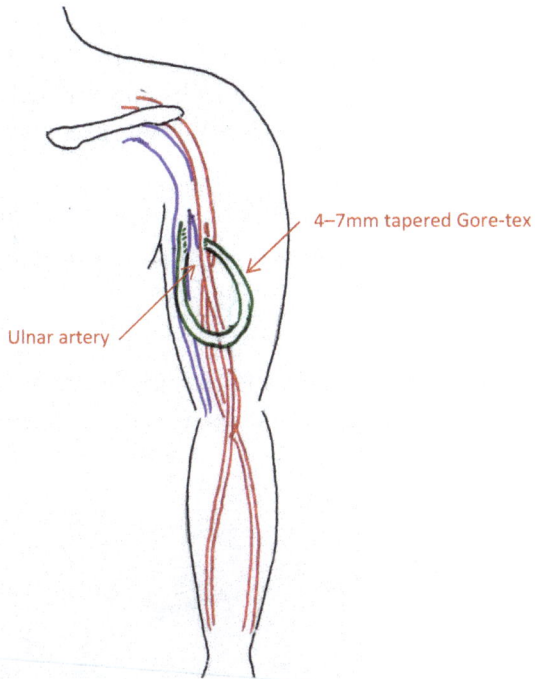

Fig. 21. Schematic diagram of left arm proximal ulnar artery to basilic vein loop AVG.

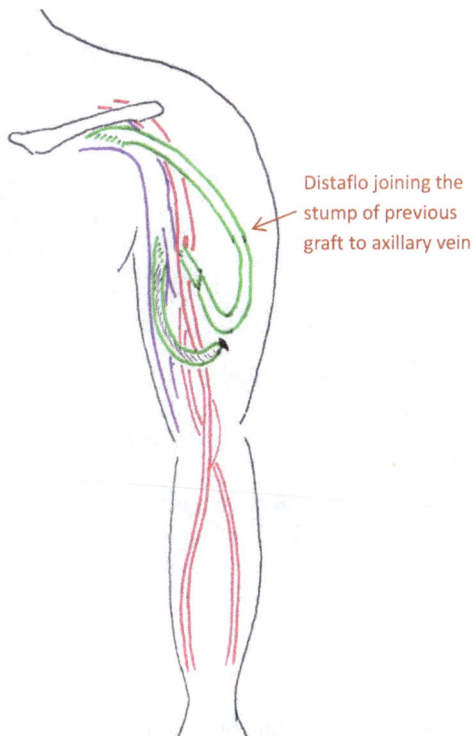

Fig. 22. Schematic diagram of jump graft converting the ulnar-basilic AVG to ulnar artery axillary vein AVG.

This graft was well until June 2013 when graft thrombosis happened again. Urokinase thrombolysis and aspiration of thrombus was performed followed by angioplasty of the in-graft and vein-graft anastomosis stenosis. Difficulty encountered on hemodialysis again in September 2013 and a fistulogram was performed (Fig. 23).

> *Suspect early bifurcation of brachial artery when an exceptionally small brachial artery is found over elbow region. Pre-operative duplex study may also miss this situation.*
> *Taking proximal artery as inflow lowers the risk of DASS.*

Two sheaths were inserted antegradely and retrogradely to tackle the arterial inflow, in-graft and vein-graft anastomosis. The AVG was successfully salvaged (Fig. 24).

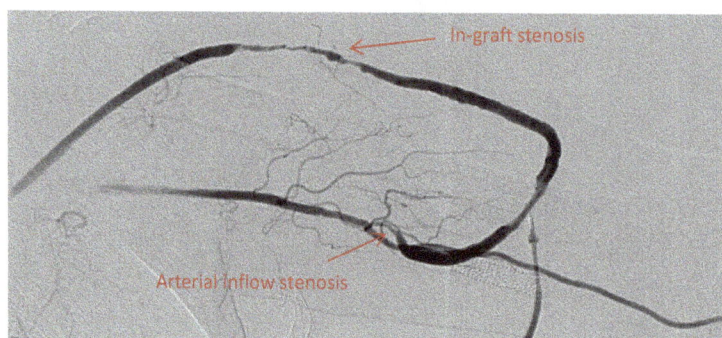

(a)

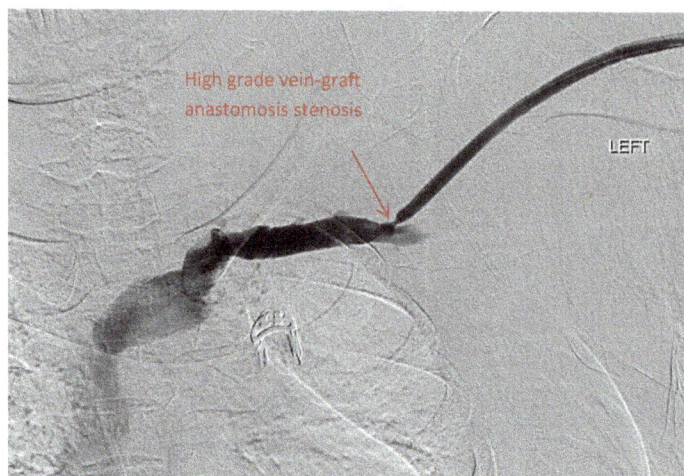

(b)

Fig. 23.   Fistulogram showing multiple sites of stenosis of patient's AVG.

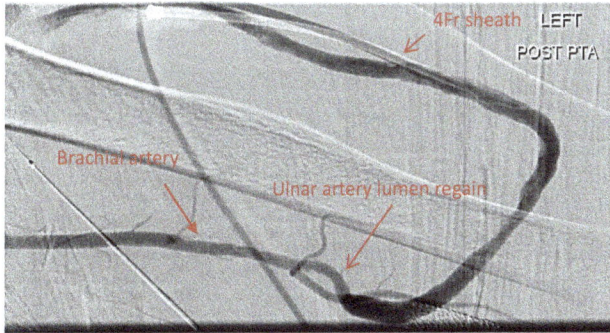

(a)

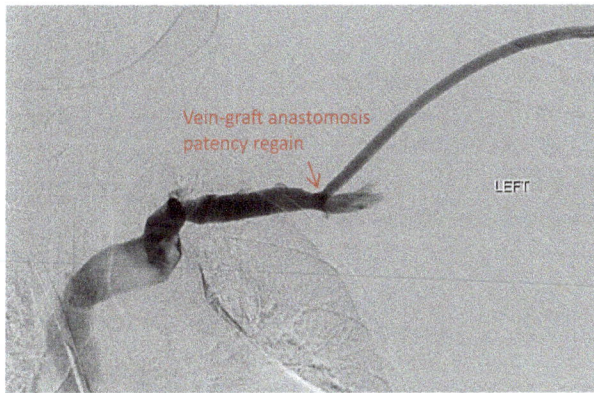

(b)

Fig. 24. Completion angiogram of Madam A after balloon angioplasty.

The graft blocked again in December 2013. At that time, she also suffered chest pain and dyspnoea. Coronary evaluation showed acute coronary event. Graft thrombolysis or thrombectomy procedure was therefore withheld. A tunnelled CVC was inserted through her left IJV for hemodialysis. She was converted to peritoneal dialysis afterwards.

References

1. DeCaprio JD, Valentine RJ, Kakish HB, *et al.* Steal syndrome complicating hemodialysis access. *Cardiovasc Surg.* 1997; **5**: 648–653.
2. Sidawy AN, Gray R, Besarab A, *et al.* Recommended standards for reports dealing with arterio-venous hemodialysis access. *J Vasc Surg.* 2002; **35**: 603–610.
3. Lazarides MK, Staramos DN, Kopadis G, *et al.* Onset of arterial 'steal' following proximal angioaccess: immediate and delayed types. *Nephrol Dial Transplant.* 2003; **18**: 2387–2390.

4. Rocha A, Silva F, Queirós J, *et al.* Predictors of steal syndrome in hemodialysis patients. *Hemodial Int.* 2012; **16**(4): 539–544.

5. Davidson D, Louridas G, Guzman R, *et al.* Steal syndrome complicating upper extremity hemoaccess procedures: incidence and risk factors. *Can J Surg.* 2003; **46**: 408–412.

6. Report on International ESRD trend. *US renal data system* 2011.

7. Miller GA, Khariton K, Kardos SV, *et al.* Flow interruption of the distal radial artery: treatment for finger ischemia in a matured radiocephalic AVF. *J Vasc Access.* 2008; **9**: 58–63.

8. Papasavas PK, Reifsnyder T, Birdas TJ, *et al.* Prediction of arteriovenous access steal syndrome utilizing digital pressure measurements. *Vasc Endovascular Surg.* 2003; **37**: 179–184.

9. Valentine RJ, Bouch CW, Scott DJ, *et al.* Do preoperative finger pressures predict early arterial steal in hemodialysis access patients? A prospective analysis. *J Vasc Surg.* 2002; **36**: 351–356.

10. Jennings W, Brown R, Blebea J, *et al.* Prevention of vascular access hand ischemia using the axillary artery as inflow. *J Vasc Surg.* 2013; **58**: 1305–1309.

11. Miller GA, Khariton K, Kardos SV, *et al.* Flow interruption of the distal radial artery: treatment for finger ischemia in a matured radiocephalic AVF. *J Vasc Access.* 2008; **9**(1): 58–63.

12. Zanow J, Petzold K, Petzold M, *et al.* Flow reduction in high-flow arteriovenous access using intraoperative flow monitoring. *J Vasc Surg.* 2006; **44**: 1273–1278.

13. Miller GA, Goel N, Friedman A, *et al.* The MILLER banding procedure is an effective method for treating dialysis-associated steal syndrome. *Kidney Int.* 2010; **77**(4): 359–366.

14. Scheltinga MR, van HF, Bruyninckx CM. Surgical banding for refractory hemodialysis access-induced distal ischemia (HAIDI). *J Vasc Access.* 2009; **10**: 43–49.

15. Minion DJ, Moore E, Endean E. Revision using distal inflow: a novel approach to dialysis-associated steal syndrome. *Ann Vasc Surg.* 2007; **19**: 625–628.

16. Thermann F, Wollert U. Proximalization of the arterial inflow: new treatment of choice in patients with advanced dialysis shunt-associated steal syndrome? *Ann Vasc Surg.* 2009; **23**: 485–490.

17. Schanzer H, Schwartz M, Harrington E, *et al.* Treatment of ischemia due to "steal" by arteriovenous fistula with distal artery ligation and revascularization. *J Vasc Surg.* 1988; **7**: 770–773.

18. Walz P, Ladowski JS, Hines A. Distal revascularization and interval ligation (DRIL) procedure for the treatment of ischemic steal syndrome after arm arteriovenous fistula. *Ann Vasc Surg.* 2007; **21**(4): 468–473.

19. Huber TS. Treatment strategies for access-related hand ischemia. *Semin Vasc Surg.* 2011; **24**: 128–136.

20. Scali ST, Chang CK, MD, Raghinaru D, *et al.* Prediction of graft patency and mortality after distal revascularization and interval ligation for hemodialysis access-related hand ischemia. *J Vasc Surg.* 2013; **57**: 451–458.

CHAPTER 13

Management of Aneurysm, Pseudoaneurysm and Infective Complications of Hemodialysis Access

Jackie P. Ho

Vascular Access Aneurysm and Pseudoaneurysm

The definition of aneurysm and pseudoaneurysm in native vein fistula is rather vague. Aneurysmal change of AVF may occur spontaneously due to increased pressure and flow (Fig. 1). More commonly, area of the AVF with repeated cannulation may have general structural weakness of the vessel wall and develop dilatation (Fig. 1). Strictly speaking, this is pseudoaneurysm but clinically it behaves more like an aneurysm. Its expansion is relatively slow and risk of rupture is low unless it continues to be repeatedly cannulated and develops skin erosion. It can be described as aneurysmal degeneration of AVF. There is another situation where a definitive defect of the fistula wall is present and a pseudoaneurysm is identified over the defect (Fig. 2). The defect is induced by either traumatic needling or intervention procedure. Very often, it is associated with underlying fistula stenosis or deeply located fistula which makes cannulation particularly difficult. The tendency for this kind of pseudoaneurysm to expand and rupture is high without proper treatment.

Most of the AVF true aneurysm do not rupture or thrombose and, as such, no intervention is needed. In some situations, the presence of aneurysm may impose a challenge to fistula cannulation. Clinicians need to communicate with the dialysis nurses to determine suitable sites for cannulation.

Even though aneurysmal degeneration of AVFs rarely rupture, with continuous repeated needling around the same region and expanding aneurysm size, thinning of the overlying skin or even ulcer (Fig. 3) may occur. Caution should be exercised to avoid cannulation to poor skin quality areas. Failure to do so, infection, torrential bleeding and rupture may develop subsequently. Sometimes, there could be associated stenosis of AVF near the aneurysmal change region and thrombus accumulation inside the aneurysm sac. The aneurysmal degenerated region could be a leading point of thrombosis in some matured AVF after prolong use. Salvaging thrombosed AVF with multiple aneurysmal

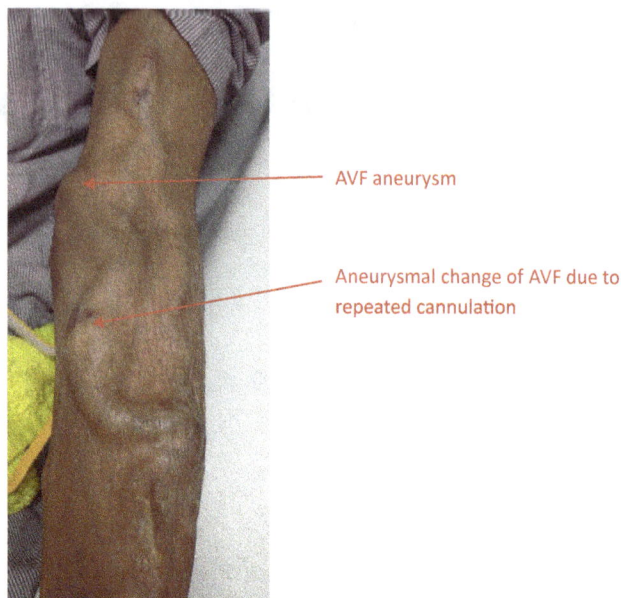

Fig. 1.   True aneurysm of AVF and aneurysmal degeneration of AVF due to repeated cannulation.

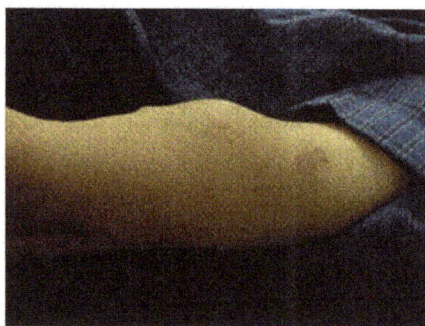

Fig. 2.   Pseudoaneurysm of left BC AVF due to traumatic cannulation.

degeneration is particularly challenging as the thrombus load is high and the content is a mix of fresh and chronic adherent clot.

Symptomatic aneurysmal AVF are mostly treated by open surgical method and, rarely, endovascular method. Conventional surgical treatment involves resection of the aneurysmal part and restoring the flow with an interposition graft. Woo *et al.*[1] reported their experience of surgical repair and revision of 19 symptomatic aneurysmal AVF size 4–7 cm without interposition graft. The criteria for symptomatic AVF aneurysm include thrombosis, skin breakdown, infection, bleeding, and/or poor flow. The surgical repair involves resection of redundant length, reduction of diameter and reconstruction of the fistula. All but one AVF maintained patency and the mean primary patency after repair was 14 months.

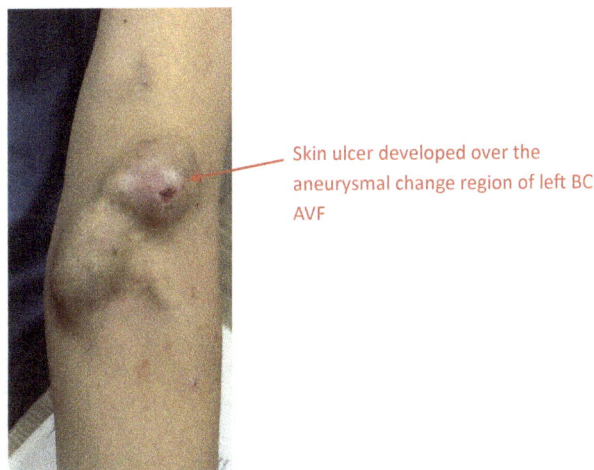

Skin ulcer developed over the aneurysmal change region of left BC AVF

Fig. 3. Ulceration of skin overlying an aneurysmal change region of left BC AVF due to repeated cannulation.

Hossny[2] also reported his series of partial aneurysmectomy for 29 symptomatic AVF aneurysm in 14 patients. The AVF salvage rate was 100% but 9 patients required temporary central venous catheter to tie over a short down time. Peden[3] suggested to perform aneurysmorrhaphy in stages for patients with more than one large aneuerysm so as to preserve an adequate length of access for cannulation and avoid the need of central venous catheter. A detailed duplex study or diagnostic fistulogram is recommended before or in the same surgical session of the open repair to rule out underlying stenotic lesion. There are also small case series[4] reporting the use of self expandable covered stent (stent-graft or endograft) for treatment of aneurysmal degeneration of AVF. The technical challenges of using covered stent in AVF aneurysm are mainly due to the variable, generally bigger size of venous fistula, potential of further dilatation of landing zone and size discrepancy between the two landing zones.

Pseudoaneurysm of AVF with a definite wall defect of the fistula has a much higher risk of expansion, bleeding and rupture. It usually develops with difficulty in cannulation. Definitive treatment is required in nearly all cases. Similar to aneurysmal degeneration of AVF, open surgical repair is the main state of treatment. Imaging to detect any underlying stenosis is a must. Treatment should be directed to the pseudoaneurysm as well as the stenosis if the access is deemed salvageable. AVF pseudoaneurysm also frequently associates with inflammation and maybe infection. If suspicious, antibiotic treatment, resting of fistula and close monitoring are adopted before any surgical procedure. If the size of vessel wall defect is small and condition of surrounding fistula is healthy, surgical exploration and simple plication of the wall defect can be performed under tourniquet or digital pressure on both inflow and outflow. If the defect is large but surrounding vessel wall is healthy, patch repair of fistula using a vein or allograft peri-cardial patch could be considered. The incision made should be perpendicular to the fistula path to avoid occupying the future potential cannulation sites over the fistula. In situations where the pseudoaneurysm is infected with abscess development, ligation of the fistula together with incision and drainage of the abscess is required. Whether an interposition graft or a secondary fistula can be created in the same session depends on the severity and extent of sepsis, quality of skin over the surrounding area and the availability of suitable vessels. The following case illustrates an alternative endovascular method to treat a non-infective pseudoaneurysm of an AVF without the implantation of covered stent.

Case 1

Mr J has DM, ESRF with hemodialysis using his newly matured left BC AVF. Left arm swelling and difficulty of cannulation was reported by the dialysis nurse. Clinical and ultrasound examination revealed a pseudoaneurysm of the fistula over his mid-arm. There was no feature of infection. Mr J was admitted for hemodialysis through a temporary femoral vein catheter. Fistulogram was performed via an antegrade brachial artery (4 Fr sheath) access. The fistulogram showed a dumbbell shape pseudoaneurysm (Fig. 4) arising from the cephalic vein fistula with narrowed pseudoaneurysm neck and a stenosis was present over the outflow vein. A 0.018" guidewire was used to cross the stenosis. Ultrasound assessment of the pseudoaneurysm sac was performed at the same time (Fig. 5). An angioplasty balloon of the diameter of the venous fistula was placed (Cook LP18 7 mm/40 Cook Medical Inc. Bloomington, IN, USA) next to the pseudoaneurysm neck and inflated with low pressure (4 atm). Immediately after inflation of the angioplasty balloon, thrombin was injected into the pseudoaneurysm sac under ultrasound guidance (Fig. 6). Compression over the sac was maintained for 2 mins together with low pressure inflation of the angioplasty balloon. The stenotic segment was then treated with balloon angioplasty as usual.

Post-thrombin injection, fistulogram showed occlusion of the pseudoaneurysm sac (Fig. 7a) and patent cephalic vein fistula. Duplex showed no flow into the pseudoaneurysm sac (Fig. 7b). Swelling on his left arm gradually subsided after the procedure. Hemodialysis was conducted through the temporary femoral vein catheter for 1 week and then shifted back to the left arm AVF.

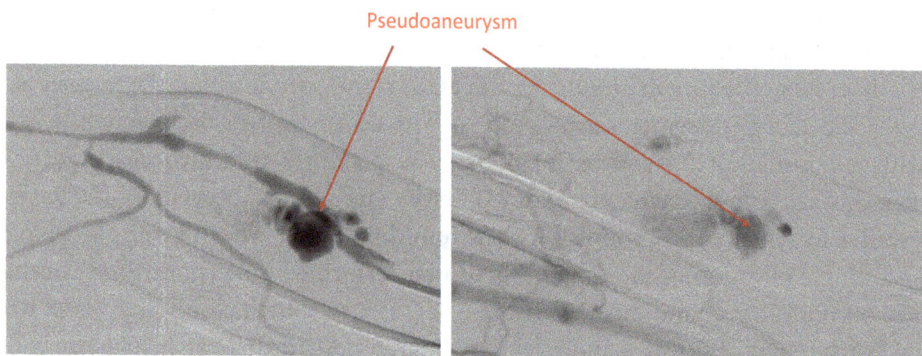

Fig. 4.   Fistulogram showing the dumbbell shape pseudoaneurysm of left BC AVF.

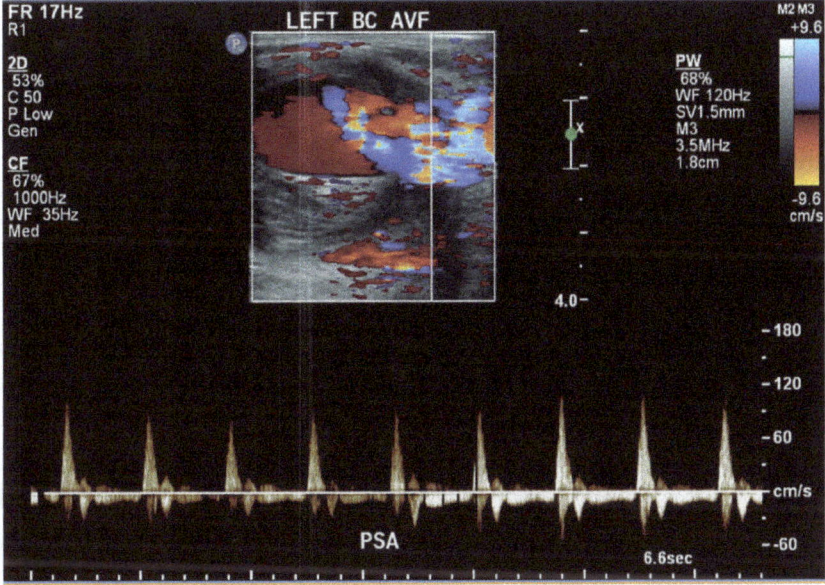

Fig. 5. Duplex ultrasound showing the pseudoaneurysm sac located superficial to the venous fistula with flow present.

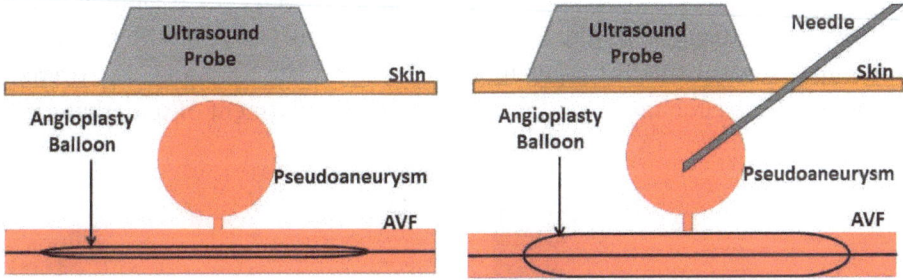

Fig. 6. Schematic diagram showing the inflation of angioplasty balloon inside the venous fistula to reduce flow and ultrasound guided injection of thrombin into the sac.

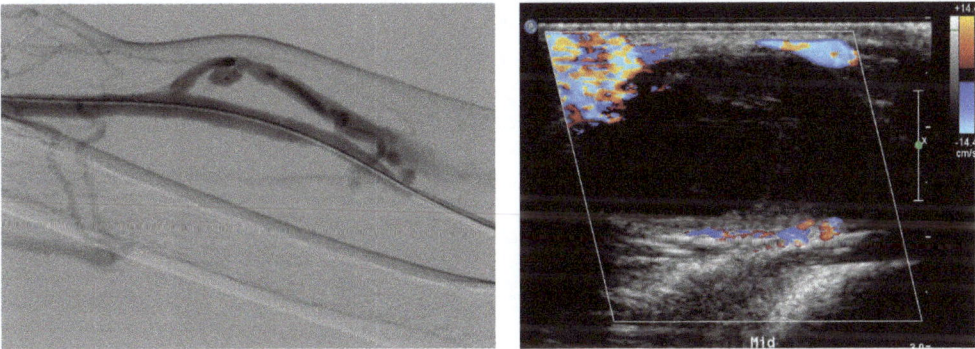

(a)   (b)

Fig. 7. Post-treatment angiogram showed patent fistula with no contrast flow into the pseudoaneurysm (a). Duplex assessment showed no flow inside the pseudoaneurysm sac (b).

For AVG, as the graft itself does not develop aneurysm, all abnormal dilatation of AVG are pseudoaneurysm. Nonetheless, the condition of pseudoaneurysm could be due to repeated cluster cannulation resulting in generalized weakening of the graft wall over a particular region, aneurysmal change of AVG (Fig. 8), or a particular identifiable significant defect of the graft wall with pseudoaneurysm overlying the defect (Fig. 9).

Aneurysmal change of AVG without feature of skin ulceration or infection can be managed with avoidance of further cluster cannulation. Nonetheless, it is not easy to change people's practice. The initiation from both patient and dialysis nurses are key to a successful change. Education and detailed explanation are needed to enhance their initiation. Without infection, aneurysmal AVG with skin ulceration can be managed by elective resection of the aneurysmal segment together with interposition graft to preserve the access as well as to avoid future infective complication.

For non-infective AVG pseudoaneurysm with a definitive wall defect, treatment to repair the wall defect is required to prevent further sac expansion and bleeding. The AVG wall defect can be repaired using open approach[5] or endovascular-covered stent.[3] Similar to AVF pseudoaneurysm repair, the incision made should be perpendicular to the graft pathway to minimize disturbance of future potential cannulation sites. The method of wall defect repair depends on the size of the defect and the quality of surrounding graft. Endovascular repair using covered stent[6-8] (or stent-graft) (Fig. 10) is gaining popularity. Most of the covered stents used were self-expandable, with PTFE or ePTFE as the covering. They are either straight-tube or taper in configuration. The advantage of covered stent treatment for AVG pseudoaneurysm include concurrent treatment of underlying vascular

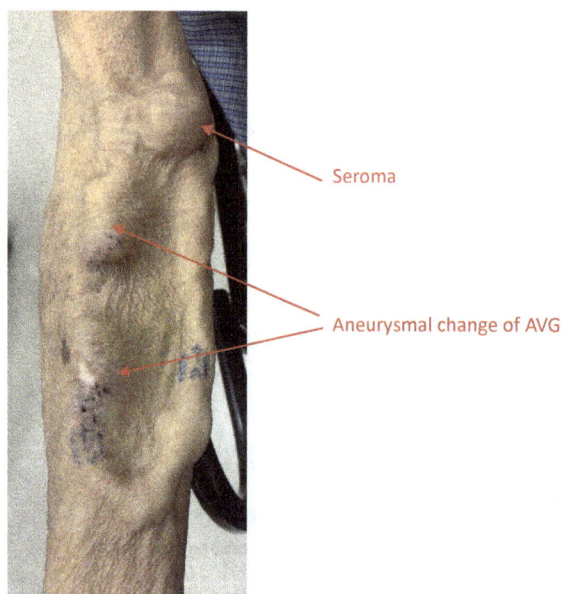

Seroma

Aneurysmal change of AVG

Fig. 8.   Aneurysmal change of BB AVG over venous limb of the graft.

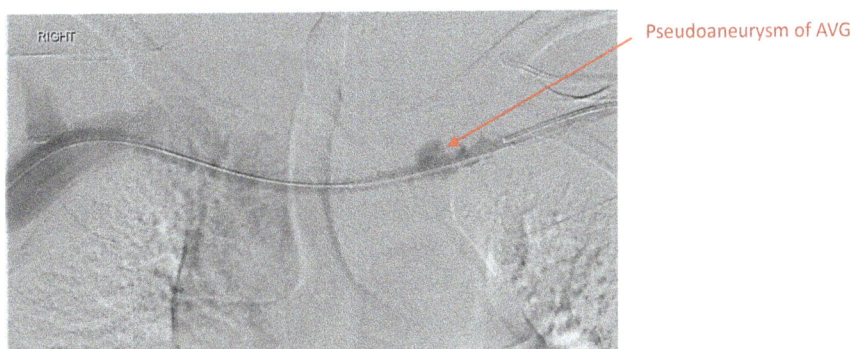

Fig. 9. Angiographic image of a pseudoaneurysm present over a cross axillary artery (left) to axillary vein (right) AVG.

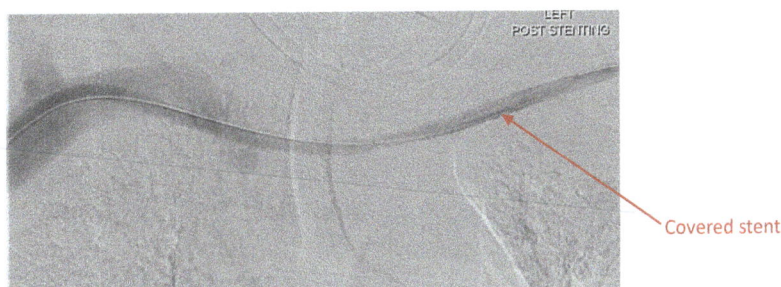

Fig. 10. Same patient with cross axillary artery to vein AVG pseudoaneurysm post-covered stent (Viabahn, W. L. Gore & Associates, Inc. Flagstaff, AZ, US) implantation.

access stenosis, minimize bleeding, and reduced infective risk associated with surgical incision. The primary patency of AVG after open and covered stent repair is comparable (highly variable, averaged 53–57% by 12 months).[3,5,9] Nonetheless, there is no consensus whether covered stents deployed inside the AVGs are safe to cannulation,[6,10] or when they will be safe to cannulate. It is generally believed to avoid cannulating the covered stent if there is plenty AVG length good for cannulation. If not, consider starting cannulation after 3 weeks of deployment. Other complications reported with covered stents include infection, migration, stent rupture, erosion, and bleeding.[7,9,11] The presentations could be access thrombosis, recurrence of pseudoaneurysm or bleeding. The risk of infective complication of covered stent was noted to be higher when skin erosion is present over the pseudoaneurysm.[7] Close monitoring of patients after covered stent treatment for pseudoaneurysm is recommended.

Infective Complications of AVF and AVG

Infection of vascular access is a common indication for hospital admission of hemodialysis patients. Various studies confirmed the infection rate of AVF is lower than AVG, which is further lower than tunnelled CVC and temporary CVC.[12] Taylor *et al.*[13] reported a 10-fold difference of HD-related bacteremia between AVF and AVG (AVF 0.2 episodes/1000 dialysis procedures versus AVG 2.5/1000 dialysis procedures). Fysaraki *et al.*[12] reviewed the blood stream infection episodes of hemodialysis patients over 6 years and reported the infection incidence for AVF, AVG, tunnelled CVC and temporary CVC as 0.18, 0.39, 1.03, 3.18 per 1000 patient-days respectively. Among AVGs, femoral AVG has higher infection risk than upper limb AVGs.[14,15] Most common pathogens are gram positive cocci (*Staphylococcus aureus*, *Staphylococcus epidermidis*, coagulase negative *Staphylococci*, and *Enterococci*) followed by gram negative rods (*Escherichia coli*).[12,16] Some are polymicrobial and sometimes no pathogen can be identified.

Infective complications of vascular accesses could be due to surgical wound infection, seedling from infected CVC, or infection introduced on repeated cannulation, often complicating hematoma formed around the needling sites.

The management goals are to eradicate the infection, prevent more sinister complications and maintain the hemodialysis access as far as possible.

Surgical wound infection

The risk of surgical wound infection can be minimized by careful operative decision, meticulous tissue handling and stringent aseptic field of surgery. The spectrum of infection may range from superficial cellulitis, superficial abscess, infection involving deep tissue and vessels. The most disastrous situation is blow out of the anastomosis due to infection of the anastomotic vessels. Therefore if wound infection happens, one should promptly commence antibiotic treatment, perform thorough assessment to detect any superficial or deep abscess (drainage if any), wound swab and blood culture to reveal pathogen identity, and close monitoring for deep spreading of the infection to vascular anastomosis.

If surgical drainage of superficial abscess is necessary, ensure adequate soft tissue coverage of the anastomosis vessel or graft. If deep infection is detected, especially in AVG situation, one should consider early surgical explantation and take down the anastomosis. It is preferable to prepare a tourniquet placed proximally to control excessive bleeding as getting an adequate length of artery proximally and distally for clamping could be challenging in an infective situation. If the inflow artery is not sacrificable, one will have to plan the strategy of arterial repair. Very often, primary repair is not feasible as the wall condition is unhealthy and may also cause stenosis or thrombosis. Patient's own vein or allograft, e.g. peri-cardial patch, is required for patch repair of the arterial defect. In the worst scenario, arterial bypass procedure might be needed. Simple ligation for the outflow vein is usually harmless unless the anastomosis is made to a deep vein. Primary wound closure may not be possible in all situations. Proper soft tissue coverage for repaired vessel and cautious negative pressure wound management are needed.

Seedling from infected CVC

The existing tunnelled or temporary CVC inserted into ESRF patient while waiting for an AVF or AVG to be ready for use can be a source of systemic sepsis, causing seedling of bacteria to the surgical wound or the graft implanted. It is a good practice to examine the CVC exit site for any feature of infection immediately before vascular access surgery, and also during follow-up assessment. Once the newly created vascular access is good for cannulation, the CVC should be removed as early as possible. Occasionally, it could be due to seedling of bacteria from a source of sepsis in other parts of the body, e.g., peri-anal abscess.

Infection introduced during access cannulation

Infection rarely happens in relatively healthy native vein fistula. The infection may occur over the vascular access frequent cannulation site, especially in AVG. More commonly, infection happens in vascular access with cannulation related hematoma, pseudoaneurysm or aneurysmal change region with poor skin condition. The spectrum of infection ranges from localized cellulitis, infected hematoma, infected expanding pseudoaneurysm, exposed graft, localized abscess formation, to pus tracking along the pathway of the vascular access. The infective complication may or may not lead to systemic sepsis. Infected pseudoaneurysm of AVF or AVG may also present as torrential bleeding.

It is important to review the infective situation promptly and thoroughly to determine the severity, and manage accordingly. On clinical assessment, localized abscess around the vascular access may present as swelling with fluctuance, or pus discharge from skin punctum. These clinical signs may evolve with time. In situations where clinical sign is equivocal, or diffuse soft tissue swelling renders definitive diagnosis difficult, ultrasound study (Fig. 11) and CT scan (Fig. 12) may help detect abscess and diffuse graft infection. The typical imaging features of graft infection is a rim of fluid surrounding the graft. Air pocket inside the fluid signifies presence of gas forming bacteria. Persistent bacteremia of blood culture despite appropriate antibiotic treatment may also signify graft infection. All exposed graft is considered as infected. However, the infection could be localized or spreading.

If only cellulitis or infected hematoma is detected without abscess or graft infection, medical treatment with potent antibiotic (preferably based on antibiotic sensitivity of cultured bacteria) will suffice. The infected site should be rested and avoided from further cannulation until infection is subsided. Depending on the location and extent of infective complication over the AVF or AVG, one may continue to use the non-infective regions for dialysis, or one may require a temporary CVC to rest the vascular access properly. For subcutaneous hematoma, it may liquefy with time and drain through the skin puncture site. Manual expression of the liquefied thrombus may help speed up the healing and surgical drainage may not be necessary. Continuous close monitoring of the infective condition (clinical, blood culture, white cell count and markers like C reactive protein, procalcitonin

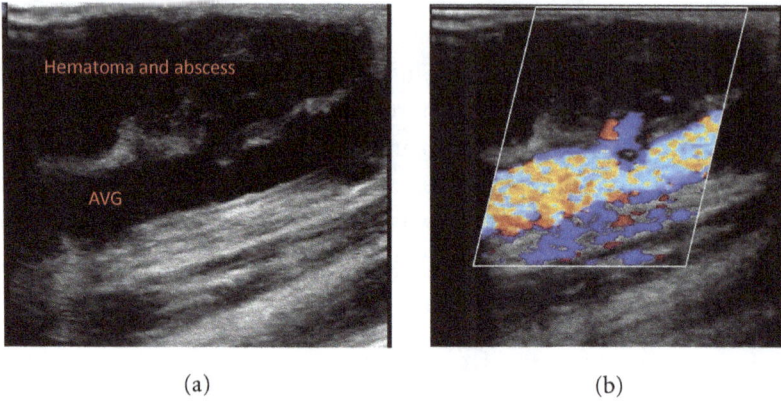

Fig. 11.   (a) Ultrasound image of an infected forearm AVG with hematoma and abscess surrounding the graft. (b) Duplex showing flow inside the graft.

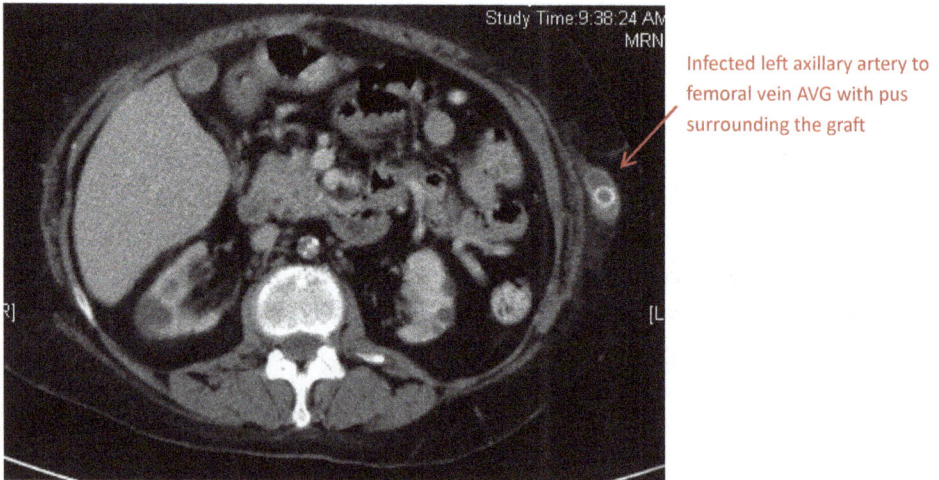

Fig. 12.   CT image of infected left axillary artery to femoral vein AVG with surrounding pus collection.

etc.) is needed and treatment strategy should change according to clinical response. Surgical treatment is necessary if there is abscess formation, graft exposure, definitive infection of the graft or progression of infection. In particular for infected AVG, there are three major surgical strategies to remove the infected graft depending on the extent of infection, patient's general medical condition, surrounding tissue condition and the availability of central vein for CVC placement:

(1) Partial or subtotal explantation of the graft;
(2) Partial explantation of the graft together with rerouting a new graft over the surrounding sterile field;
(3) Total explantation of the graft.

There are two important principles of graft explantation surgeries. The first is to avoid excessive bleeding and the second is to minimize contamination of the surgical field made to take down the graft from the artery (in total explantation) or near the arterial anastomosis (partial/subtital explantation). If the graft infection is localized, the first incision should be made over a clinically non-infective part towards the arterial end well away from infected region to divide and ligate the graft so as to cut away the arterial inflow. The graft is then dissected away from surrounding soft tissue towards both sides of the wound. The arterial end of the graft requires oversew with non-absorbable suture to secure hemostasis. If a subtotal or total explantation is planned, the initial incision should be made close to the arterial anastomosis so as to facilitate the dissection around the anastomotic site. If there is a punctum over the abscess area with pus oozing out, one may cover this infective part with a water proof adhesive dressing before the incision make over non-infective region. The clean wounds can be closed primarily and should be before incision to infected region. The contaminated wounds usually require healing with secondary intention.

In patients with poor medical health, severe regional graft infection (but not involving the arterial or venous anastomosis), partial or subtotal (leaving a short stump of graft close to the vascular anastomotic sites) would be a safe approach. The patient will require CVC placement for HD. If patient's medical condition is relatively stable, clinicians should also look out for possibility of secondary fistula creation once the initial sepsis is eradicated, e.g., look for arterialized arm basilic vein for single stage BB AVF and BBT after explantation of a forearm loop BB AVG.

If the infection has spread to involve the vascular anastomotic site, then total explantation of through graft is required. One should prepare proximal tourniquet for control of arterial bleeding and autogenous/allograft for arterial repair as mentioned above. Soft tissue coverage to protect the arterial repair site is important.

In relatively fit patients with localized graft infection and healthy surrounding tissue, without better secondary AVF option, one may consider construction of a new route of graft through the surrounding non-infective tissue by joining with the divided arterial and venous ends of the existing graft. Temporary CVC placement might be necessary if the new graft is not a rapid access type. Negative pressure wound therapy for graft explantation site(s) promotes wound healing and also help to isolate the explantation wound from the surrounding clean areas to avoid introducing infection to the new graft during cannulation.

After infected graft explantation with all the above strategies, meticulous wound management, long period antibiotic cover (usually 6 weeks or more) and close monitoring of persistent infection (clinical and blood parameters) are essential. Progression of infection may lead to arterial blow out or new graft infection. Based on the study of Ryan *et al.*, the risk of persistent infection is higher with partial explantation and simultaneous re-routing of new graft compared to the other two strategies.[17] However, this approach minimizes central catheter time and provides a new access to the patient without a second operation.

One should also look out for occult lung abscess if patient develops persistent systemic sepsis after graft explantation and without evidence of wound or central catheter infection, especially in medically frail patients.

Infective complications of hemodialysis accesses are common clinical occurrences. Proper diagnosis, prompt antibiotic ± surgical treatment, and close monitoring for response to treatment are the key elements of management.

Case 2

Mr S has DM, ESRF, failed RC AVF and hemodialysis via a left forearm loop BC AVG created in another hospital 2 years ago. He carried a right IJV tunnelled CVC for more than 1 year before the left forearm AVG was created. He presented with a swelling over the venous limb of the graft about 5 cm away from the cubital fossa surgical scar. It is also associated with decrease in access flow and increase in venous pressure. A quick ultrasound scan of the swelling showed pseudoaneurysm of the AVG. There was no feature of infection. Mr S was brought in for a fistulogram (Fig. 13) through the graft, which showed graft-vein anastomotic stenosis. The pseudoaneurysm was mostly thrombosed with only a small stump demonstrated in the fistulogram. This was confirmed by Duplex ultrasound study.

Graft-vein anastomosis stenosis was treated with balloon angioplasty (Fig. 14). The neck of the pseudoaneurysm was about 3 mm on ultrasound assessment. Since it is mostly

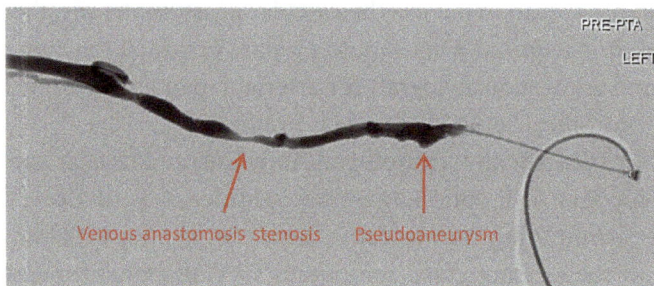

Fig. 13.   Fistulogram via the left forearm BC AVG showing graft-vein anastomosis stenosis and a small stump over the pseudoaneurysm.

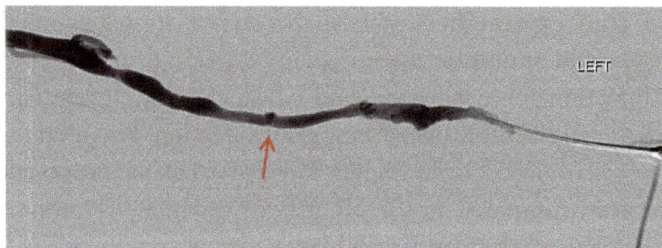

Fig. 14.   Patent graft-vein anastomosis after angioplasty.

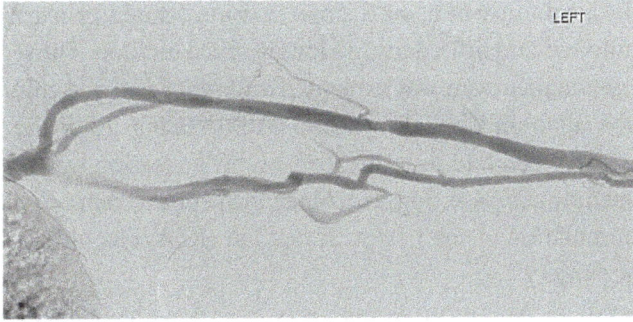

Fig. 15. Except a short segment 40% stenosis, the rest of the arm cephalic vein was of good size on fistulogram.

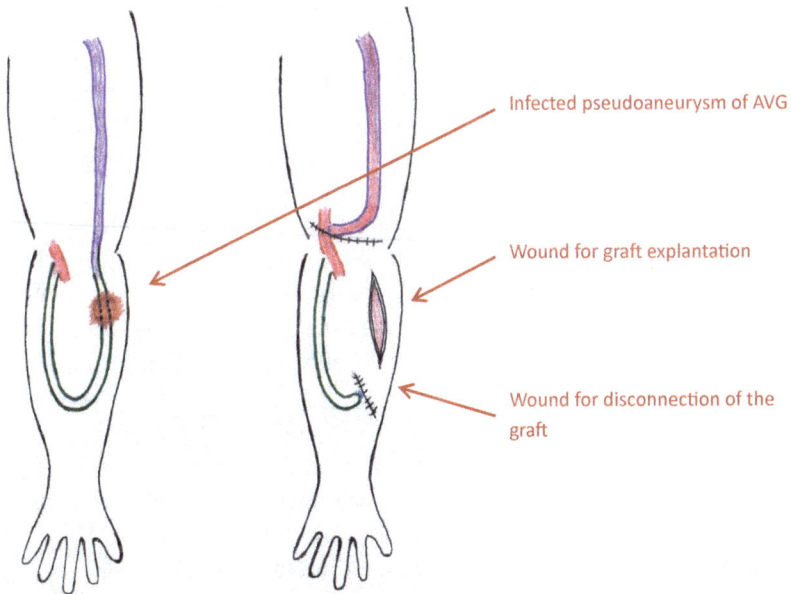

Infected pseudoaneurysm of AVG

Wound for graft explantation

Wound for disconnection of the graft

Fig. 16. Schematic diagram of left BC AVG partial graft explantation and concurrent secondary left BC AVF creation.

thrombosed, the decision was to manage it conservatively and avoid any needling over the pseudoaneurysm region. The arm cephalic vein was of good size (~6 mm) and generally healthy (Fig. 15). No arterial limb lesion was detected.

Mr S was discharged after the fistuloplasty. The dialysis center staff was liaised to avoid needling the pseudoaneurysm. Unfortunately, 4 weeks later, Mr S was admitted to Emergency Department due to fever, chills and rigor together with oozing from the left forearm swelling over the AVG. The size of swelling was much bigger compared to 4 weeks ago. Clinically, the left forearm BC AVG psuedoaneurysm was infected. Ultrasound scan revealed localized infection around the pseudoaneurysm. To avoid sudden rupture of the AVG pseudoaneurysm, partial explantation of the graft is performed. The graft was disconnected and ligated over the distal forearm using a small incision (closed primarily). The graft-vein junction was disconnected and ligated using a longitudinal incision. The segment of the graft together with the pseudoaneurysm was explanted and sent for tissue culture. The explantation wound was laid open and initially packed with dressing. After the explantation, the upper limb was cleaned again and drapped. Left BC AVF creation was performed immediately in the same operating session (Fig. 16). Since the cephalic vein was already arterialized and pumped up, cannulation of the newly created left BC AVF could be performed on the second day after the surgery.

The forearm graft explantation wound was healed with secondary intention using negative pressure wound therapy. The graft culture grew non-tuberculous mycobacterium. Anti-TB medication was prescribed by the infectious disease physician for a total of 6 months. Both the surgical and the explantation wounds healed well. There was no further infection of the remaining graft in the left forearm.

Please also see Chapter 14 Case 1.

References

1. Woo K, Cook PR, Garg J, *et al.* Midterm results of a novel technique to salvage autogenous dialysis access in aneurysmal arteriovenous fistulas. *J Vasc Surg.* 2010; **51**(4): 921–925.
2. Hossny A. Partial aneurysmectomy for salvage of autogenous arteriovenous fistula with complicated venous aneurysms. sites. *J Vasc Surg.* 2014; **59**(4): 1073–1077.
3. Peden EK. Role of stent grafts for the treatment of failing hemodialysis accesses. *Semin Vasc Surg.* 2011; **24**(2): 119–127.
4. Quinn SF, Kim J, Sheley RC. Transluminally placed endovascular grafts for venous lesions in patients on hemodialysis. *Cardiovasc Intervent Radiol.* 2003; **26**(4): 365–369.
5. Georgiadis GS, Lazarides MK, Panagoutsos SA, *et al.* Surgical revision of complicated false and true vascular access-related aneurysms. *J Vasc Surg.* 2008; **47**(6): 1284–1291.
6. Barshes NR, Annambhotia S, Bechara C, *et al.* Endovascular repair of hemodialysis graft-related pseudoaneurysm: An alternative treatment strategy in salvaging failing dialysis access. *Vasc Endovascular Surg.* 2008; **42**: 228–234.
7. Shah AS, Valdes J, Charlton-Ouw KM, *et al.* Endovascular treatment of hemodialysis access pseudoaneurysms. *J Vasc Surg.* 2012; **55**(4): 1058–1062.
8. Pandolfe LR, Malamis AP, Pierce K, *et al.* Treatment of hemodialysis graft pseudoaneurysms with stent grafts; institutional experience and review of the literature. *Semin Intervent Radiol.* 2009; **26**: 89–95.
9. Kinning AJ, Becker BW, Fortin GJ, *et al.* Endograft salvage of hemodialysis accesses threatened by pseudoaneurysms. *J Vasc Surg.* 2013; **57**(1): 137–143.
10. Niyyar ND, Moossavi S, Vachharajani TJ. Cannulating the hemodialysis access through a stent graft — is it advisable? *Clin Nephrol.* 2012; **77**(5): 409–412.
11. Zink JN, Netzley R, Erzurum V, *et al.* Complications of endovascular grafts in the treatment of pseudoaneurysms and stenoses in arteriovenous access. *J Vasc Surg.* 2013; **57**(1): 144–148.

12. Fysaraki M, Samonis G, Valachis A, *et al.* Incidence, clinical, microbiological features and outcome of bloodstream infections in patients undergoing hemodialysis. *Int J Med Sci.* 2013; **10**(12): 1632–1638.

13. Taylor G, Gravel D, Johnston L, *et al.* Prospective surveillance for primary bloodstream infections occurring in Canadian hemodialysis units. *Infect Control Hosp Epidemiol.* 2002; **23**: 716–720.

14. Miller CD, Robbin ML, Barker J, *et al.* Comparison of arteriovenous grafts in the thigh and upper extremities in hemodialysis patients. *J Am Soc Nephrol.* 2003; **14**(11): 2942–2947.

15. Akoh JA, Patel N. Infection of hemodialysis arteriovenous grafts. *J Vasc Access.* 2010; **11**(2): 155–158.

16. Gupta V, Yassin MH. Infection and hemodialysis access: an updated review. *Infect Disord Drug Targets.* 2013; **13**(3): 196–205.

17. Ryan SV, Calligaro KD, Scharff J, *et al.* Management of infected prosthetic dialysis arteriovenous grafts. *J Vasc Surg.* 2004; **39**: 73–78.

CHAPTER 14

Vascular Access for Desperate Situations

Jackie P. Ho

When common sites of upper limb vascular access (RC AVF, BC AVF, BBT, forearm loop AVG, arm AVG, BA AVG) are exhausted, clinicians will need to thoroughly evaluate the patient and consider more complex vascular access.

Factors to Consider

(1) Any central vein obstruction — Patients running out of upper limb access options mostly had placement of tunnelled CVC at different sites for various durations. It is important to know the patency of the central vein for planning of access.

(2) Any arterial insufficiency — Majority of Asian ESRF patients are diabetic and have multiple medical co-morbidities associating with atherosclerosis. Peripheral arterial disease is common among them. Multiple access procedures may further affect the patency of the limb arteries.

(3) The locations of accessible patent arteries and outflow veins — Good inflow artery and outflow vein are essential components of vascular access. Reasonable graft length of AVG is needed for arterial and venous site cannulation. However, if the inflow artery and outflow vein are situated too far apart, the risk of graft compression will be increased while crossing a joint or over other pressure points. The tract of the complex vascular access has to be carefully planned.

(4) Personal hygiene, body habitus and social culture — One should look out for the skin condition over groin and axillary regions, especially in obese or poor self-care patients, and avoid placing a vascular access near the un-hygienic or fungal infected area. The tract of the vascular access requires repeated exposure and repeated needling. Planning of the complex access should take into consideration of patient's body habitus, ease of cannulation by the dialysis nurse and psycho-social factors. For example, an axillo-femoral AVG would be difficult to needle in grossly obese patients with thick layer of truncal subcutaneous fat. Exposing certain regions of body, especially in female patients (e.g. groin or chest region), may be socially embarrassing to the patient. Thorough discussion and proper patient counselling is needed before proceeding for these access surgeries.

Do Not Take It for Granted

Each patient's arterial and venous status changes with time, especially in the presence of a hemodialysis access over the same limb. Patients with initially small superficial upper limb vein and no AVF option may have his/her vein enlarged progressively after an AVG is created. Whereas with time, peripheral arterial disease may set in. An artery previously patent may become stenotic and not suitable for hemodialysis access creation. To plan a new access for patients desperate for hemodialysis access, clinicians have to review their previous access history as well as thoroughly assess the current arterial and venous condition using clinical examination and imaging studies.

Strategies for Desperate Access

Utilization of deep vein

In the absence of central vein obstruction, the brachial vein together with proximal radial or ulnar vein of upper limb, and superficial femoral vein of the lower limb can be utilized for AVF creation. Upper limb deep vein is preferred more than lower limb as lower limb is more prone to dependent edema. A detailed ultrasound assessment of the course and size of the deep vein is an essential step of pre-operative planning. Unless the deep vein is originally big in size (diameter >4 or 5 mm), two-stage procedure is preferred.[1] In the upper limb, select the larger size and straighter pathway of the two brachial veins or proximal radial, proximal ulnar vein will be chosen to anastomose with distal brachial artery or proximal radial artery respectively. An ultrasound evaluation of the deep vein fistula would be preferred before proceeding to second stage transposition 4–6 weeks after the first procedure. Connecting veins between the two deep veins might be short and wide. Care has to be exercised to ensure proper control of bleeding and avoid narrowing of either deep veins.

In patients with troublesome CVC and prefer to have functional vascular access early, a graft may alternatively be used to connect to the distal brachial vein of reasonable calibre. Diffuse limb swelling would be common even though it is usually not severe and temporary. Compressive bandage for the hand and forearm is preferred after AVF or AVG creation utilizing the deep vein.

Complex AVG

Complex AVG can be constructed over shoulder (and upper chest), lower limb or trunk regions. Shoulder AVG is preferable to the lower limb and trunk because of the ease of care, better skin hygiene, less embarrassment on exposure and ease of cannulation.

In patients with patent internal jugular vein over the neck, brachial artery to IJV AVG or axillary artery to IJV loop AVG[2] (Fig. 1) can be considered. A careful review of any history of long-term tunnelled CVC insertion using the same IJV is important. A stenosis or occlusion may present over the proximal IJV behind the clavicle or inside the chest due to previous CVC while the neck IJV is patent and of good size. Conventional venogram or CT

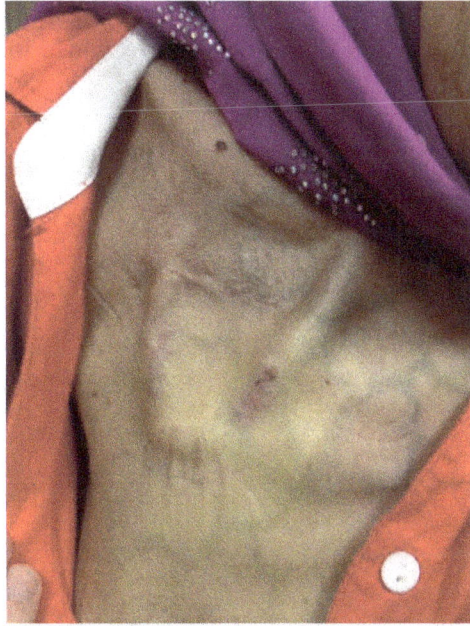

Fig. 1. This lady had exhausted her left upper limb accesses, failed right BC AVF, BB AVF and arm BB AVG. She had left IJV tunnelled CVC *in situ*. Right IJV was patent. Right shoulder loop axillary artery to IJV AVG was constructed.

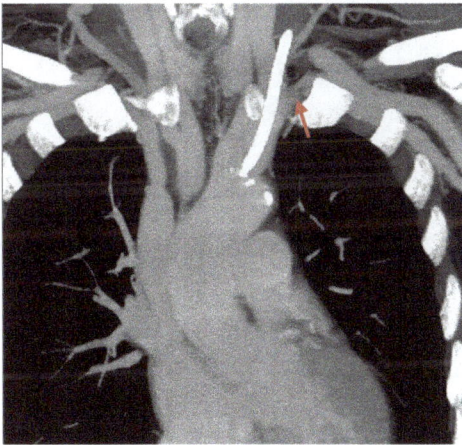
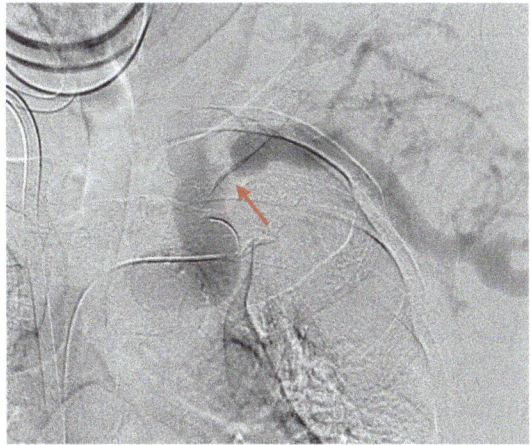

(a)   (b)

Fig. 2. (a) CT venogram showing mild disease of the left subclavian vein but subsequently, (b) venogram showed high grade focal stenosis of the subclavian vein.

venogram of the jugular and central vein should be performed before the procedure. Sometimes, CT venogram could be misleading if the stenosis is of a short segment (Fig. 2). Alternatively, if fluoroscopy facility is readily available in the operating theatre, an on-table venogram can be performed once the IJV is dissected out. As the AVG will cross the clavicle, ring supported graft is preferred. The length of the loop axillary artery to IJV AVG should be tailored to be adequate for cannulation and not extended too caudally to the breast region for female patient.

"Necklace" AVG can be constructed between the axillary artery on one side and contralateral axillary/subclavian vein or IJV (Fig. 3). This AVG is preferred if the more favorable artery inflow and venous outflow are on different side. Again, ring-supported graft is required for this type of AVG. Morsy *et al.* reported low complication rate and reasonable patency in a series of 18 patients with necklace AVG.[3] Some dialysis nurses may worry about accidentally needling into the chest wall and causing pneumothorax, especially for slim and frail patients. Marking out selected location for cannulation and thorough communication with the dialysis nurses would help to relieve the worry.

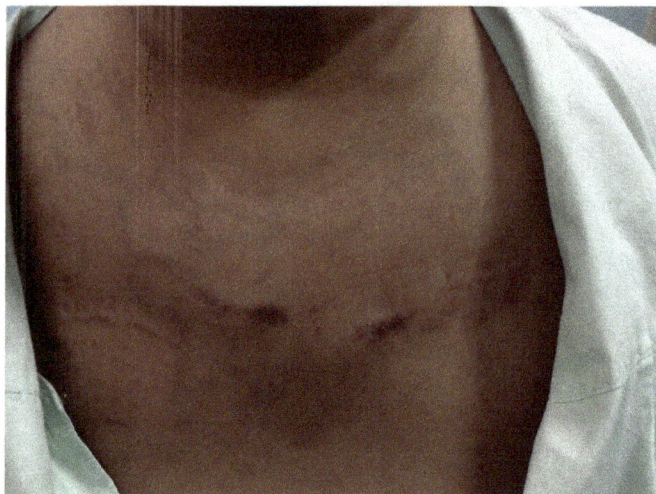

Fig. 3.   Obese patient with "necklace" left axillary artery to right axillary vein AVG.

In patients who have exhausted both upper limb common AVF and AVG sites and with stenosis of the central vein but treatable by endovascular intervention, one may consider implantation of a prosthetic graft with an implantable segment inside the central vein and the graft segment anastomose to the arterial inflow,[4,5] e.g. HeRO graft (CryoLife, Inc. Kennesaw, GA, US) and Hybrid vascular graft (W. L. Gore & Associates, Inc. Flagstaff, AZ, US).

For hemodialysis patients with extensive central vein obstruction (bilateral brachioce-phalic vein or SVC occlusion) that is not amenable by endovascular intervention, and not suited for peritoneal dialysis or transplantation, lower limb and truncal vascular access or accessing the right atrium for venous outflow could be the options.

Various options of lower limb vascular access

- Saphenous vein to femoral artery AVF with loop transposition;
- Saphenous vein to popliteal artery AVF with transposition;
- Superficial femoral vein to superficial femoral artery AVF with transposition;
- Femoral artery to saphenous or femoral vein AVG;
- Popliteal artery to saphenous or femoral vein AVG;
- Suprapublic cross femoral artery to femoral vein AVG.

The most important prerequisite of creating a lower limb vascular access is the absence of significant peripheral arterial disease. In Asia, majority of ESRF patients are diabetic. Prevalence of peripheral arterial disease is high especially in elderly patients. Lower limb steal syndrome DASS is less forgiving than upper limb and the consequence could be major limb loss. All the lower limb pulses have to be examined carefully. Only bouncing pulse of the dorsalis pedis and posterior tibial artery can be accepted as normal. Even palpable, if the pulse is weak or of doubtful quality, a formal toe pressure index study is warranted (normal ≥0.64). In diabetic patients, ankle brachial index can be falsely high and misleading. However, ankle brachial index together with its waveform is also an useful study of the arterial system for younger non-diabetic ESRF patients. Ultrasound assessment of the size of the common femoral, superficial femoral and popliteal artery should also be performed to help making the clinical decision of which target arterial to use and the size of arteriotomy for anastomosis during operation.

Skin and hygiene condition of the lower limb particular groin region should be well assessed before planning a lower limb vascular access. Septic complications of the lower limb vascular accesses are far more common than the upper limb. Vascular access infection spreading to the arterial anastomosis may also jeopardize lower limb circulation and lead to major limb loss. In grossly obese patients, the groin area is usually folded. One should avoid creating AVF or AVG over the groin but may consider using the upper and mid-thigh.

Venous system study is also important in lower limb vascular access. The size and course of great saphenous vein and femoral vein can be easily assessed by simple bedside ultrasound. However, if there is suspicious history of deep vein thrombosis, history of kidney transplantation or previous history of CVC insertion via the femoral vein of the planned operation side, a detailed venous duplex study together with conventional venogram of the iliac vein and IVC will be warranted.

A) *Saphenous vein to femoral artery AVF with loop transposition and saphenous vein to popliteal artery AVF with transposition*

Great saphenous vein GSV, a commonly used conduit for femoro-popliteal bypass in peripheral arterial disease, performs less well as a dialysis conduit when compared to cephalic vein in the upper limb. It is a thick walled vein with limited dilatation ability.[6] Usually, clinicians would prefer a saphenous vein size $\geq$ 4 mm for consideration as a conduit. Secondly, even though it is a superficial vein, it is usually deeply embedded in the subcutaneous tissue underneath the superficial fascia and medially situated. Whether the saphenous vein is connected to popliteal or femoral artery, the whole course of the venous fistula should be mobilized, transposed and superficialized to facilitate future cannulation. GSV is medially situated. Transposition of the GSV to a much more lateral position is important in facilitating subsequent cannulation. It is advisable to draw out the intended transposition location over the thigh with the patient's lower limb at neutral position and hip joint in a non-abducted position before the surgery. This is the posture of future cannulation in hemodialysis center. Different from the position during operation with the hip joint abducted and externally rotated.

The great saphenous vein is mobilized and then either turned into a loop connecting to common femoral artery or proximal superficial femoral artery,[7] or transposed into a superficial plane connecting to distal superficial femoral artery[8] or proximal popliteal artery. The wound of GSV mobilization can be reduced with laparoscopic technique.[9] Superficial femoral artery is prone to calcification and atherosclerosis in diabetic and elderly patients. Thus, saphenofemoral AVF connecting to distal superficial femoral artery may not be a good choice in elderly diabetic patients.

B) *Great saphenous vein semipanel graft (Tagliatelle technique[10])*

C) *Femoral vein transposition AVF[11]*

Systematic review by Antoniou *et al.*[12] showed a better primary and secondary patency in transposed femoral vein AVF compared to loop femoral AVG. However, the ischemia complication was more frequent then AVG. Rueda *et al.*[13] reported a high harvest site and ischemia complications with femoral vein transposition. It is certainly not an entry level operation.

D) *Upper thigh loop AVG connecting common or superficial femoral artery to proximal saphenous vein or common femoral vein*

If the proximal GSV size is satisfactory, graft anastomosis to the proximal GSV is preferred to the femoral vein. When vein-graft anastomosis stenosis occurs, the chance of developing venous hypertension and deep vein stenosis/thrombosis is less with proximal GSV anastomosis than that of femoral vein.

Again during operation, patient will be positioned with hip joint abducted and externally rotated like a frog leg. In the dialysis center, the size of the dialysis chair usually cannot accommodate the patient to assume frog leg posture. An AVG created over the thigh can look central in the operation but becomes too medial in the dialysis center and poses a challenge to cannulation. At the beginning of the operation, it is better to position the hip in neutral position and mark the preferred course of the AVG before proceeding to the actual operation. Of course, make sure an adequate length of straight pathway of AVG was made.

E) *Mid-thigh loop AVG connecting distal superficial femoral artery to superficial femoral vein*

This AVG is away from the less hygienic groin region. Nonetheless, superficial femoral artery, prone to have disease in peripheral arterial disease, is not the preferred inflow artery in elderly and diabetic patients.

F) *Cross femoral AVG connecting femoral artery and femoral vein (or proximal saphenous vein)*

Cross femoral artery to femoral vein (or proximal GSV) AVG is constructed when the patient has one side femoral or iliac vein occlusion and atherosclerotic disease of the other side lower limb arteries. Again, in constructing the pathway of this AVG, ensure adequate length of straight pathway of graft for needling. The AVG pathway should be kept at the suprapubic region and not going proximally into the lower abdominal wall so that the graft will not get exposed in case patient requires midline laparotomy for any acute abdominal condition. The arteriotomy and venotomy are preferably in an oblique rather than straight longitudinal manner to allow better configuration of the graft vessels complex (Fig. 4). A thorough patient counselling is needed prior to making a cross femoral AVG as patient may feel embarrassed exposing the pubic area for needling on every hemodialysis session.

Uzun et al.[14] compared the clinical course of 29 saphenous vein AVF and 25 lower limb AVG showed significantly higher primary and secondary patency rate of the saphenous vein AVF. A systemic review[12] on lower limb vascular accesses reported the weighted mean primary patency rates for upper thigh AVG, mid-thigh AVG and the femoral vein transposition AVF at 12 months were 48%, 43% and 83% respectively. Most commonly associated complications are infection and limb ischemia. The incidence of infection ranged

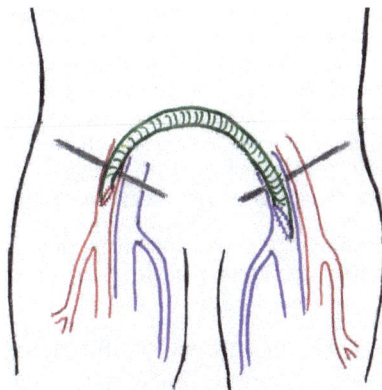

Fig. 4. Oblique arteriotomy and venotomy for vascular anastomosis in supra-pubic AVG to avoid kinking.

from 0% to 41%, whereas the incidence of lower limb ischemia ranged between 0% and 33%. Major limb amputation ranged between 0% and 9%. Infection was more common in AVG and ischemic complication was more in femoral vein transposition AVF. According to author's experience, lower limb edema due to venous hypertension, or outflow vein stenosis/thrombosis is not uncommon.

Truncal AVG

This can be constructed using axillary/subclavian artery as inflow, iliac,[15] femoral, saphenous or popliteal vein as outflow. However, as the AVG crosses longer tract or over a joint, the chance of compression and thrombosis also increases. Ring-supported ePTFE graft is required for truncal AVG. In patients where the skin condition is not favorable over the groin area, external iliac vein can be used (retroperitoneal dissection) as the alternative outflow.

During the operation, one may consider elevating the shoulder and loin with folded drape or towel and abduct slightly the shoulder joint. To facilitate proper surgical judgement, clean, prepare and expose a large area of patient's body from the neck all the way down to mid-thigh or leg (depending on the target vein), from mid-line of the chest and abdomen to beyond mid-axillary line laterally. If deep vein (iliac vein and IVC) patency is not certain, one has to prepare the position of the trunk to be screened by image intensifier or theatre mounted fluoroscopy unit, so that intraoperative venogram can be performed. Over the truncal region, the graft is tunnelled along the antero-lateral side of the body to minimize compression when the patient turns laterally. During subcutaneous tunnelling, the operator has to ensure a large proportion of the AVG is close to the skin (especially obese patients) and superficial enough for easy cannulation in the future, as the subcutaneous tissue over the body can be rather thick. The bony prominence of iliac crest should be avoided. Often, it is difficult to get an optimal subcutaneous tract from the upper chest incision all the way to iliac fossa or groin wound. An additional skin incision over the mid-way of the subcutaneous tract may help to keep the graft in a superficial layer of the subcutaneous tissue.

Other rarely used prosthetic vascular accesses include accessing right atrium as the target venous outflow for AVG creation[16] and arterial-arterial AVG[17] in patients running out of all other dialysis options. Exposure of the right atrium requires highly specific surgical skill and is suitable only for relatively young surgically fit patients. Such procedure should only be carried out in tertiary referral centers and with cardiac surgery backup. Certainly, a concern is that if infection of the AVG happens, it may easily spread to the right atrium and explantation surgery would then be extensive. Alternatively, in surgically fit patients with bilateral brachiocephalic vein occlusion, surgical bypass of one side short segment brachiocephalic vein occlusion can be considered. This enables further vascular access options in the ipsilateral upper limb. Arterial-arterial bypass usually only applies in desperate situations. Embolization of small air bubbles or clots to the end organ supplied by the artery is a main concern.

Care, Surveillance, and Maintenance

The complex vascular accesses are usually the last few access options for the patients to maintain their renal replacement therapy. Therefore, up-keeping of these accesses is important. As these accesses are less often encountered by the dialysis nurses, a diagram to illustrate the pathway of the vascular access and preferred site of cannulation would be helpful to the nurses. Clinician may let the patient keep a picture of preferred "A" and "V" cannulation sites drawn on their accesses to show the dialysis center staff. The vascular access should only be needled by a few experienced dialysis nurses. Patient should be referred back to the clinicians with much lower threshold if any problem with the dialysis access happens. A vigorous follow-up program for this group of hemodialysis patients is also required so that problems can be recognized earlier and rectified. Minimal invasive secondary intervention can be performed to preserve longer secondary patency of these accesses.

Case 1

Madam J, 37 years old, suffered renal failure since childhood. She had failed kidney transplant due to rejection. She had multiple previous upper limb access and bilateral brachiocephalic vein occlusion. There were two times right lower limb AVG implantation, complicated by infection and then explanted. She was using left femoral vein tunnelled catheter for dialysis at time of presentation. Previously there were multiple times of right femoral vein tunnelled CVC insertion (Fig. 5).

Clinically right radial and ulnar pulses, and bilateral DP and PT pulses were palpable and strong. There was scar over the right groin due to previous tunnelled CVC insertion. Ultrasound showed right axillary artery 5 mm diameter. Although only 4–6 mm in size, right femoral vein was patent. The distal external iliac vein EIV was patent and of good size, 9 mm. Proximal iliac vein cannot be assessed. Right axillary artery to right distal external iliac vein AVG was planned together with on-table venogram of the right iliac vein. During the operation, there was moderate stenosis of the right common iliac vein with angioplasty performed. Right axillary artery to distal EIV was constructed with ring-supported ePTFE graft.

She was able to use this AVG for dialysis two weeks after the creation and was followed up regularly in clinic monthly. One year later, she was referred from the dialysis center to emergency unit for painful swelling around the graft and fever. An infected hematoma was diagnosed. Intravenous antibiotic was started immediately (Fig. 6). Ultrasound study at the time showed fluid and air inside the 3 cm hematoma next to the graft. Abscess over the AVG diagnosed. In order to salvage this dialysis access, partial

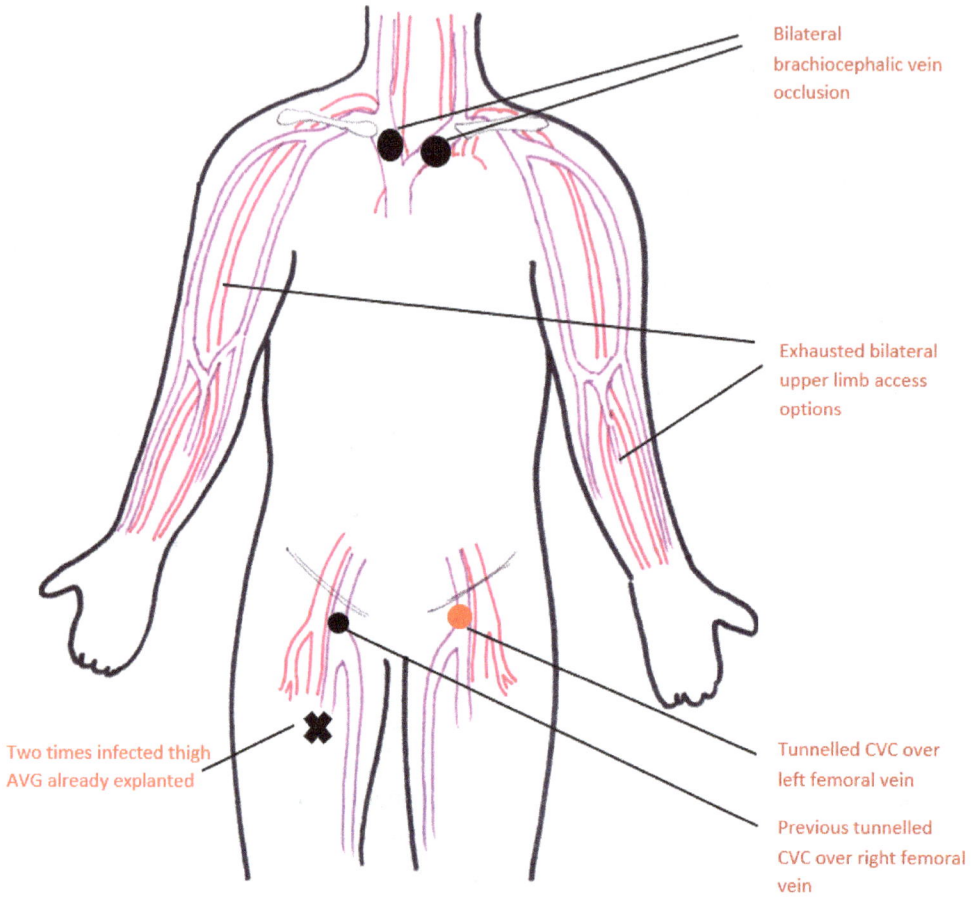

Fig. 5.   Summary of vascular history and vessel condition of Madam J.

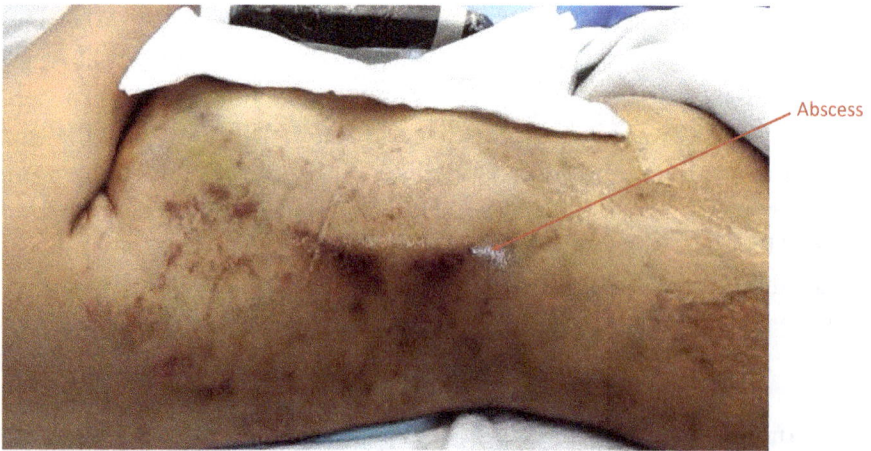

Fig. 6.   Clinical photo showing right axillary artery to external iliac vein AVG with localized abscess formation.

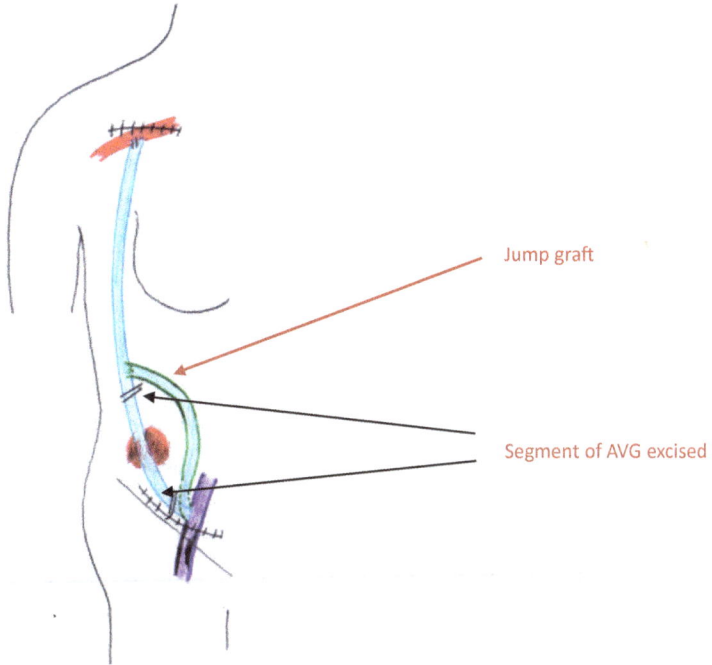

Fig. 7. Schematic diagram of the operation with excision of the segment of infected graft and implantation of a jump graft.

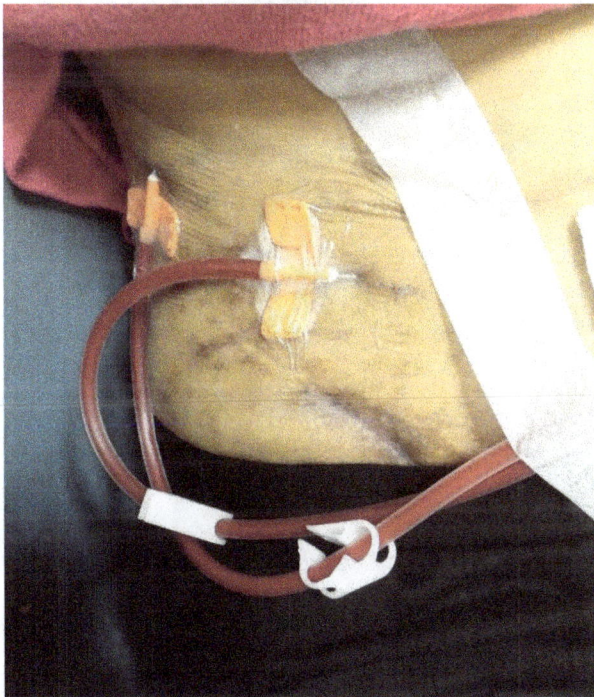

Fig. 8. Clinical photo showing the cannulation of the original AVG and the jump graft for hemodialysis.

explantation of the infected graft and jump graft to maintain the dialysis access was performed. (Fig. 7)

The wound for hematoma, abscess and graft explantation site was treated with negative pressure wound therapy. The patient was put back to left femoral vein tunnelled CVC dialysis. After four weeks, the explantation wound shrunk in size. Needling of the new jump graft and the existing AVG resumed (Fig. 8). Eventually, the left femoral CVC was removed and she was back to hemodialysis using the AVG.

Case 2

Madam L, 43 years old, had systemic lupus induced renal failure for two years. She also had depression and refused permanent vascular access creation. She had multiple tunnelled CVC insertion for a long time and was diagnosed to have bilateral brachiocephalic vein occlusion (Fig. 9). She also had bilateral iliac vein stenosis with stenting of the right common and external iliac vein performed. She was using a tunnelled CVC through the right femoral vein at the time of presentation. She was lately diagnosed to have triple vessel coronary artery disease and planned for coronary bypass surgery. Bilateral upper limb pulses were good and bilateral median cephalic vein size was about 2.8 mm.

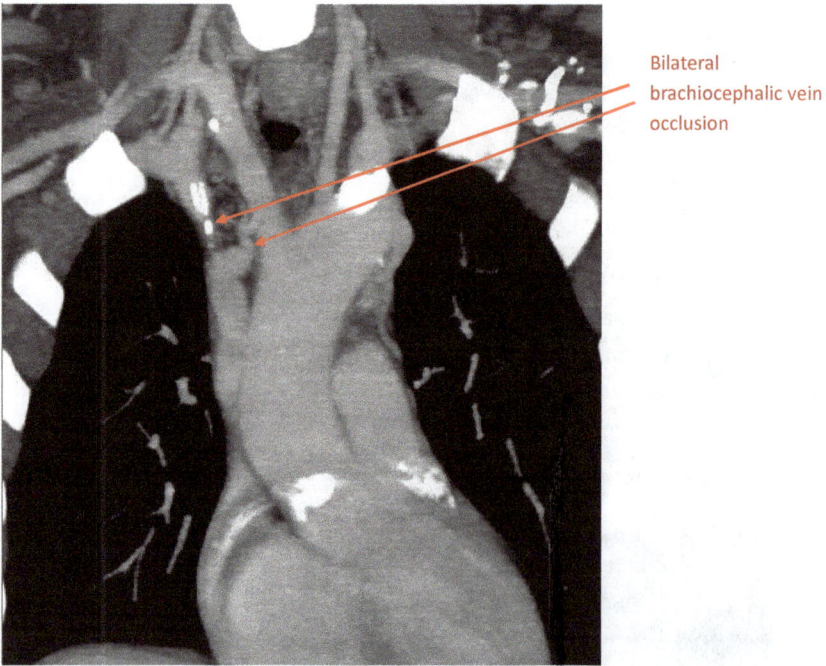

Bilateral brachiocephalic vein occlusion

Fig. 9.   Coronal reformat of CT thorax venogram showing bilateral brachiocephalic vein occlusion.

After a lengthy and detailed discussion with the patient of her future dialysis plan, she finally agreed for interposition graft between the SVC and distal right brachiocephalic vein at the same session of coronary bypass surgery. However, she preferred right BC AVF creation in a later stage.

During the coronary bypass surgery, sternotomy was performed. The SVC and the patent segment of distal right brachiocephalic vein were dissected. the right brachiocephaic vein occlusion was bypassed with an ePTFE (9 mm) graft between the distal right brachiocephalic vein and the SVC. Anticoagulation therapy was commenced after the surgery and maintained for three months.

CT scan performed two weeks after the bypass surgery showed patent bypass graft (Fig. 10).

After recovering from her coronary bypass procedure, one of her family decided to donate a kidney to her. She underwent kidney transplantation three months after the coronary bypass. The patent right brachiocephalic vein was used for central venous catheter insertion to monitor her condition (Fig. 11) and provide infusion during and after the surgery. Vascular access creation was not required.

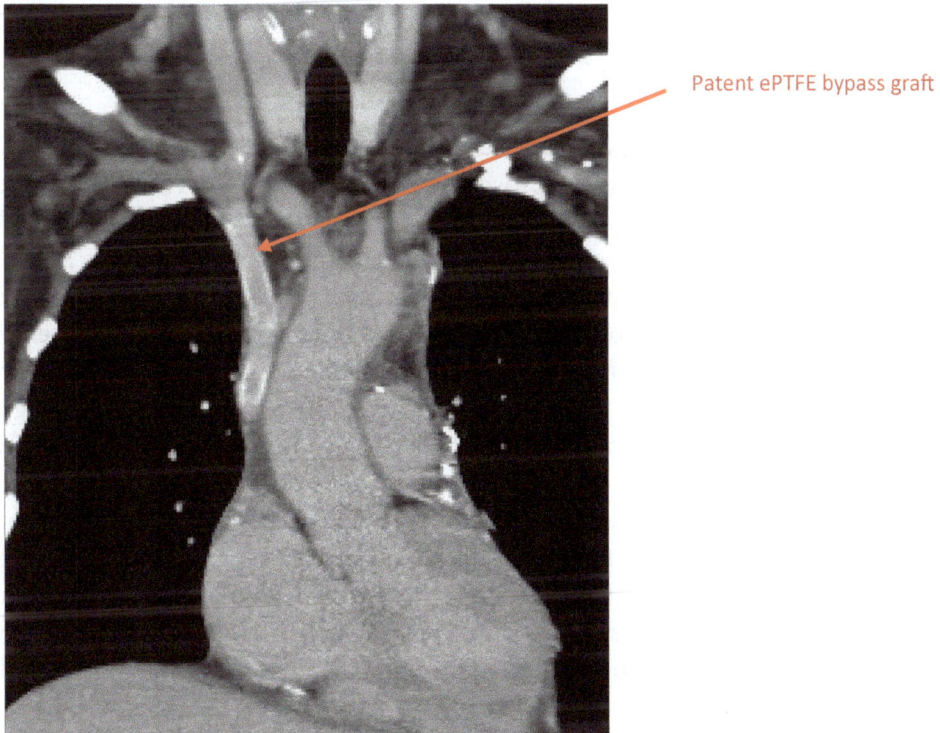

Patent ePTFE bypass graft

Fig. 10. Coronal reformat of CT thorax showed patent bypass ePTFE graft between the SVC and right brachiocephalic vein.

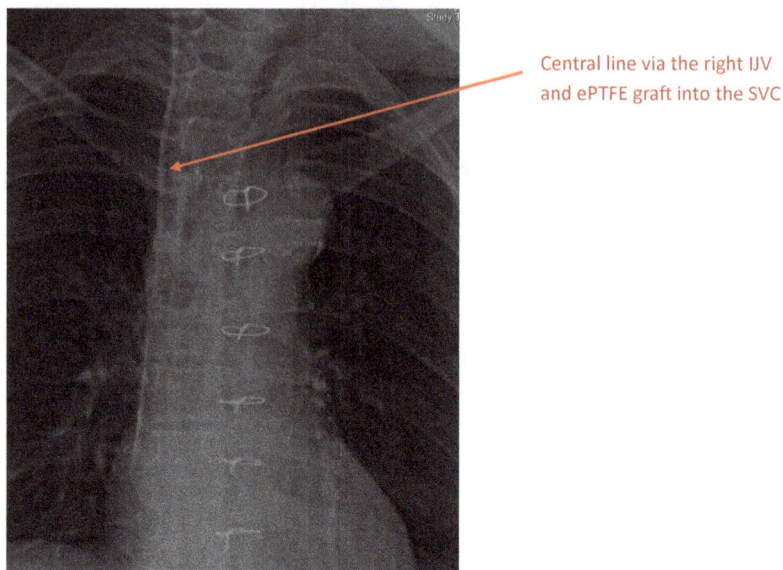

Central line via the right IJV
and ePTFE graft into the SVC

Fig. 11.   CXR showing the central venous catheter inside the bypass graft.

Case 3

A 35-year-old man, Mr H, has ESRF since childhood. He had been using tunnelled CVC via both sides of his IJV for a long time. Two years ago, he was referred for vascular access creation. He has a small body build with body weight of 40 kg. Ultrasound study of his both upper limbs showed small size (<2 mm) forearm and arm cephalic vein as well as distal arm basilic vein. The left proximal basilic vein size was 3.5 mm. His left radial and ulnar artery were patent but both were small in size (<2 mm). His brachial artery size was reasonable (~3.5 mm). Left brachial artery to proximal basilic vein C shape AVG was created for him at that time. He had in-graft as well as graft vein anastomosis high grade stenosis, requiring three times of fistuloplasty to salvage the AVG within the two years. Unfortunately, the AVG became thrombosed again two weeks after the last fistuloplasty. In view of rapid onset thrombosis after fistuloplasty, endovascular salvage was considered to have a high chance of failure. A new hemodialysis access was considered. Ultrasound study (Fig. 12) of his both upper limb showed small sized cephalic and basilic vein over the right upper limb. However, the cephalic vein of the left arm was found to be of a good size (3.0–3.8 mm), although it is rather far away from brachial artery over the antecubital region (Fig. 13).

As patient was temporarily running out of hemodialysis access and his right IJV was also thrombosed, a tunnelled CVC was inserted through the left IJV after a central

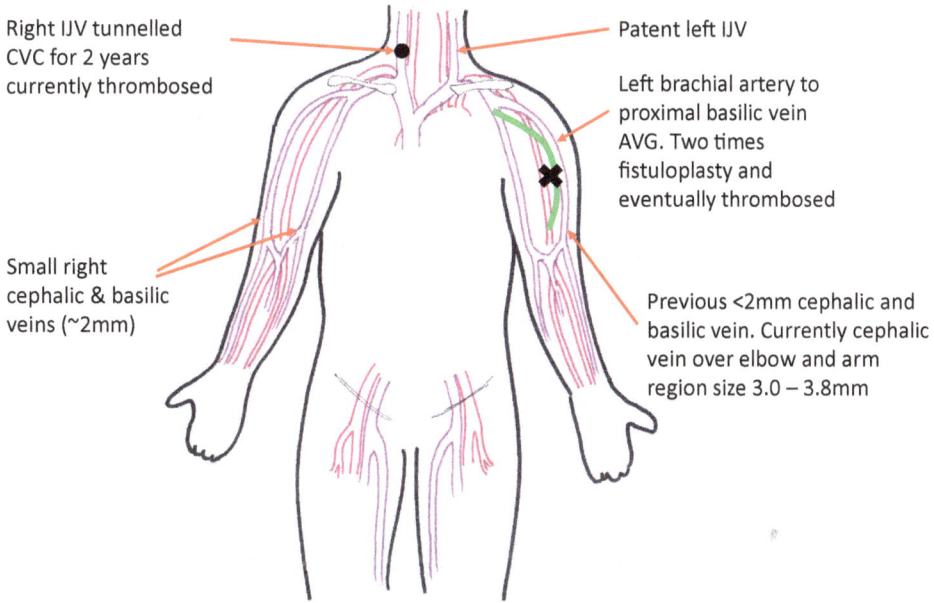

Right IJV tunnelled CVC for 2 years currently thrombosed

Patent left IJV

Left brachial artery to proximal basilic vein AVG. Two times fistuloplasty and eventually thrombosed

Small right cephalic & basilic veins (~2mm)

Previous <2mm cephalic and basilic vein. Currently cephalic vein over elbow and arm region size 3.0 – 3.8mm

Fig. 12. Pictorial illustration of Mr. H's vascular conditions relating to hemodialysis access.

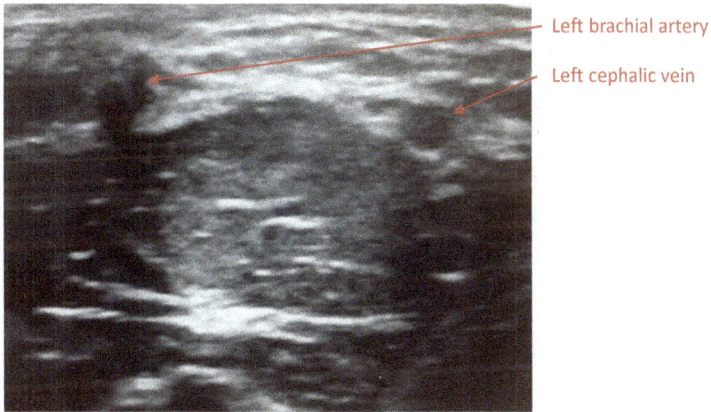

Left brachial artery

Left cephalic vein

Fig. 13. Ultrasound imaging showing good sized left cephalic vein and brachial artery over the antecubital region but separate wide apart.

venogram clear of any obstruction. Then, left cephalic vein to brachial artery AVF was created (Fig. 14).

Four weeks after the AVF creation, the left BC AVF was matured and ready for cannulation. Figure 15 shows marking over the matured AVF for cannulation. His left IJV CVC was removed two weeks later.

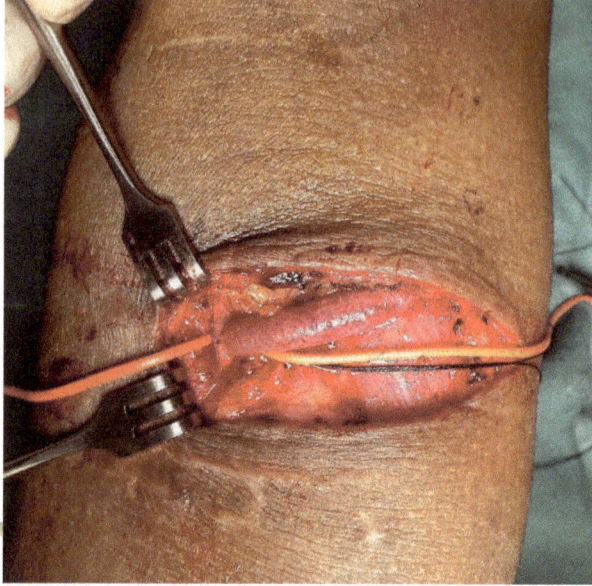

Fig. 14.   Intra-operative picture after left cephalic vein to brachial artery anastomosis created.

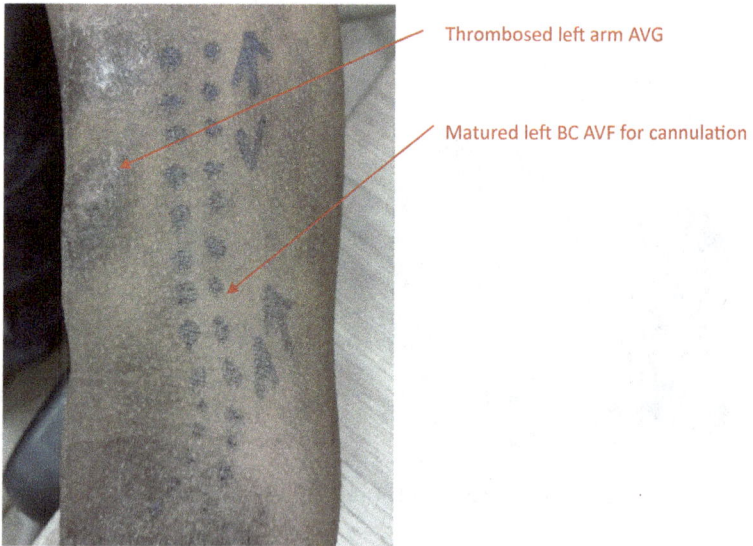

Fig. 15.   Marking of the matured left BC AVF for cannulation.

Case 4

Madam H, 64 years old, DM, HT, morbid obesity (body mass index 39), developed ESRF in 2008. Tunnelled CVC was inserted multiple times to both sides IJV and bilateral femoral vein during the first year of her HD, and during the down time of her vascular accesses. Figure 16 summarize the vascular access history of madam H. Left and right BC AVF were created sequentially but all failed to mature. Eventually left arm BB AVG was successfully created in June 2009 but was complicated with steal syndrome. Left proximal to distal brachial artery bypass with ePTFE graft (without interval ligation) was performed to reverse the steal syndrome. This AVG was used for HD between June 09 to January 2011 with 2 episodes of graft thrombosis salvaged by thrombectomy and fistuloplasty. Third episode of AVG thrombosis occurred again 10 days after the graft salvage. Unable to set up a temporary CVC, and a plasma potassium level of 5.7 mmmol/L, an emergency left BA AVG using the arterial end of the old graft as inflow was created for madam H. After the BA AVG creation and proper hemodialysis, a CT venogram was performed to assess her central vein which showed bilateral IJV and iliac vein occlusion (Fig. 17).

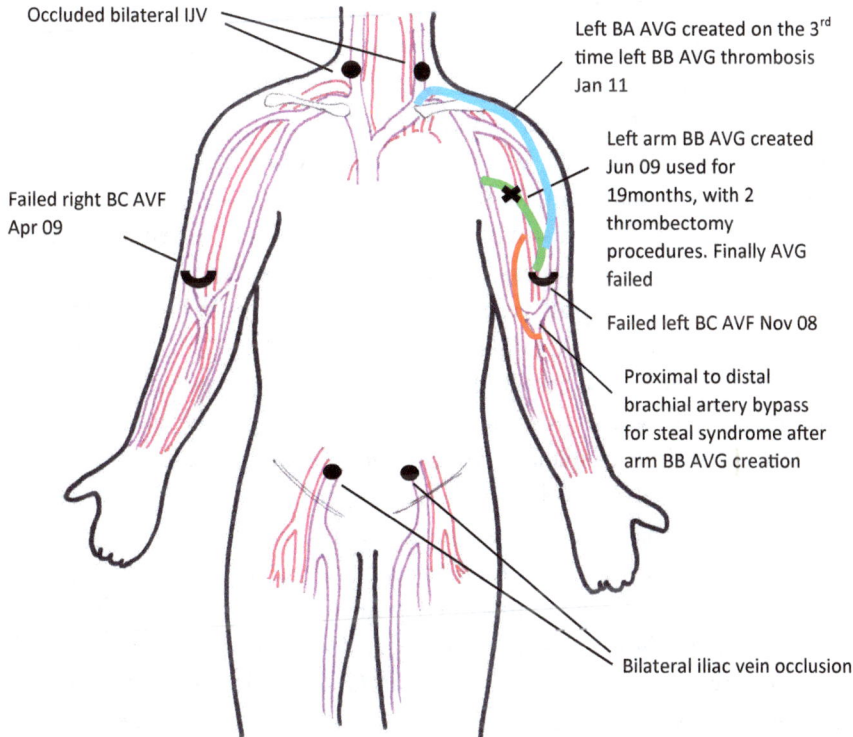

Fig. 16.   Summary of madam H's vascular access history.

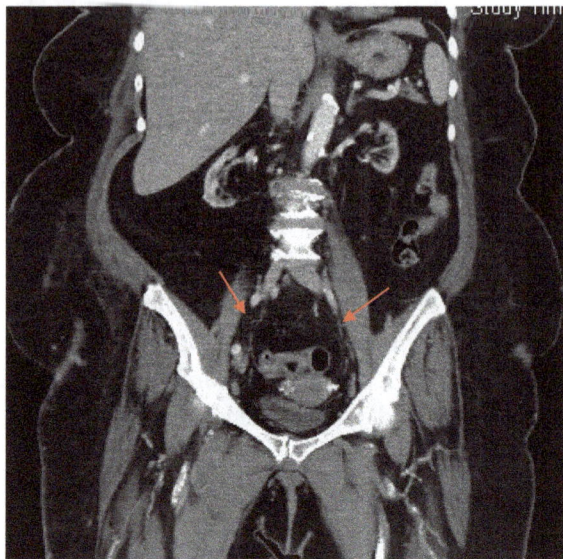

Fig. 17.   CT venogram showing bilateral iliac vein stenosis and occlusion.

The venous pressure during HD raised to >160mmHg persistently 6 months after the BA AVG creation. Madam H was being brought back for fistulogram which showed graft-vein anastomosis high grade stenosis (80%). Fistuloplasty was performed and the venous pressure returned to below 160mmHg. Subsequently, repeated fistuloplasty was required for graft-vein anastomosis re-stenosis and occasionally also for in-graft stenosis approximately every 3–4 months to maintain the AVG patency.

The patency of this BA AVG was maintained until May 2015 when she developed 2 episodes of sudden AVG blockage, 2 weeks apart, both times managed with emergency graft thrombectomy, fistulogram and graft-vein anastomosis angioplasty by the on-call team. However, the venous pressure during HD remained high (~180–200mmHg) after the second thrombectomy. In view of no access for temporary CVC placement, the plan was to preserve the left BA AVG as long as possible and establish a secondary vascular access preferably AVF at the soonest time. In madam H's case, the secondary AVF would have to be on the right upper limb. Ultrasound study of the right upper limb revealed a laterally located (2.5 cm lateral to the brachial artery) cephalic vein 3 mm diameter over the elbow level despite previous failed BC AVF using the median cephalic vein. However, the right cephalic vein over the proximal arm was deeply embedded in the thick subcutaneous fat and the size was small (1.8 mm) with slightly thickened wall. The basilic vein was slightly smaller of 2.5 mm diameter over elbow level. Although previous CT central venogram did not suggest any right subclavian or brachiocephalic vein obstruction, a venogram will be needed to rule out central vein lesion in view of her history of long period tuennlled CVC through the right IJV. Her right brachial artery measured 4 mm over the elbow level. Both right radial and ulnar artery were small (< 2 mm diameter) but with patent lumen. Clinically a weak right radial pulse was palpable but the ulnar pulse was absent. Allen's test was negative.

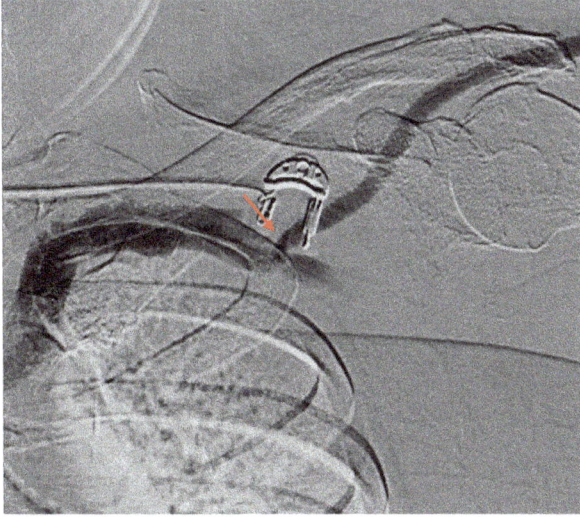

Fig. 18. Fistulogram of the venous limb of the left BA AVG showing re-stenosis (arrow) of the graft-vein anastomosis.

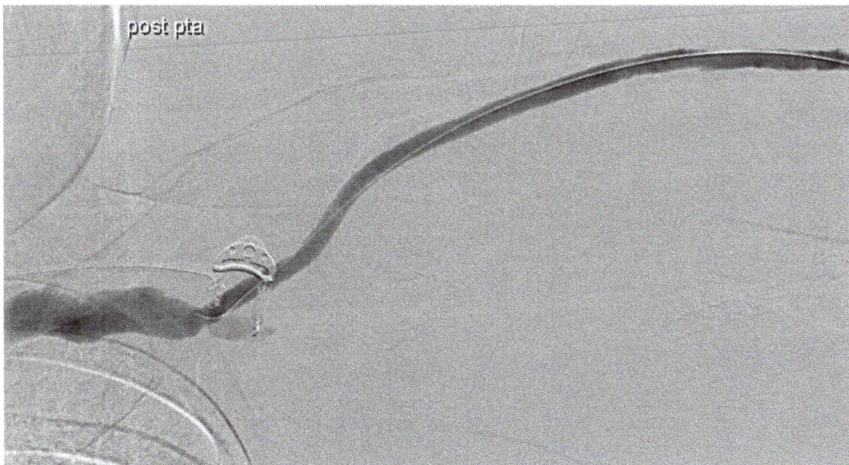

Fig. 19. Graft-vein anastomosis on fistulogram after focal pressure balloon angioplasty with Enforcer balloon 6 mm/40.

An early procedure was arranged 2 days after the second thrombectomy. The left BA AVG fistulogram and intervention was done together with the establishment of right upper limb secondary AVF in the same operative session. The left BA AVG fistulogram showed a 60% re-stenosis of the graft-vein anastomosis (Fig. 18). This lesion was first treated with a focal pressure angioplasty balloon Enforcer 6mm/40 (Cook Medical Inc. Bloomington, IN, US) (Fig. 19), followed by stent deployment from the proximal end of the graft extending 1cm into the axillary vein using Complete SE 7mm/40 (Medtronic Inc. Minneapolis, MN, US) (Fig. 20). Then the arterial anastomosis of the graft was also being interrogated. A mild

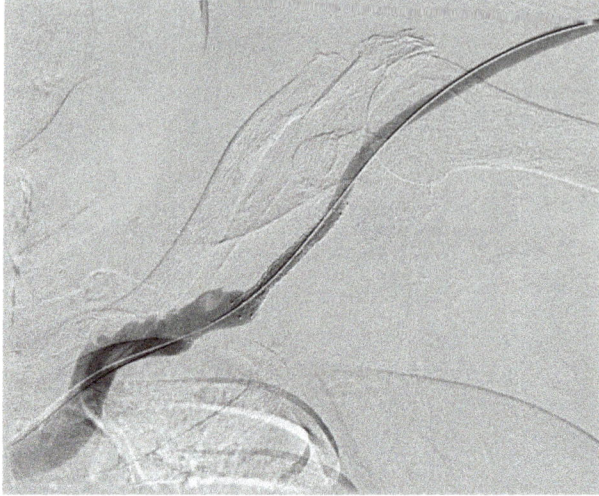

Fig. 20.   Angiogram of the venous anastomosis after stenting with Complete SE 7 mm/40.

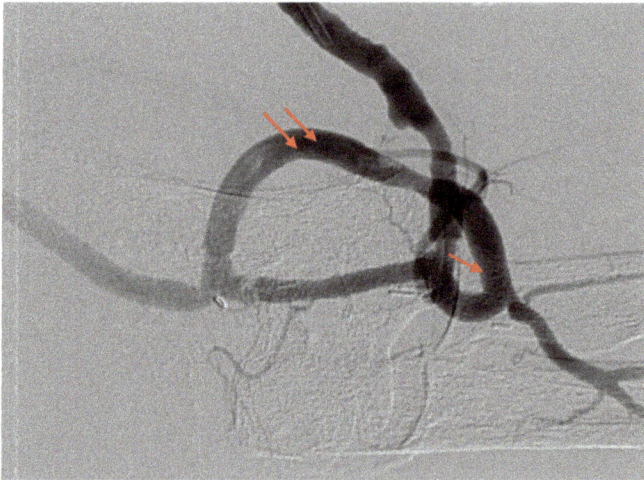

Fig. 21.   Fistulogram of the arterial limb of the BA AVG revealed mild juxta-anastomostic stenosis (single arrow) and patent proximal to distal brachial artery bypass graft (double arrow).

(~30%) stenosis was noted over the graft just next to the anastomosis, and the proximal to distal brachial artery bypass graft was patent (Fig. 21). As the thrill over the AVG was strong at that juncture, no intervention was deemed necessary for the arterial end.

For the secondary AVF on the right side, we decided to proceeded with the BC AVF under brachial plexus block and through the cephalic vein perform a venogram. Right BC

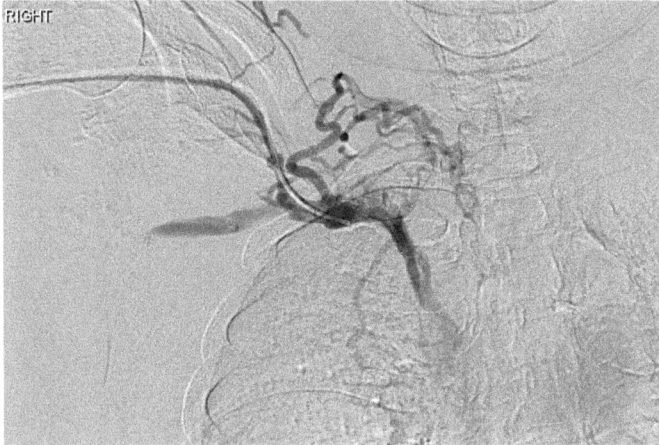

Fig. 22. Venogram of the right central vein showing short segment high grade stenosis of the juncture between subclavian vein and brachiocephalic vein. Enlarged collateral vessels were noted around the stenotic segment.

AVF was more preferable to BB AVF and BBT because cephalic vein size is bigger and the maturation time for BC AVF is usually shorter than BBT. We were not certain when the left BA AVG might thrombose again. On surgical exploration, the cubital fossa cephalic vein was of good size about 3mm. However, on venogram, there was a segment of sclerotic cephalic vein over the proximal arm with diameter only about 1.5mm. There was also a short segment high grade stenosis over the juncture between right subclavian and brachio-cephalic vein. Enlarged collaterals vessels (Fig. 22) were noted around the right subclavian vein as well as proximal cephalic vein.

Balloon angioplasty for both proximal right cephalic vein sclerotic segment and the central vein was performed using Mustang 5mm/60 (Boston Scientific Co. MA, US) and Mustang 7mm/40 (Boston Scientific Co. MA, US) respectively. Stenting of the right central vein was performed using Zilver stent 8mm/60 (Cook Medical Inc. Bloomington, IN, US). Completion angiogram showed good lumen regain of the central vein and reduced collateral flow (Fig. 23). There was re-coil of the proximal arm cephalic vein lesion. Nonetheless, the flow of the contrast was smooth. Then, BC AVF was created. By July 2015, left BA AVG was functioning well with a venous pressure between 120–150 mmHg. Simple ultrasound examination showed her right distal and mid-arm cephalic vein size range between 5.0 mm and 5.5 mm. Proximal arm cephalic vein remained sclerotic with diameter only 1.5 mm. Large size collaterals were noted branching out from mid-arm cephalic vein. As the thickness of her right mid-arm subcutaneous fat was about 7–8 mm, we plan to start cannulation only when the cephalic vein fistula is of larger size (about 7 mm).

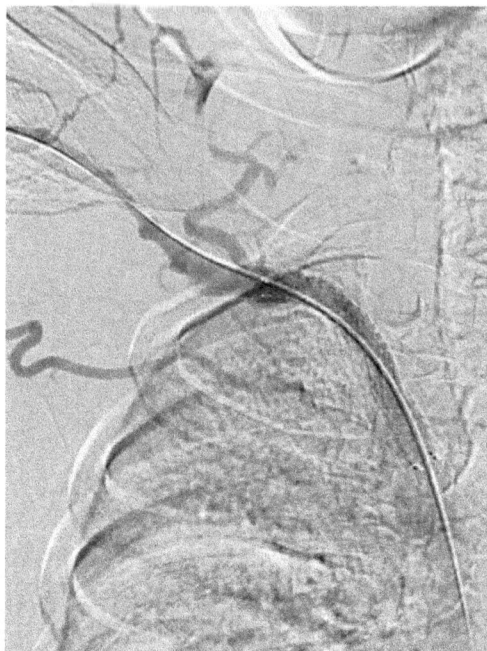

Fig. 23.   Venogram of the right central vein after balloon angioplasty and stenting.

References

1. Jennings WC, Taubman KE. Alternative autogenous arteriovenous hemodialysis access options. *Semin Vasc Surg.* 2011; **24**: 72–81.
2. Jean-Baptiste E ,Hassen-Khodja R, Haudebourg P, *et al.* Axillary loop grafts for hemodialysis access: midterm results from a single-center study. *J Vasc Surg.* 2008; **47**: 138–143.
3. Morsy MA, Khan A, Chemla ES. Prosthetic axillary-axillary arteriovenous straight access (necklace graft) for difficult hemodialysis patients: A prospective single-center experience. *J Vasc Surg.* 2008; **48**: 1251–1254.
4. Katzman HE, McLafferty RB, Ross JR, *et al.* Initial experience and outcome of a new hemodialysis access device for catheter-dependent patients. *J Vasc Surg.* 2009; **50**: 600–607.
5. Jones RG, Inston NG, Brown T. Arteriovenous fistula salvage utilizing a hybrid vascular graft. *J Vasc Access.* 2014; **15**(2): 135–137.
6. Shenoy S. Innovative surgical approaches to maximize arteriovenous fistula creation. *Semin Vasc Surg.* 2007; **20**: 141–147.
7. Daphne Pierre-Paul D, Williams S, Lee T, *et al.* Saphenous vein loop to femoral artery arteriovenous fistula: a practical alternative. *Ann Vasc Surg.* 2004; **18**: 223–227.
8. Correa JA, Abreu LC, Pires AC, *et al.* Saphenofemoral arteriovenous fistula as hemodialysis access. *BMC Surg.* 2010, **10**: 28.
9. Oto T. Endoscopic saphenous vein harvesting for hemodialysis vascular access creation in the forearm: A new approach for arteriovenous bridge graft. *J Vasc Access.* 2003; **4**(3): 98–101.

10. Alomran F, Boura B, Mallios A, *et al.* Tagliatelle technique for arteriovenous fistula creation using a great saphenous vein semipanel graft. *J Vasc Surg.* 2013; 58: 1705–1708.
11. Bourquelot P, Rawa M, Van Laere O, *et al.* Long-term results of femoral vein transposition for autogenous arteriovenous hemodialysis access. *J Vasc Surg.* 2012; **56**: 440–445.
12. Antoniou GA, Lazarides MK, Georgiadis GS, *et al.* Lower-extremity arteriovenous access for haemodialysis: a systematic review. *Eur J Vasc Endovasc Surg.* 2009; **38**: 365–372.
13. Rueda CA, Nehler MR, Kimball TA, *et al.* Arteriovenous fistula construction using femoral vein in the thigh and upper extremity: single-center experience. *Ann Vasc Surg.* 2008; **22**(6): 806–814.
14. Uzun A, Diken AI, Yalçınkaya A, *et al.* Long-term patency of autogenous saphenous veins vs. PTFE interposition graft for prosthetic hemodialysis access. *Anadolu Kardiyol Derg.* 2014; **14**(6): 542–546.
15. Hamish M, Shalhoub J, Rodd CD, *et al.* Axillo-iliac conduit for haemodialysis vascular access. *Eur J Vasc Endovasc Surg.* 2006; **31**(5): 530–534.
16. Kopriva D, Moustapha A. Axillary artery to right atrium shunt for hemodialysis access in a patient with advanced central vein thrombosis: a case report. *Ann Vasc Surg.* 2006; **20**(3): 418–421.
17. Moncef G. Arterio-arterial graft interposition and superficial femoral vein transposition: an unusual vascular access. *Saudi J Kidney Dis Transpl.* 2005; **16**(2): 171–175.

CHAPTER 15

Multi-Disciplinary Team Approach for Optimal Hemodialysis Access Care

Jackie P. Ho

Multi-Disciplinary Team Approach

The needs of ESRF patients on hemodialysis are multi-faceted. Each healthcare provider can only address a component of their needs. The creation and assessment of hemodialysis accesses are usually managed by surgeons in a hospital or clinic setting. Medical problems and adequacy of dialysis are taken care of by the nephrologists, while the regular cannulation and care of vascular accesses are handled by the dialysis nurses at the dialysis centers.

In addition to medical and surgical needs, many patients may have financial and psychosocial issues. Therefore to ensure comprehensive and effective care of hemodialysis patients, a multi-disciplinary approach service has to be established.[1-4] Essential members of this multi-disciplinary team should include nephrologists, vascular access surgeons (general or vascular surgeons), interventionists, dialysis nurses, vascular access nurse specialists, medical social workers and, preferably, infectious disease specialists. As the healthcare professionals involved usually based across different locations and facilities, a consensus on protocols and effective communication are essential to establish a successful program. Berdud *et al.*[3] suggested the establishment of a common treatment protocol between the dialysis center and the referred hospital. The protocol should cover dialysis access management including referral for suspected access failure and infection. Curtis *et al.*[4] evaluated clinical outcomes of patients who are managed in a formalized multi-disciplinary clinic program compared to only with a nephrologist. They reported an improved survival with the formalized multi-disciplinary approach.

In our setting, the nephrologists determine when a patient is suitable for consideration of vascular access creation (both for pre-emptive and those already on hemodialysis), provide counsel on modality of dialysis and optimize the medical health. Vascular surgeons are responsible for the creation of vascular access, monitoring, maintenance and salvage of failing or failed accesses. Nephrologists also share the monitoring task for vascular access. To communicate with dialysis centers, we utilize a standard memo. Illustrations of the configuration and condition of the vascular accesses are frequently used. In addition, we also mark on the patient's limb to indicate suggested sites for "A" and "V" cannulation.

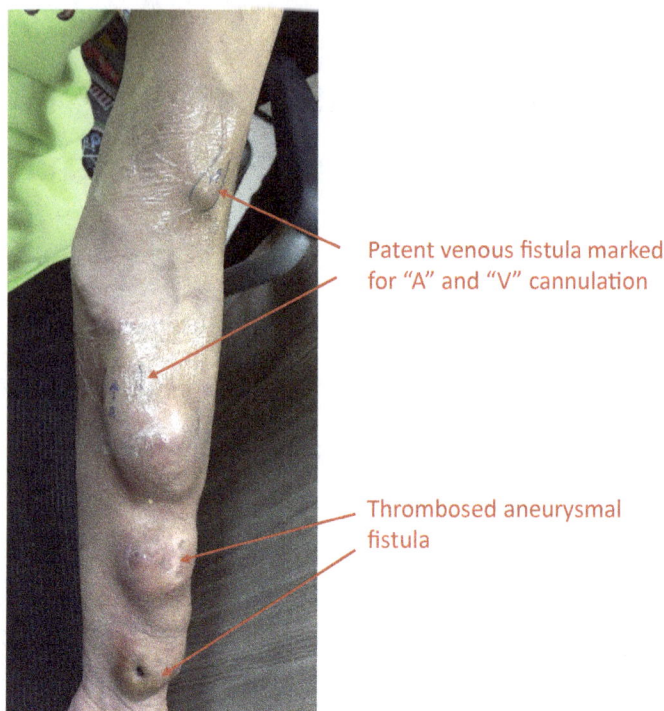

Fig. 1. Suggested "A" and "V" site cannulation drawn on patient's upper limb and the image captured on patient's handphone to communicate with dialysis nurses. This patient had ligated left RC AVF due to bleeding. Secondary AVF with side-to-side anastomosis between left antebrachial vein and distal brachial artery was performed. The distal forearm aneurysmal fistula was thrombosed. Only the proximal forearm aneurysmal area and the arm cephalic and basilic vein are patent.

Since most patients carry their personal handphone, photographs indicating the cannulation sites are stored in their handphone to effectively direct the dialysis nurses (Figs. 1 and 2) on cannulation locations. The dialysis center will also provide a summary of dialysis parameters for patients to attend clinic follow-up with the surgical team. Usually the Qb, arterial and venous pressure recorded during dialysis will be provided (Fig. 3a). In some centers, access flow of the vascular access will be monitored monthly and plotted for review (Fig. 3b). The workload relating to salvage of failing and thrombosed vascular access is extensive. The endovascular interventions are performed by surgeons, intervention radiologists and intervention nephrologists, depending on the earliest availability of both clinician and treatment facility.

The vascular access nurse specialist forms a very important bridge between surgeons, dialysis nurses, nephrologists, patients and other healthcare professionals (Fig. 4). The nurse specialist is often the first point of contact for patients and dialysis center staff when issues of dialysis access occur. The nurse specialist is also able to help patients on their financial and social issues by liaising with the medical social workers and patients' family or caregiver.

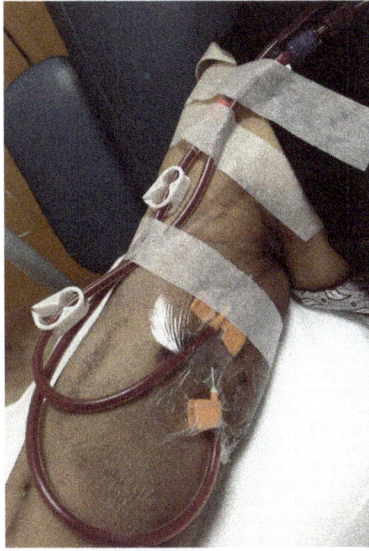

Fig. 2. Picture taken in hospital out-patient dialysis center before transferral of patient to community hemodialysis center, indicating site and direction of cannulation on a loop basilic vein to proximal brachial artery AVF.

Preservation of Forearm and Arm Vein for Future AVF Creation

The service of vascular access for CKD patients actually starts well before the patient develops stage 4 or 5 disease. Many renal impaired patients might require hospital admission for various reasons including chest or urinary tract infection, cardiovascular diseases, diabetes-related problems (e.g. diabetic foot ulcer or skin infection). Intravenous fluid infusion, antibiotic therapy and the administration of other medications are commonly required. Repeated intravenous cannulation to forearm or antecubital superficial veins will induce stenosis and fibrosis of those salient veins (Fig. 5) and affect their eligibility for AVF creation in the future. In many healthcare facilities, the junior frontline doctors, nurses and phlebotomists perform most of the IV cannula insertions. In situations (e.g. severe sepsis, trauma or bleeding) where patients might require large volume transfusion, a large bore cannula insertion to the distal forearm cephalic vein is required. Apart from that, usually a 20G or 22G cannula over the dorsal vein of hand should suffice. Dorsal vein of the hand, although visually prominent, might be challenging to cannulate because the skin and connective tissue are loose over the veins making the veins more "mobile." In some centers, a bracelet reminding healthcare professionals "not to insert IV cannula" is issued and to be worn at all times on the non-dominant hand or the upper limb bearing the vascular access of renal impairment or ESRF patients respectively. However, non-dominant upper limb may not be suitable for pre-emptive AVF creation due to various arterial, venous or skin conditions in some patients. In these situations, the dominant limb may be an important alternative. With the same token, when existing vascular access failed over one limb, second access might have to be placed onto the opposite limb. Therefore, education should be provided to all the healthcare professionals to use the dorsal veins of the hand as far as possible,[5] and to the kidney disease

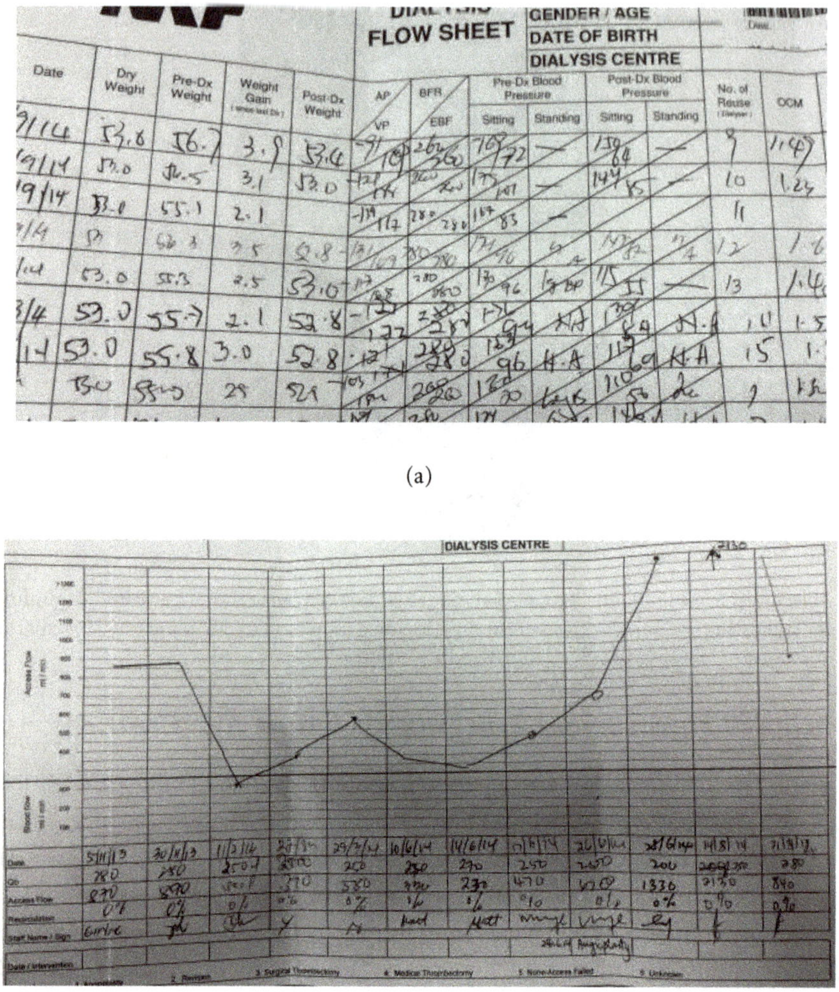

(a)

(b)

Fig. 3. Hemodialysis parameter charts provided by dialysis centers show the recorded arterial and venous pressure during dialysis (a) and access flow plot (b).

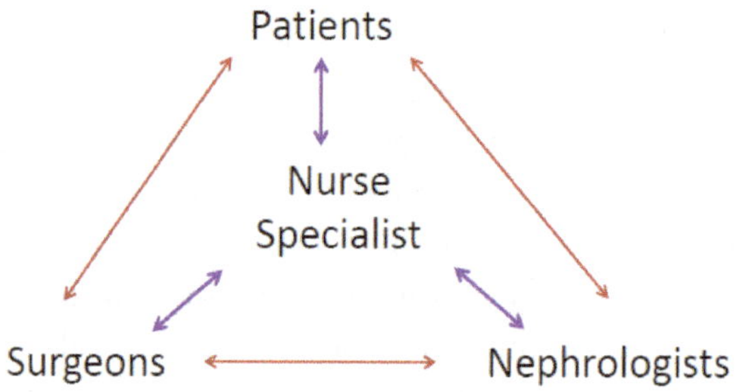

Fig. 4. Vascular access nurse specialist's role within the multi-disciplinary team.

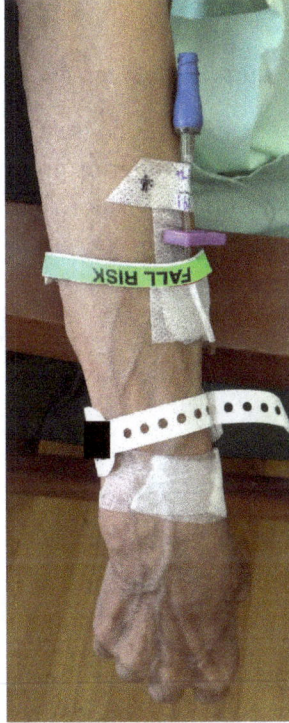

Fig. 5. An intravenous cannula was inserted into the cephalic vein of a patient with stage 4 CKD. This patient has a good sized dorsal vein of the hand but was not being utilized.

patients (including those at early stages of renal insufficiency) alerting healthcare personnel of their kidney disease whenever they are being admitted to hospital. Certainly, no IV cannula should be inserted into existing AVF except in life saving situations.

Patients as a Member of the Vascular Access Care Team

Educating kidney disease patients to actively take part in vascular access care is important. KDOQI (2006)[6] recommends that all hemodialysis patients should be educated on: the basics of assessing flow in the vascular access; prompt report of any loss of thrill; personal and access hygiene; avoidance of excessive pressure on the access; compression for mild bleeding; recognition of infection; monitoring aseptic process of cannulation; and encouraging alternation of cannulation sites. Indeed, vascular access education for kidney disease patients should commence well before the start of dialysis therapy.

Since kidney disease patients come from a wide spectrum of age, ethnicities, cultures and education levels, their ability to understand, recall and use of information provided will also vary. An effective patient education program should therefore contain distinct messages expressed in simple language, lots of pictorial illustrations, with maximal use of audio-visual and electronic media. Furthermore, the education material must be readily available and widely disseminated, with the content fully endorsed by all members of the multi-disciplinary team.

Method of Cannulation

Area technique, Rope-ladder and Buttonhole strategy

In general, there are three different cannulation strategies used by dialysis nurses. In the area technique, dialysis nurses perform cannulation around the same region of the vascular access (Figs. 6 and 7). After the initial few cannulations, adhesion will form between the vascular access and its overlying skin. The chances of successful cannulation will

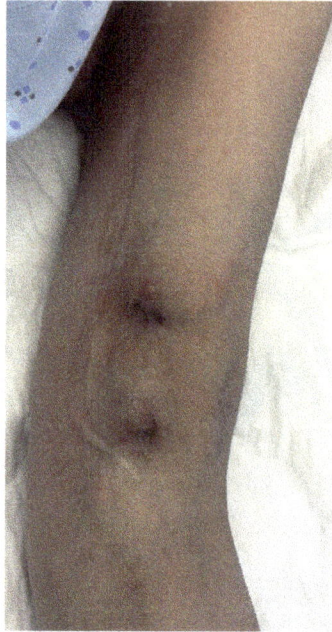

Fig. 6.   Area technique cannulation of left BC AVF resulting in aneurysmal changes of AVF.

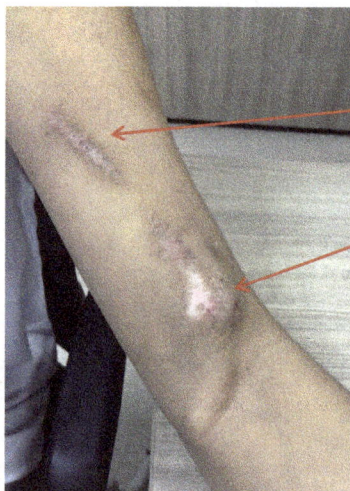

Subsequent rope-ladder cannulation of AVG

Initial area technique cannulation resulting in aneurysmal change of AVG and thinning of overlying skin

Fig. 7.   Area technique and subsequent rope-ladder technique cannulation of left arm BB AVG.

improve over that region and the sensation of skin after repeated cannulation is also reduced. Many a time, the area technique is mistakenly considered as buttonhole technique. The downsides of this strategy are: a higher risk of causing aneurysmal change of access; possibility of skin erosion in the long term (Fig. 8); and the development of access stenosis.

Rope-ladder (or step-ladder) strategy involves sequentially shifting the site of cannulation gradually along the pathway of the vascular access a short distance cranially and then moving caudally again after a period of time (Figs. 7 and 9). As the cannulation sites are not clustered within the same region, there is less structural damage to the vascular access (either native vein or graft). In situations where the depth of the vascular access is uneven, or the straight part of the access is of limited length, the application of rope-ladder technique will be challenging.

Buttonhole strategy involves the creation of a tract between the overlying skin and the underlying fistula. This method is only applicable to native vein fistula and not AVG. A trained cannulator (clinician or nurse, preferably the same cannulator) is required to create the buttonhole tract during the initial few cannulations of the AVF over the same site, following the same angle. Special devices are available at the beginning to assist the formation of the tract. Ordinary sharp needles are used at the beginning for fistula cannulation until the tract has been well formed. Blunt needles can then be used for subsequent cannulations. The buttonhole strategy is particularly useful in AVFs with only a short segment available for cannulation,[7] with loose skin overlying, and for self-cannulation in home dialysis. Currently in Singapore, there are only a limited number of clinicians and

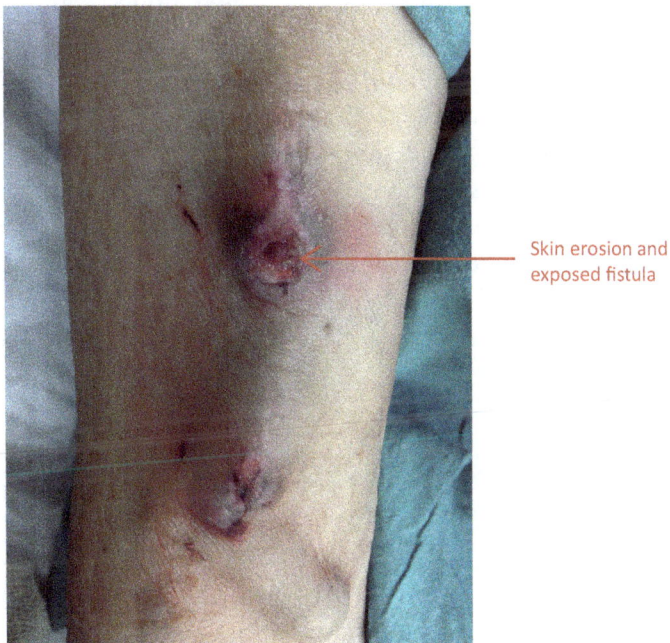

Skin erosion and exposed fistula

Fig. 8.   Right BC AVF with erosion of skin and exposure of the underlying fistula over the cluster needling site.

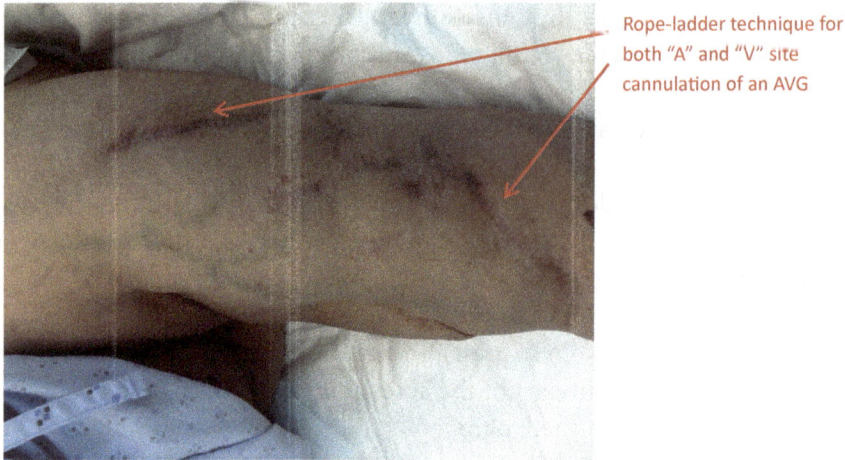

Fig. 9.   Rope-ladder technique to cannulate left BA AVG.

nurses that have mastered the buttonhole technique. Only a small percentage of hemodi-alysis patients are on the buttonhole cannulation strategy.

A cohort study[8] of 7,058 hemodialysis patients in 171 dialysis centers in Germany showed that area technique (65.8%) was the most frequently adopted method, followed by rope-ladder (28.2%) and then buttonhole (6.0%). Area technique was associated with a significantly higher risk of access failure compared to rope-ladder and buttonhole tech-nique. A randomized controlled study comparing buttonhole and rope-ladder technique in 140 hemodialysis patient in Calgary[9] showed a similar median access survival and median fistulogram rate in both groups. However, the buttonhole group was associated with a higher rate of access-related infection. The general impression is that the button-hole is associated with less cannulation pain compared to rope-ladder method but might be associated with a higher infection rate. A systemic review[10] of the literature comparing buttonhole and rope-ladder reported that a statistically significant cannulation pain reduction was found only in observational studies, not in randomized controlled trials. The rate of local and systemic infection was higher with buttonhole than rope-ladder technique.[10,11] Nonetheless, there are vascular access or patient conditions where one technique may be preferred over the other. Clinicians must consider the pros and cons when making a decision on cannulation strategy and communicate their selection clearly to the dialysis center.

Prevention and Control of Infection in Hemodialysis Unit

Dialysis infection and bloodsteam infection are the most common causes for hospital admission among hemodialysis patients. The care and manipulation of tunnelled CVCs and vascular accesses for hemodialysis patients are occurring daily in hemodialysis cent-ers. Hemodiaysis centers' environment, setups, staffing, infection control protocol, staff and patient education and monitoring measures all have a strong influence on the risk of access related infections.[12–14] Higgins *et al.*[15] performed a survey on the knowledge and practice of infection control among dialysis nurses in the Republic of Ireland. They found that even though infection control policy is established, the knowledge and compliance to infection control still have significant scope for improvement. Development of practice guidelines, regular review and update of infection control policy as well as surveillance of staff compliance were recommended to ensure high quality infection prevention. Individual centers may derive their own guidelines and policies based on international guidelines, such as:

- APIC (Association for Professionals in Infection Control and Epidermiology) Guide to the Elimination of Infections in Hemodialysis;
- CDC (Centers for Disease Control and Infection) Health-Associated Infections.

By adopting proper infection control guidelines and performing vigilant surveillance of infection through a quality improvement program, Patel *et al.*[16] reported a successful reduction in overall and access-related bloodstream infection.

Hemodialysis regimen — intermittent dialysis versus frequent short day or nocturnal dialysis

In recent years, there has been a surge of interest in more frequent (either short day or nocturnal hemodialysis) hemodialysis rather than the current two or three sessions per week intermittent dialysis. The concept is to reduce fluctuation of the body's metabolic and physiological stress in-between hemodialysis sessions. From the perspective of quality of life considerations, frequent nocturnal dialysis will minimize disruption to patients' day time work and activities. Observational studies showed that frequent home hemodialysis reduced all cause mortality.[17] A small scale RCT study[18] showed that frequent nocturnal hemodialysis significantly reduced left ventricular hypertrophy, improved BP and phosphate control. Another RCT, however, reported no significant difference between the two hemodialysis strategies in terms of death and left ventricular mass.[19] Ongoing RCTs will provide more evidence on which is the better hemodialysis regimen.

Currently in Singapore, only a small number of patients are practising self-cannulation and frequent home hemodialysis. For frequent hemodialysis, the patient or everyday caregiver would be the person to perform cannulation of vascular access and commence the dialysis circuit. Cafazzo *et al.*[20] reported that barriers for patients and caregivers to conduct frequent home hemodialysis were fear of self-cannulation, lack of confidence in conducting hemodialysis and fear of a catastrophic event. A well-designed education and training program with good backup support will better equip patients and their caregivers to carry out home dialysis, if such a regimen is to be promoted in the region.

The creation, care, monitoring and maintenance of hemodialysis access for kidney disease patients involves a wide spectrum of services and treatments. Different expertise and skill sets are required. The quality of dialysis access care has a strong impact on the well-being of patients as well as on overall healthcare cost. A multi-disciplinary approach with concerted efforts from various healthcare professionals to provide a comprehensive hemodialysis access program is the way to go.

References

1. Santoro D, Benedetto F, Mondello P, *et al.* Vascular access for hemodialysis: current perspectives. *Int J Nephrol Renovasc Dis.* 2014; **8**(7): 281–294.
2. Medkouri G, Aghai R, Anabi A, *et al.* Analysis of vascular access in hemodialysis patients: a report from a dialysis unit in Casablanca. *Saudi J Kidney Dis Transpl.* 2006; **17**(4): 516–520.
3. Berdud I, Arenas MD, Bernat A, *et al.* Appendix to dialysis centre guidelines: recommendations for the relationship between outpatient haemodialysis centres and reference hospitals. Opinions from the Outpatient Dialysis Group. Grupo de Trabajo de Hemodiálisis Extrahospitalaria. *Nefrologia.* 2011; **31**(6): 664–669.
4. Curtis BM, Ravani P, Malberti F, *et al.* The short- and long-term impact of multi-disciplinary clinics in addition to standard nephrology care on patient outcomes. *Nephrol Dial Transplant.* 2005; **20**(1): 147–154.
5. Akoh JA, Dutta S. Autogenous arteriovenous fistulas for haemodialysis: a review. *Niger Postgrad Med J.* 2003; **10**(2): 125–130.
6. KDOQI 2006 Updates Clinical Practice Guidelines and Recommendations. Clinical Practice Recommendations for Vascular Access. Clinical Practice Recommendations for Guideline 4: Detection of Access Dysfunction: Monitoring, Surveillance, and Diagnostic Testing.
7. Jennings WC, Galt SW, Shenoy S, *et al.* The venous window needle guide, a hemodialysis cannulation device for salvage of uncannulatable arteriovenous fistulas. *J Vasc Surg.* 2014; **60**(4): 1024–1032.
8. Parisotto MT, Schoder VU, Miriunis C, *et al.* Cannulation technique influences arteriovenous fistula and graft survival. *Kid Int.* 2014; **86**: 790–797.
9. Macrae JM, Ahmed SB, Hemmelgarn BR, *et al.* Arteriovenous fistula survival and needling technique: long-term results from a randomized buttonhole trial. *Am J Kidney Dis.* 2014; **63**(4): 636–642.
10. Wong B, Muneer M, Wiebe N, *et al.* Buttonhole versus rope-ladder cannulation of arteriovenous fistulas for hemodialysis: a systematic review. *Am J Kidney Dis.* 2014; **64**(6): 918–936.
11. Grudzinski A, Mendelssohn D, Pierratos A, *et al.* A systematic review of buttonhole cannulation practices and outcomes. *Semin Dial.* 2013; **26**(4): 465–475.
12. Arenas MD, Sanchez-Paya J, Barril G, *et al.* A multicentric survey of the practice of hand hygiene in haemodialysis units: Factors affecting compliance. *Nephrol Dial Transplant.* 2005; **20**: 1164–1171.
13. Cimiotti JP, Aiken LH, Sloane DM *et al.* Nurse staffing, burnout, and healthcare-associated infection. *Am J Infect Control.* 2012; **40**: 486–490.
14. Karkar A, Bouhaha BM, Dammang ML. Infection control in hemodialysis units: A quick access to essential elements. *Saudi J Kidney Dis Transpl.* 2014; **25**(3): 496–519.

15. Higgins M, Evans DS. Nurses' knowledge and practice of vascular access infection control in haemodialysis patients in the Republic of Ireland. *J Ren Care.* 2008; **34**(2): 48–53.

16. Patel PR, Yi SH, Booth S, *et al.* Bloodstream infection rates in outpatient hemodialysis facilities participating in a collaborative prevention effort: a quality improvement report. *Am J Kidney Dis.* 2013; **62**(2): 322–330.

17. Georgianos PI, Sarafidis PA. Pro: Should we move to more frequent haemodialysis schedules? *Nephrol Dial Transplant.* 2015; **30**(1): 18–22.

18. Culleton BF, Walsh M, Klarenbach SW, *et al.* Effect of frequent nocturnal hemodialysis vs conventional hemodialysis on left ventricular mass and quality of life. A randomized controlled trial. *JAMA* 2007; **298**: 1291–1299.

19. Rocco MV, Jr Lockridge RS, Beck GJ, *et al.* The effects of frequent nocturnal home hemodialysis: the Frequent Hemodialysis Network Nocturnal Trial. *Kidney Int.* 2011; **80**(10): 1080–1091.

20. Cafazzo JA, Leonard K, Easty AC, *et al.* Patient-perceived barriers to the adoption of nocturnal home hemodialysis. *Clin J Am Asoc Nephrol.* 2009; **4**: 784–789.

Index

Blossom in the spring of 2014 Singapore — an exceptional drought period at that time.